Springer Tracts in Additive Manufacturing

* Indexed in Scopus 2024 *

The book series aims to recognise the innovative nature of additive manufacturing and all its related processes and materials and applications to present current and future developments. The book series will cover a wide scope, comprising new technologies, processes, methods, materials, hardware and software systems, and applications within the field of additive manufacturing and related topics ranging from data processing (design tools, data formats, numerical simulations), materials and multi-materials, new processes or combination of processes, new testing methods for AM parts, process monitoring, standardization, combination of digital and physical fabrication technologies and direct digital fabrication.

Dietmar Drummer · Michael Schmidt
Editors

Progress in Powder Based Additive Manufacturing

 Springer

Editors
Dietmar Drummer
Institute of Polymer Technology
Friedrich-Alexander-Universität
Erlangen-Nürnberg
Erlangen, Bayern, Germany

Michael Schmidt
Institute of Photonic Technologies
Friedrich-Alexander-Universität
Erlangen-Nürnberg
Erlangen, Bayern, Germany

ISSN 2730-9576   ISSN 2730-9584   (electronic)
Springer Tracts in Additive Manufacturing
ISBN 978-3-031-78349-4   ISBN 978-3-031-78350-0   (eBook)
https://doi.org/10.1007/978-3-031-78350-0

Preface

In modern manufacturing, the advent of additive manufacturing (AM) processes has revolutionized how we conceive and create complex components. Unlike traditional methods, reliant on shaping tools, AM offers unmatched geometric design freedom thanks to its layered approach to component fabrication. This remarkable capability has positioned AM as the method of choice for producing intricately designed components optimized for form and function.

At the forefront of this transformative field lay the Collaborative Research Center 814 (CRC 814), which operated within the dynamic interplay of innovation and necessity from 2011 to 2023. Focused on laser-based powder bed fusion of polymers and metals as well as electron beam powder bed fusion of metals, the CRC 814 is dedicated to realizing a vision of producing multi-material components with precisely defined, reproducible, and graded properties. This ambitious endeavor necessitates comprehensively examining the entire manufacturing process chain, from powder production to final component realization, accompanied throughout by simulation.

Chapter 1 briefly describes the field of AM, its benefits, and its applications. Furthermore, the used processes are introduced and explained in detail.

Part I, Powder Materials, focuses on different ways of powder production, modification, and process-adapted powder characterization.

Notable innovations included the integration of functional fillers, enabling the fabrication of magnetically responsive components, among other applications. Strategies for surface functionalization were devised, allowing precise control over particle surface coverage. Integrating nanoparticles in laser-based powder bed fusion of metals broadened the process window and enhanced reproducibility while fostering grain formation and mitigating hot cracking susceptibility. Insights into polymers' mechanical and optical behavior during melt crystallization revealed significant effects achieved by minor alterations. These findings formed adaptable process strategies and expanded simulation model parameters.

Part II, Process and Exposure Strategies, explores monitoring methods essential for gaining insights into processes and adapting process management accordingly.

Noteworthy advancements include a framework for defect-free production of complex components in electron beam powder bed fusion of metals and geometry-independent exposure strategies derived from temperature field analysis during laser-based powder bed melting of polymers. This chapter also addresses overcoming one of the limiting constraints for efficient production of multi-material components through electrophotographic powder deposition. This innovative technique enables selective, uniform, and precise material deposition, facilitating locally adjustable material composition, reducing material usage, and processing poorly flowable powder materials. Additionally, insights into process strategies for hybrid metal components are presented, focusing on understanding the interactions between additively manufactured parts and hybrid components during forming.

Part III, Process Strategies for Multi-material and Multi-functional Parts, provides comprehensive insights into the expanded strategies for multi-material components.

This involved hybrid components were produced by combining forming technology and AM. A primary goal was to thoroughly investigate the interactions during the forming of additively manufactured components. The consistent further development of process strategies has led to remarkable successes, including the pioneering production of thermoset–thermoplastic combinations using additive processes. This means that complex multi-material components can now be produced in a single process, significantly expanding the possibilities of AM.

Part IV, Simulation Techniques, contains profound insights into in-process simulation for metal and polymer powders.

Simulation models for powder deposition have evolved to include already fused layers to clarify the intricacies of the process, such as the correlation between deposition rate and powder bed integrity, as well as the influence of particle size, shape, and electrostatic forces. Optimization models were refined to integrate lattice structures into solid components, optimize buckling stability, and create graded lattice structures to reduce micro-buckling. In process simulation, particularly in laser-based powder bed fusion of metals, significant progress has been made in predicting thermomechanical properties, reducing calculation time, and improving the prediction of residual stresses and distortion. Laser-based powder bed fusion of polymers advances in crystallization models enable a differentiated analysis of temperature and crystallization development. Temperature fields derived from process simulations form the basis for modeling the microstructure and facilitate the prediction of grain growth and the resulting microstructure. Micromechanical models and homogenization approaches refine the prediction capabilities and enable the representation of elastic and elastoplastic behavior.

Part V, Quality Control, examines the metrological challenges in sample detection for non-destructive component analysis. This leads to the development of a methodology for measuring and classifying pores in additively manufactured components that provide information about their formation.

As we embark on this journey through the pages of this book, we invite you to explore the cutting-edge research, innovative methodologies, and groundbreaking insights cultivated at CRC 814. Together, let us dive into the intricate world of AM and chart a course toward a future of limitless possibilities.

Erlangen, Germany   Dietmar Drummer
                                                                                      Michael Schmidt

Acknowledgment

We thank the German Research Foundations and their respective contacts, Dr. Ursula von Gliscynski, Dr. Thomas Münker, and Ursula Hüllen, for their support since 2011.

We want to express our special thanks to everyone who has contributed in any way to the publication of this book in its present form. In addition to the mentioned authors, we would also like to mention other project contributors over the past 12 years. Our heartfelt thanks go to:

Philipp Amend, Andreas Bauereiß, Maximilian Dechet, Maximilian Drexler, Thomas Frick, Bogdan Galovskyi, Harald Helmer, Florian Huber, Andreas Jaksch, Vera Jüchter, Michael Karg, Stephanie Kloos, Johannes Köpf, Florian Kühnlein, Lydia Lanzl, Martin Lerchen (née. Heinl), Fuad Osmanlic, Thomas Papke, Christoph Pobel, Daniel Riedlbauer, Dominik Rietzel, Thorsten Scharowsky, Adam Schaub, Christian Scheitler, Meng Zhao.

We would also like to thank the two most recent managing directors, Simon Cholewa and Andreas Jaksch, for their organization and coordination of this book.

Contents

1 Introduction to Powder and Beam Based Additive Manufacturing 1
Andreas Jaksch, Simon Cholewa, Dominic Bartels,
Matthias Markl, and Dietmar Drummer

Part I Powder Materials

2 Production of Functional Polymer Particles via Liquid-Based Top-down Processes 15
Florentin Tischer, Nicolas Hesse, Wolfgang Peukert,
and Jochen Schmidt

3 Gas Phase Functionalization of Polymer and Metallic Materials for Powder-Based Additive Manufacturing 37
Björn Düsenberg, Juan Sebastian Gomez Bonilla, Marius Christ,
Jochen Schmidt, Karl-Ernst Wirth, Wolfgang Peukert,
and Andreas Bück

4 Process Adapted Multiscale Material Characterization 57
Simon Cholewa, Maximilian Marschall, Michael Schmidt,
and Dietmar Drummer

Part II Processes and Exposure Strategies

5 New Process Strategies for Laser Powder Bed Fusion of Polymers 83
Sandra Greiner, Samuel Schlicht, and Dietmar Drummer

6 Three-Dimensional Multi-material Parts 107
Sebastian-Paul Kopp and Stephan Roth

7   Processing Strategies for Electron Beam Based Powder Bed
     Fusion 127
     Christoph Breuning, Jakob Renner, Matthias Markl,
     and Carolin Körner

8   Laser Beam Melting of Metals 149
     Florian Nahr and Michael Schmidt

Part III   Process Strategies for Multi-material and
           Multi-functional Parts

9   Selective Laser Beam Sintering of Multiphase Systems 175
     Matthias Lindbüchl and Dietmar Drummer

10   Laser Beam Melting of Plastics with Reactive Liquids 197
     Robert Setter and Katrin Wudy

11   Additive and Formative Manufacturing of Hybrid Parts
     with Locally Adapted, Tailored Properties 219
     Jan Hafenecker, Richard Rothfelder, Michael Schmidt,
     and Marion Merklein

Part IV   Simulation Techniques

12   Robust Structure-Process-Optimization 243
     Daniel Huebner, Jannis Greifenstein, Fabian Wein,
     and Michael Stingl

13   DEM Simulation of the Powder Application in Powder Bed
     Fusion 263
     Vasileios Angelidakis, Michael Blank, Eric J. R. Parteli,
     Sudeshna Roy, Daniel Schiochet Nasato, Hongyi Xiao,
     and Thorsten Pöschel

14   Macroscopic Modeling, Simulation, and Optimization 285
     Christian Burkhardt, Dominic Soldner, Paul Steinmann,
     and Julia Mergheim

15   Mesoscopic Modeling and Simulation of Properties
     of Additively Manufactured Metallic Parts 309
     Zerong Yang, Ludwig Herrnböck, Matthias Markl,
     Julia Mergheim, Paul Steinmann, and Carolin Körner

Part V   Quality Control

16   Geometric Measurement and Testing Technology for Additive
     Manufacturing 333
     Benjamin Baumgärtner and Tino Hausotte

About the Editors

Prof. Dietmar Drummer succeeded Prof. Ernst Schmachtenberg as Head of the Institute of Polymer Technology at Friedrich-Alexander-Universität Erlangen-Nürnberg on May 1, 2009. He was the spokesperson for the Collaborative Research Center 814 Additive Manufacturing, which focused on process understanding, development, and process-adapted material characterization. Furthermore, he heads the Polymer Group of Neue Materialien Fürth GmbH and 2 Keylabs at the Bavarian Polymer Institute.

Prof. Michael Schmidt has headed the Institute of Photonic Technologies since its founding in 2009 at Friedrich-Alexander-Universität Erlangen-Nürnberg. His research interests include laser application from micro- to macroscopic scales within industrial manufacturing, additive manufacturing, and medical engineering. He was the vice spokesperson of the Collaborative Research Center 814 Additive Manufacturing.

Chapter 1
Introduction to Powder and Beam Based Additive Manufacturing

Andreas Jaksch, Simon Cholewa, Dominic Bartels, Matthias Markl, and Dietmar Drummer

Abstract This chapter briefly describes the field of Additive Manufacturing, its benefits, and its applications. Furthermore, the used processes are introduced and explained.

1.1 Introduction

1.1.1 Additive Manufacturing

Additive manufacturing (AM) is often referred to as the disruptive technology of the 21st century, primarily due to its ability to produce and functionalize geometrically complex components and even assemblies. A faster time-to-market can be achieved through tool-free production, which can generate economic advantages compared to conventional manufacturing processes. AM can also help to increase flexibility in product development, improve the CO_2 balance, and strengthen the innovative power of companies. In addition, the layered structure makes it possible to produce complex components with undercuts, internal structures, etc., for the first time. AM enables the layer-by-layer production of components using Computer-aided design data, a feature that clearly distinguishes AM from subtractive or shaping manufacturing processes. In contrast to conventional manufacturing processes, the manufacturing

A. Jaksch · S. Cholewa · D. Drummer (✉)
Friedrich-Alexander-Universität Erlangen-Nürnberg, Institute of Polymer Technology, Am Weichselgarten 10, 91058 Erlangen-Tennenlohe, Germany
e-mail: dietmar.drummer@fau.de

D. Bartels
Friedrich-Alexander-Universität Erlangen-Nürnberg, Institute of Photonic Technologies, Konrad-Zuse-Straße 3/5, 91052 Erlangen, Germany

M. Markl
Friedrich-Alexander-Universität Erlangen-Nürnberg, Chair of Materials Science and Engineering for Metals, Martensstraße 5, 91058 Erlangen, Germany

© The Author(s) 2025
D. Drummer and M. Schmidt (eds.), *Progress in Powder Based Additive Manufacturing*, Springer Tracts in Additive Manufacturing,
https://doi.org/10.1007/978-3-031-78350-0_1

complexity of an AM-produced component is primarily influenced by its volume and, above all, its height rather than its complexity. The following schematic diagrams illustrate the differences between the three main manufacturing processes according to DIN EN ISO/ASTM 52900 [1] (Fig. 1.1).

The upper row illustrates subtractive manufacturing, where parts are formed or modified by removing material through milling, turning, cutting, grinding processes, electrical discharge machining, et cetera. Tools such as molds or tool sets are used to form parts in the shaping manufacturing processes shown in the middle row. The desired shape is achieved by applying pressure to a body of raw material, e.g., injection molding, die casting, and compaction of green bodies in conventional powder metallurgy or ceramic processing, among others. Additionally, shaping processes in this category employ tools to shape the initial material, such as thermoforming, forging, rolling, extrusion, and stamping. Finally, the bottom row illustrates AM processes, which involve building up parts layer-by-layer by adding material.

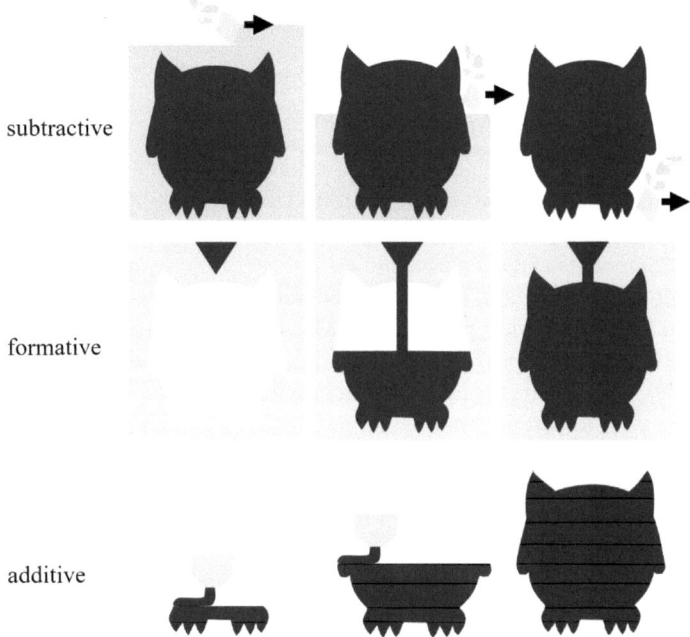

Fig. 1.1 Schematic illustration of subtractive (top), formative (middle), and additive manufacturing (bottom) [1]

Application of Additive Manufacturing

Additive manufacturing processes can be used in a wide range of different industries. Accordingly, mechanical engineering, automotive engineering, aerospace, and (dental) medical applications are the main application areas. Initially commercialized as rapid prototyping, the processes were so named to reflect its main application. Over the last two decades, however, AM has seen a significant increase in its use for tooling applications and final manufacturing (see Fig. 1.2). Nevertheless, prototyping remains an essential application for cosmetic, optical, and presentation models, as well as visual aids and functional parts for fit and function testing. Tooling includes polymer and sand models, cores and molds, and metal molds produced directly on metal AM systems, fixtures, and assembly aids. In recent years, there has been a significant increase in end part production, in which parts are manufactured for direct use by end customers [12].

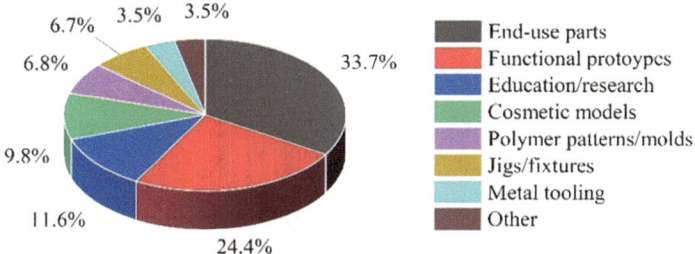

Fig. 1.2 Applications fields of AM from Wohlers Report [12]

Development of Additive Manufacturing

In the past three decades, the AM market has demonstrated nearly consistent double-digit growth, surpassing the $18 billion mark for the first time in 2022 (Fig. 1.3). Forecasts for the forthcoming years remain optimistic, projecting sustained double-digit expansion and an anticipated doubling of the market size by 2028–2030. This expansion may be fueled by the rise of economical and streamlined manufacturing methods across diverse sectors such as aerospace, automotive, healthcare, and consumer goods. AM offers remarkable benefits such as minimized material waste, rapid prototyping, and adaptable design, driving its widespread adoption.

The Wohlers reports show that powder-based manufacturing processes are of fundamental importance given the increasing industrial significance of functional components. Therefore, this book focuses on research activities and findings on powder-based process beam technologies. These technologies are the most promising and economically relevant for technical parts today and in the future. The following chapters cover the fundamentals and specific knowledge of polymers and various metals, highlighting the familiar settings and effects and their key differences.

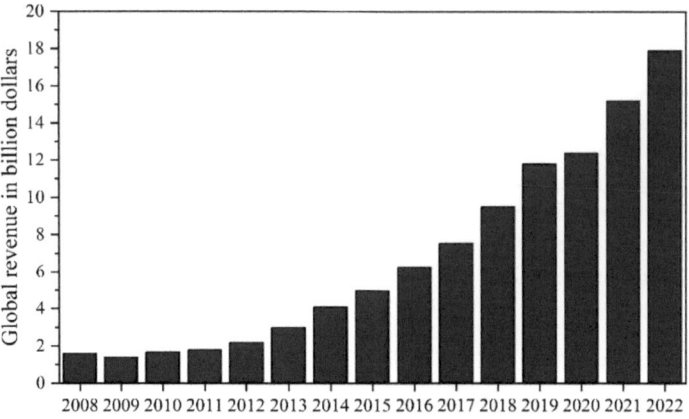

Fig. 1.3 Global revenue for additive manufacturing from Wohlers Report [11]

1.1.2 Laser Based Powder Bed Fusion of Polymers

The most significant industrial class comprises powder-based additive manufacturing processes. The fabrication of the component can be achieved through either melting and solidification or by bonding a fine, thermoplastic powder. Semi-crystalline thermoplastics are generally of industrial importance, as they lead to a significant reduction in viscosity after melting and thus achieve higher densities than amorphous systems. Common to all processes within this class is the absence of support structures required to generate components with undercuts. The first Laser Based Powder bed fusions of Polymer (PBF-LB/P) systems were commercially available as early as the early nineties, rendering PBF-LB/P an established and mature technology.

Set up: Laser Based Powder Bed Fusion of Polymers

Figure 1.4 illustrates a schematic depiction of the standard configuration PBF-LB/P machine. Variations in the configuration of these systems may arise depending on the manufacturer. Broadly, the system can be delineated into three functional domains: the optics area, the process chamber, and the powder area. The optics area encompasses critical components such as the laser source, the scanning apparatus comprising galvanic mirrors, and supplementary optical elements responsible for directing the laser beam. The processing activities occur predominantly within the process chamber. Here, the powdered material is metered from a reservoir—bottom or top feeder—via a dosing unit and applied onto the powder bed using a coating mechanism, for instance, a counter-rotating roller or a blade. Infrared radiant heaters are employed to elevate the applied powder to the requisite processing temperature. The powder area of the laser sintering system includes the build platform, the component fabricated via laser irradiation, and the surrounding powder cake. In commercial systems, the removal chamber is heated as well [9].

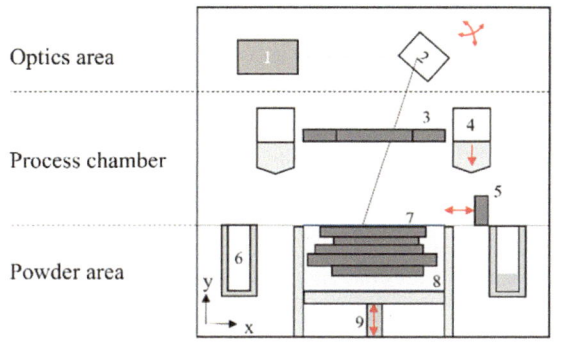

1	Laser
2	Galvano mirror / scanner
3	IR –heater
4	Powder reservoir with dosing unit
5	Coating mechanism
6	Overflow container
7	Component
8	Powdercake
9	Building platform

Fig. 1.4 Set up PBF-LB/P machine according to Greiner [5]

Process: Laser Based Powder Bed Fusion of Polymers

The overall PBF-LB/P process consists of preheating, building, and cooling phases (see Fig. 1.5). In the preheating phase, polymer powder provided by the powder reservoir is deposited in the build chamber via a deposition system and preheated to a temperature just below the material's melting range using infrared heating elements. Following the preheating phase, additional layers, called base layers, are applied. These layers are necessary to reduce the heat transfer between the exposed layers and the metal frame.

During the building phase, the desired geometry is selectively exposed to a focused laser beam through a mirror system (galvanometer mirrors). Typically, a CO_2 laser is used for energy input, as its emitted wavelength of around 10.6 μm exhibits high absorption for most polymers [6]. After exposure, the powder bed is lowered according to the set layer thickness, usually between 40 μm and 150 μm. A new layer of powder is applied, and the repetitive geometry exposure ensures not only

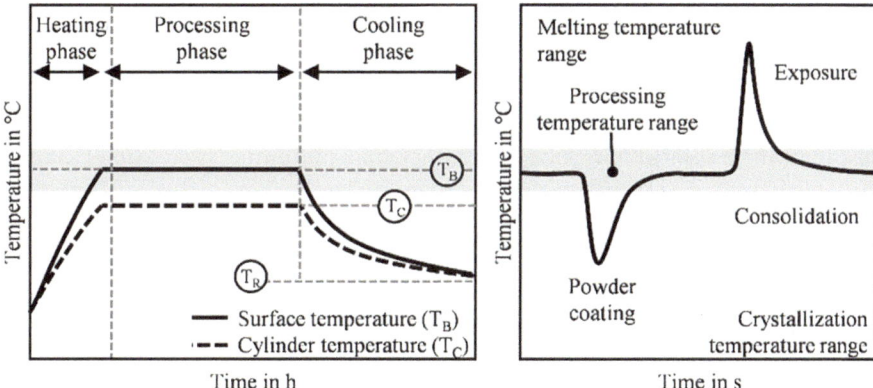

Fig. 1.5 Temperature profile during processing [5]

the particles of the current layer fuse but also the individual layers bond together in the z-direction or build direction. The process repeats until all components are completed. The unmelted powder serves as a support structure for the fabricated components. A defined number of empty powder layers are applied at the end of the building phase to provide thermal insulation.

Subsequently, the heating systems are turned off, and the build chamber cools slowly and uniformly. All three phases occur under continuous inert gas flushing to prevent powder oxidation and thus, aging processes within the material. The components can be removed once the powder cake has cooled to room temperature.

1.1.3 Laser Based Powder Fed Fusion of Metals

Laser-based powder bed fusion of metals (PBF-LB/M) is the most established additive manufacturing technology for the fabrication of metal parts. The material portfolio ranges from low density materials like aluminum- or titanium-based alloys to high density materials such as iron-, cobalt-, and nickel-based alloys. In all cases, the main prerequisite is good weldability of the base material to assure defect-free processing in the absence of pores and cracks (Fig. 1.6).

Set up: Laser Based Powder Bed Fusion of Metals

As for PBF-LB/P, the component is manufactured from a powder bed by placing single weld tracks next to each other in a layer-by-layer process. However, the main difference lies in the energy source used. Whereas polymeric materials are typically processed using CO_2 or CO lasers with rather low powers in the range of a few watts, PBF-LB/M requires hundreds of watts for melting the materials. Consequently, fiber lasers are often used.

The powder is supplied using a recoating system. Since the raw materials are typically atomized from the molten state under a shielding gas atmosphere, the powders possess good sphericity and, thus, good flowability. The high-power laser is scanned over the part's surface using a scanning system. A shielding gas flow (e.g., argon, nitrogen) is used to remove spatters and fumes from the process zone. After the layer is illuminated where needed, the build platform is lowered. The process steps are then repeated iteratively for manufacturing three-dimensional parts (Fig. 1.6). A platform heating can be applied to reduce the thermal gradients during processing [13].

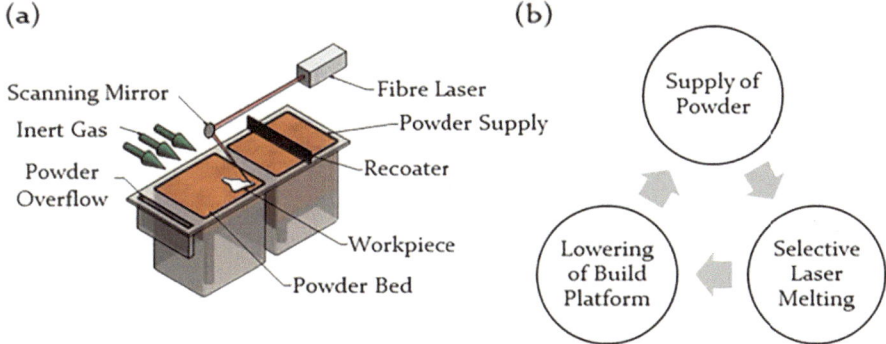

Fig. 1.6 **a** Set-up of PBF-LB/M machine and **b** iteratively repeated process steps

Process: Laser Based Powder Bed Fusion of Metals

Single-mode fiber lasers are most commonly used for PBF-LB/M since they can be focused very well onto the surface of the workpiece due to their high beam quality. Typical laser spot sizes fall in the range of 70 μm to 100 μm. These single-mode fiber lasers are characterized by output powers in the range of 400 W to 1,000 W. The surface of the powder bed is scanned with the high-power laser at high scanning speeds (vs 500 mm/s). The hatch distance between two layers depends on the average weld track width and is in the range of 80 μm to 160 μm. Established particle size distributions for PBF-LB/M range between 15 μm and 63 μm. Consequently, the average layer thickness falls between 20 μm and 100 μm, with thin layers supporting the fabrication of highly resolved components. However, a small layer thickness results in prolonged manufacturing times since the time-consuming recoating operation needs to be repeated accordingly [10].

When welding the material, two key defects need to be considered [8]. On the one hand, a lack of fusion can occur when the applied energy for welding and joining the material with the lower layer is too low. This defect can be countered by increasing e.g. the laser power. On the other hand, however, laser powers that are too high are also detrimental since gas porosity can arise. Gas porosity can be attributed to the formation of a keyhole during processing, which can collapse and lead to residual gas entrapment within the part. Other defects during processing include crack formation. This type of defect is mainly defined by the crack susceptibility of the material used. However, by adjusting the processing strategy, e.g., through tailoring of the energy input via beam shaping, the cooling conditions can be modified, and the crack sensitivity can be reduced [8].

Another aspect that needs to be considered in PBF-LB/M is associated with the high melting points of metallic materials. Due to the high melting points (e.g., steel 1,600 °C), preheating slightly below the melting point, as is done for polymers, is no longer feasible. The combination of fast heating and cooling of the part promotes the formation of internal stresses. These internal stresses lead to undesired distortion and warping of the component. To counter these effects, which could result in a failure of the build job, support structures are used to fix the component on the build plate. The mechanical connection furthermore supports the dissipation of the energy supplied by the high-power laser source. Another lever to reduce the thermal gradients and the cooling rates is achieved by heating the build plate to temperatures typically in the range of 150 °C to 250 °C [2].

The material properties of additively manufactured components in the as-built state differ from the ones of conventionally manufactured products. This difference can be attributed to the fast heating and cooling of the single weld tracks that form the final component. All these cycles affect the thermal history of the workpiece. Consequently, different regions within one component can be characterized by different microstructures. The main reasons for these differences are intrinsic effects such as in-situ tempering or annealing. However, using established heat treatment strategies, material properties comparable to those of conventional parts can be achieved. These heat treatments help to homogenize the microstructure of the additively manufactured parts [3].

1.1.4 Electron Beam Powder Bed Fusion

Electron Beam Powder Bed Fusion (PBF-EB) attracts increasing interest in research and industry. Especially the last few years have seen incredible growth in machine and process development as well as material research. In PBF-EB, components are built layer by layer within a powder bed by selectively melting the powder with a high-power electron beam. This technology is restricted to metals because electric conductivity is required. However, it shows various unique features and advantages for processing high-performance metallic materials, as described in this section. The contents of this section are collected from the review articles of Körner [7] and Fu and Körner [4].

Machine Setup

One key element of a PBF-EB machine is the electron beam gun. Electrons are generated at the cathode and are accelerated using a high voltage between the cathode and the anode. The accelerated electrons pass through a series of electromagnetic coils. Deflection coils are used to adjust the beam position, resulting in beam velocities that are up to the order of km/s. However, the beam deflection alters the circularity of the electron beam for different incident locations. Astigmatism coils recover the circularity. Finally, focus coils are used to modify the intensity distribution of the electron beam.

A second key element is the vacuum chamber beneath the electron beam column, where powder feeding, raking, and selective melting occur (see Fig. 1.7). The process operates under a technical vacuum (10^{-4} mbar to 10^{-5} mbar), which is particularly important for metals and alloys with a high affinity to gases like oxygen and nitrogen. A controlled vacuum, i.e., a small helium pressure of 10^{-3} mbar, is applied to prevent electrostatic charging. In most PBF-EB systems, the powder is delivered from

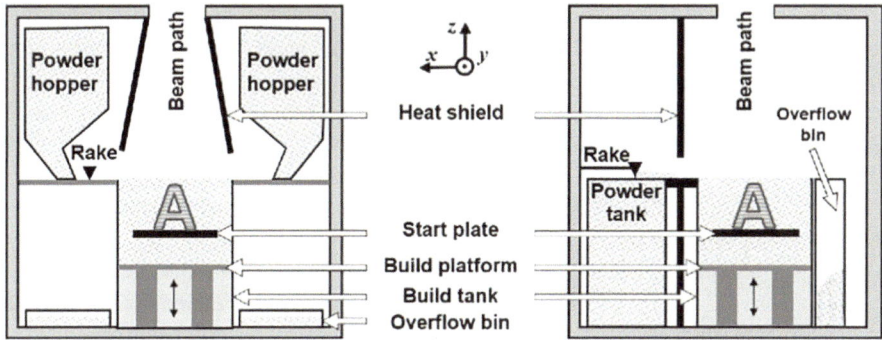

Fig. 1.7 Schematic designs of the vacuum chamber of PBF-EB systems. Adapted from Fu and Körner [4] under a CC BY 4.0 license

hoppers (see Fig. 1.7 left), requiring good powder flowability. A different approach uses an additional powder feed tank (see Fig. 1.7 right), which is mechanically lifted before powder deposition. The rake distributes the provided powder into the build tank. Excessive powder is collected inside the overflow bins. The heat shields protect the powder in its storage device from high build temperatures. The component itself is built on a start plate located inside the build tank covered by a powder bed. The build platform is mobile to allow the layer by layer manufacturing.

Process Cycle

The overall process cycle for each layer is visualized in Fig. 1.8. It consists of three manufacturing steps, namely the powder layer deposition via raking followed by preheating and melting, and two intermediate process observation steps by electron optical (ELO) imaging.

In the first manufacturing process step, a thin powder layer is applied by the rake in the build tank. The thickness of the powder layer is normally between 50 μm and 150 μm. The very first layer is applied on a preheated start plate. The second manufacturing step is the preheating step. During preheating, the applied powder is heated by scanning the electron beam several times across the layer. This step is necessary to maintain elevated build temperatures between 300 °C, e.g., for pure

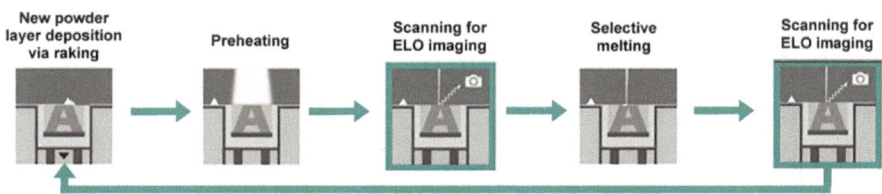

Fig. 1.8 Individual PBF-EB process steps consisting of part generation and process observation by ELO imaging. Adapted from Fu and Körner [4] under a CC BY 4.0 license

copper, and up to 1100 °C, e.g., for some nickel-base alloys. In addition, the powder particles are slightly sintered, which is essential to increase the electric conductivity of the powder to prevent process instabilities like 'smoke', where the powder gets distributed within the machine due to the repulsion of the charged powder particles. After preheating, the electron beam scans the powder layer and melts the powder particles where solid material is to form. Melting is commonly achieved by line hatching, where the direction of the beam changes in each line by 180° and also in each layer by 90°. In recent years, point melting patterns and strategies have been developed, and the high beam velocity is used to jump between individual melt spots.

In recent years, a new process monitoring method was established in PBF-EB. The process offers the possibility to measure the electrons produced as by-products of the beam-material interaction—mainly backscattered electrons—similar to a scanning electron microscopy to generate an ELO image of each layer. The signal intensity (brightness in ELO images) depends on the atomic number of the material and the surface topography. Thus, ELO imaging is well suited to detect defects and the surface topography in each layer after melting. In addition, an ELO imaging step after preheating provides information about the distributed powder layer. In comparison to other optical monitoring systems, the ELO imaging process is very robust, i.e., it is not affected by any temperature effects and is not sensitive to metal vaporization during the process.

The process cycle is embedded into pre- and post-processing tasks. The pre-processing contains the machine setup (e.g., filling of powder devices, mounting of start plate, etc.) as well as establishing the vacuum and an initial preheating to ensure the build temperature before the first process cycle. As a result of the preheating steps in each layer, the manufactured components are finally embedded within a slightly sintered powder bed. During post-processing, the powder can be removed by sand blasting using the same powder particles as the blasting abrasive. Generally, the powder can be nearly completely recycled and reused if appropriate processing conditions are applied. After sandblasting, a component is in the as-built state and can be further processed by surface or heat treatments.

References

1. ISO/ASTM 52900:2021 additive manufacturing–general principles–fundamentals and vocabulary, 2021. German version EN ISO/ASTM 52900:2021
2. Buchbinder, D., Meiners, W., Pirch, N., Wissenbach, K., Schrage, J.: Investigation on reducing distortion by preheating during manufacture of aluminum components using selective laser melting. J. Laser Appl. **26**(1), (2014)
3. Deirmina, F., Peghini, N., AlMangour, B., Grzesiak, D., Pellizzari, M.: Heat treatment and properties of a hot work tool steel fabricated by additive manufacturing. Mater. Sci. Eng. A **753**, 109–121 (2019)
4. Fu, Z., Körner, C.: Actual state-of-the-art of electron beam powder bed fusion. Eur. J. Mater. **2**(1), 54–116 (2022)

5. Greiner, S.: Bedeutung der geometrieabhängigen Belichtungstemperaturen für das Lasersintern von Kunststoffen. Ph.D. thesis, Friedrich-Alexander-Universität Erlangen-Nürnberg (2023)
6. Kruth, J.-P., Wang, X., Laoui, T., Froyen, L.: Lasers and materials in selective laser sintering. Assem. Autom. **23**(4), 357–371 (2003)
7. Körner, C.: Additive manufacturing of metallic components by selective electron beam melting—a review. Int. Mater. Rev. **61**(5), 361–377 (2016)
8. Nahr, F., Bartels, D., Rothfelder, R., Schmidt, M.: Influence of novel beam shapes on laser-based processing of high-strength aluminium alloys on the basis of EN AW-5083 single weld tracks. J. Manuf. Mater. Process. **7**(3), 93 (2023)
9. Schmid, M.: Laser Sintering with Plastics, vol. 1. Carl Hanser Verlag, München (2018)
10. Singla, A., Banerjee, M., Sharma, A., Singh, J., Bansal, A., Gupta, M., Khanna, N., Shahi, A., Goyal, D.: Selective laser melting of ti6al4v alloy: process parameters, defects and post-treatments. J. Manuf. Process. **64**, 161–187 (2021)
11. Wohlers, T.: Wohlers report 2023—3D printing and additive manufacturing 2023 (2023)
12. Wohlers, T., Campbell, I., Diegel, O., Huff, R., Kowen, J.: Wohlers report 2022 3d printing and additive manufacturing global state of the industry (2022)
13. Yadroitsev, I., Yadroitsava, I., Du Plessis, A., MacDonald, E.: Fundamentals of Laser Powder Bed Fusion of Metals. Elsevier Science (2021)

Matthias Markl completed his doctorate in 2015 in the field of simulation of electron beam powder bed fusion under the supervision of Prof. Dr.-Ing. habil. Carolin Körner. Matthias Markl then took over as head of the Numerical Simulation working group, which focuses on the simulation of metal additive manufacturing processes.

Dietmar Drummer succeeded Prof. Ernst Schmachtenberg as Head of the Institute of Polymer Technology at Friedrich-Alexander-Universität Erlangen-Nürnberg on May 1, 2009. He was the spokesperson for the Collaborative Research Center 814 Additive Manufacturing, which focused on process understanding, development, and process-adapted material characterization. Furthermore, he heads the Polymer Group of Neue Materialien Fürth GmbH and 2 Keylabs at the Bavarian Polymer Institute.

Part I
Powder Materials

Chapter 2
Production of Functional Polymer Particles via Liquid-Based Top-down Processes

Florentin Tischer, Nicolas Hesse, Wolfgang Peukert, and Jochen Schmidt

2.1 Introduction

In this chapter, top-down approaches for the development of novel functional thermoplast particle systems are discussed. The capabilities of three processes for the production of polymer powders that could act as a raw material for laser-based powder bed fusion of polymers (PBF-LB/P) feedstock production were investigated, namely cold wet comminution in stirred media mills, thermally induced precipitation, and melt emulsification. The aforementioned processes were assessed in detail.

The studied approaches for converting millimeter-sized polymer granules into functional micro-particulate polymer particle systems for PBF-LB/P via a process chain approach starting with a comminution step (left), melt emulsification (middle), and thermal precipitation (right) are outlined in Fig. 2.1.

2.2 Cold Wet Comminution in Stirred Media Mills

Comminution is one of the most important methods for the mechanical production of particles. The objectives of comminution processes are either to tailor the particle size distribution of a material for a desired application, to decompose heterogeneous substances, such as in minerals processing or recycling, or to mechanically activate a material. Comminution is carried out in mills, in which feed particles are subjected to mechanical stresses. If the stress is sufficient, cracking and subsequent breakage of the particle occurs. Mills can be classified according to the type of acting stress

F. Tischer · N. Hesse · W. Peukert · J. Schmidt (✉)
Friedrich-Alexander-Universität Erlangen-Nürnberg, Institute of Particle Technology, Cauerstr. 4, 91058 Erlangen, Germany
e-mail: jochen.schmidt@fau.de

© The Author(s) 2025

D. Drummer and M. Schmidt (eds.), *Progress in Powder Based Additive Manufacturing*, Springer Tracts in Additive Manufacturing,
https://doi.org/10.1007/978-3-031-78350-0_2

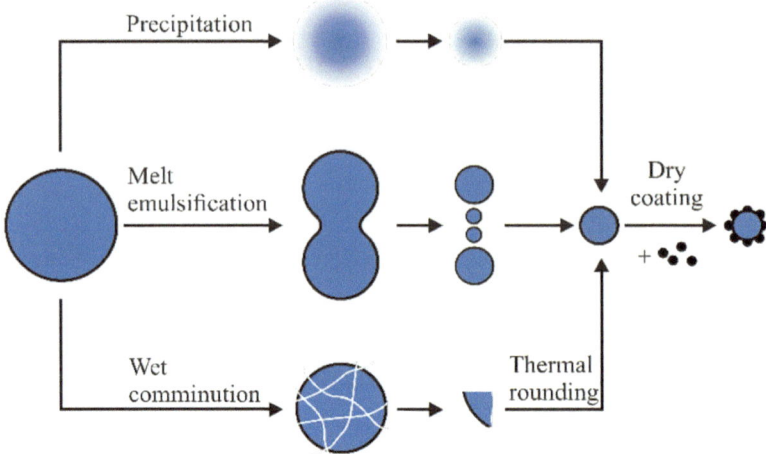

Fig. 2.1 Methods to produce spherical polymer microparticles for PBF-LB/P. Top: thermal precipitation. Middle: melt emulsification. Bottom: comminution of polymers followed by particle rounding in a heated downer reactor (see Chap. 3). Dry coating of the spherical particles obtained further improves the powder flowability

(cf. compression, impact, shear, cutting), and with respect to the stress intensity and stress frequency they provide. The visco-elastic behavior of polymers makes their comminution challenging [41]. Especially for the production of fine polymer particles, which are mandatory for the PBF process, high energy inputs and often low temperatures—for embrittlement of the thermoplast to improve its grindability—are required. Industrially, this is mostly facilitated by cryogenic impact comminution, in which liquid nitrogen or solid carbon dioxide are used to cool the polymer [4, 11, 19]. Examples of commercial PBF feedstocks produced by comminution are e.g., PA1101 (EOS) and HT-23 (EOS).

For the comminution of several thermoplastics, including polyesters, cold wet comminution in stirred media mills at reduced temperatures or even at ambient temperature was shown to be a feasible approach for producing fine polymer particles without the need for cryogenic cooling [32]. In addition, the option to produce very fine powders (<5 µm) via this process enables the production of composite particles (see Chap. 3). Cold wet comminution in stirred media mills uses grinding beads in a liquid to stress and comminute particles. In the case of polymers, the liquid is usually an organic solvent that wets the polymer well. As shown in Fig. 2.2, the energy input is realized by an agitator, which sets the grinding media in motion, resulting in collisions among the grinding beads. The stressing of particles by compression and shearing takes place through the transfer of kinetic energy from the grinding media to the feed particles in the collision zone.

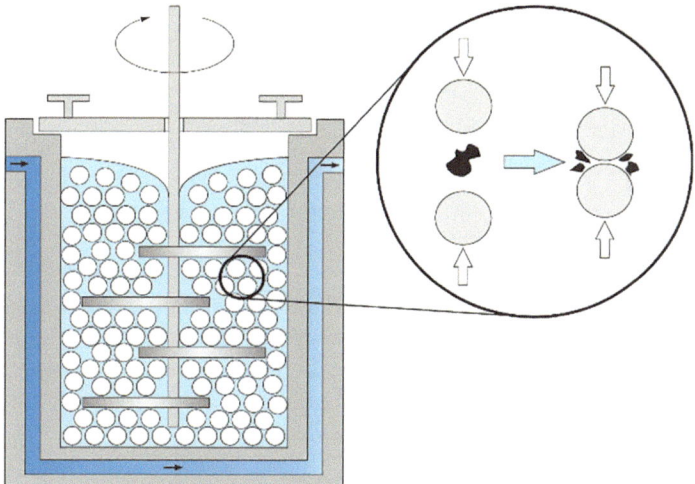

Fig. 2.2 Schematic representation of a batch stirred media mill

2.2.1   Effect of Process Parameters on Product Particle Size

According to Kwade [17, 18], the change in particle size during comminution in stirred media mills is described by the mass-specific energy input E_M, which is given by the product of the stress energy SE and stress number SN (Eq. 2.1), whereby SN describes how often stress events take place and SE describes their intensity. The stress energy transferred to a particle is proportional to the product of the maximum stress energy SE_{Max} (Eq. 2.2), an elastic factor F [3] (Eq. 2.3), and an energy transfer coefficient r_η (Eq. 2.4) [16], which can also be used to account for viscous dampening phenomena. The energy transfer coefficient is a function of the diameter d_{GM} and density ρ_{GM} of the grinding media, the agitator tip speed v_{tip}, the Young's moduli of the grinding media E_{GM} and the feed material E_{mat}, the particle size of the feed material x, and the Stokes number St_{GM} (Eq. 2.5) of the grinding media. The stress number SN depends on the filling ratio ϕ_{GM}, the porosity ϵ_{GM}, the solid's concentration c_V, the rotational speed n, the process time t, and the grinding media diameter d_{GM} (Eq. 2.6).

$$E_M = SE \cdot SN \tag{2.1}$$

$$SE \propto SE_{\text{Max}} \cdot F \cdot r_\eta = d_{GM}^3 \cdot v_{\text{tip}}^2 \cdot \rho_{GM} \cdot F \cdot r_\eta \tag{2.2}$$

$$F = \frac{E_{GM}}{E_{GM} + E_{\text{mat}}} \tag{2.3}$$

$$r_\eta = \left(1 + \ln\left(\frac{x}{d_{GM}}\right) \cdot \frac{1}{St_{GM}}\right)^2 \tag{2.4}$$

$$St_{GM} = \frac{v_{GM} \cdot d_{GM} \cdot \rho_{GM}}{9\eta} \tag{2.5}$$

$$SN \propto \frac{\phi_{GM}(1-\epsilon)}{c_V(1-\phi_{GM}(1-\epsilon))} \cdot \frac{nt}{d_{GM}} \tag{2.6}$$

The minimum achievable particle size during comminution is given by the grinding limit, which is in the order of 10 nm for crystalline inorganic materials [15]. For polymers, the extension of the plastic zone δ_C, which according to Dugdale [14] can be described with Eq. 2.7, may serve as an estimate for the grinding limit. Here, K_{Ic} is the fracture toughness, and σ_F is the yield stress.

$$\delta_C = \frac{K_{Ic}^2}{2\sigma_F E} \tag{2.7}$$

Cold wet comminution was extensively investigated as part of a process chain (Fig. 2.1) for the production of commercially not available micron-scale polymer powders used as feed material for the production of PBF feedstocks. Cold wet comminution was found to be an effective process to tailor the particle size distribution of a variety of different polymers [32]. These include polystyrene (PS), polybutylene terephthalate (PBT), polycarbonate (PC), polyoxymethylene (POM), and polyetheretherketone (PEEK), which, as shown in Table 2.1, could be comminuted to sizes well below the extent of the plastic zone according to Dugdale. The comminution took place in 650 ml batch mills in ethanol with Yttrium-stabilized zirconia grinding beads. The temperatures required to produce very fine polymer powders <5 μm are between −30 °C and −40 °C, which are significantly higher than the typical temperatures during cryogenic impact comminution of thermoplastics. To get into the size range of commercial PBT polymer powders ($x_{50.3} \sim 40 - 70$ μm), for example, it is sufficient to cool the grinding chamber to room temperature. Regarding the required cooling capacity, cold wet comminution of thermoplastics with low

Table 2.1 Expansion of the plastic zone according to Dugdale (Eq. 2.7) at 20 °C combined with experimental data of the observed minimum particle sizes $x_{50.3}$ of selected polymers during cold wet comminution at temperatures of −15 °C to −40 °C

Polymer	$K_{Ic}/MPa\sqrt{m}$	σ_F/MPa	E/MPa	$\delta_C/\mu m$	$x_{50.3\,exp.}/\mu m$
PS	0.8	55	3200	1.8	1.9
PEEK	5	100	3800	32.9	4.2
POM	4	65	2700	45.6	5.0
PBT	5	55	2500	90.9	8.0
PC	3.5	65	2500	37.7	10.0

elongation at break is significantly more effective than cryogenic comminution. The process is also feasible for polyesters, which are known to show elongations at break of >10 %. However, thermoplastics with a very high elongation at break, such as polyolefins or polyamides, cannot be processed using this approach, as the necessary elongations for fracture of the stressed particle cannot be realized in this type of mill.

In Fig. 2.3, the course of the mean particle size $x_{50.3}$ with advancing process times is shown in a double logarithmic plot, exemplified by the PS comminution [32]. Here, different stress energies were adjusted by varying the agitator speed v_{tip} (a), respectively, the diameter of the grinding media d_{GM} (b). A linear decrease in particle size is clearly visible until the grinding limit is reached. At higher stress energies, the decline of the particle size is steeper, whereby the grinding limit is reached earlier.

Thus, the course of the evolution of the particle size distribution can also be plotted against the specific energy input to describe the comminution kinetics, whereby the different temporal progressions for different stress energies are described by one function. Therefore, to achieve the desired particle size, only the appropriate energy input has to be selected. This is shown using PBT as an example in Fig. 2.4a. Figure 2.4b portrays the required mass-specific energies of different substances together with the comminution law according to Bond and Rittinger [41].

It can be seen that higher mass-specific energies are required for cold wet comminution of polymers (<10 μm) compared to cryogenic comminution of thermoplasts yielding typical product particle sizes in the range of several 10 to 1000 microns, i.e., the wet comminution approach allows for the production of very fine thermoplast particles. In addition, Fig. 2.4b does not take the required cooling capacity into account, which is orders of magnitude higher in the case of cryogenic comminution. Apart from the particle size distribution, the PBF process has further requirements concerning the polymer powder in terms of its bulk properties, such

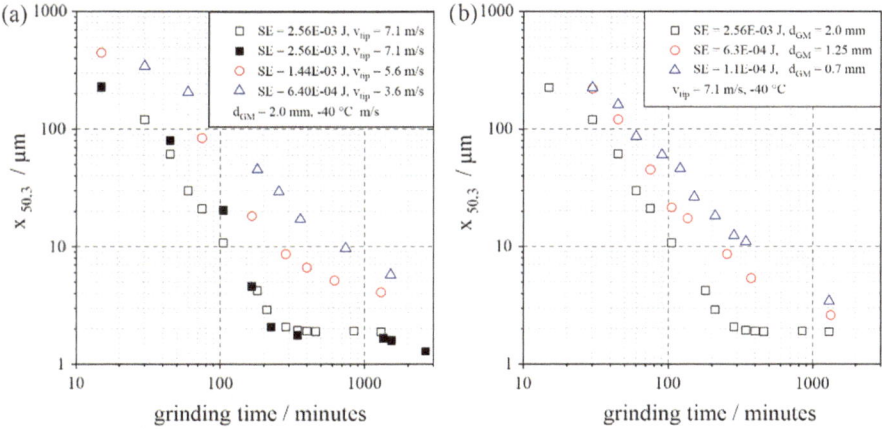

Fig. 2.3 Mean particle size $x_{50.3}$ of cold wet comminuted PS as a function of time for different **a** stirring speeds, and **b** grinding media diameters (reproduced from [32])

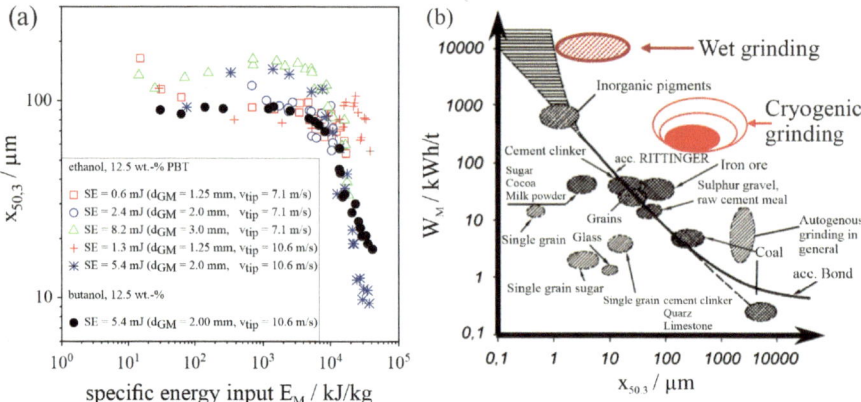

Fig. 2.4 **a** Mean particle size $x_{50.3}$ of cold wet comminuted PBT at different stress energies as a function of mass specific energy input [33]; **b** energy required to achieve the desired mean particle sizes for different materials. The areas of cold wet comminution and cryogenic comminution of polymers are marked in red (reproduced from [33, 41])

as flowability, and intrinsic properties, such as crystallization kinetics, which will be discussed in the following.

2.2.2 Particle Shape of Wet Comminuted Polymers

The particle shape is a parameter that significantly influences the flowability of powders. To be suitable for the PBF process, the powder must have good flowability at temperatures within the sintering window [12, 28, 31] in order to obtain a powder bed that is as homogeneous as possible [5]. The smaller the contact area between the particles, the lower the adhesive forces, which reduces flowability. Therefore, round particles display better flow properties than irregularly shaped particles and, thus, are to be favored for the PBF process [42].

During the comminution of polymers, apart from the particle size reduction, particles are always reshaped. Due to the brittle-to-ductile transition of polymers, polymer particles become more ductile with decreasing particle size. In the case of stirred media mills, the shape of the particles approaches that of a plate, the closer the particle size comes to the grinding limit [32], as the small particles become more ductile and thus can no longer fracture, but rather reshape. There are several options to obtain powders of suitable shape and flowability by exploiting cold wet comminution. For instance, a promising method was developed and studied in which highly spherical polypropylene (PP), PBT/PC, or PBT/glass powders with a suitable size range for the PBF process were produced using a combination of cold wet comminution and thermal rounding or spray drying [30, 32, 33]. More details on the latter processes are provided in Chap. 3.

Another possibility to tune product properties is the separation of particles during the comminution process, which have reached a particle size of 50 μm and are only slightly deformed. This is part of ongoing work to set up and characterize a wet grinding and classification circuit in which the fine particles are continuously separated using a combination of hydrocyclones and filters.

2.2.3   Crystallization Behavior of Wet Comminuted Polymers

The crystallization and solidification of the plastic component plays an important role in the PBF process. Excessively rapid crystallization kinetics can lead to component distortion (curling) during the process which, in the worst case, can lead to the process being aborted. In general, slow crystallization kinetics and a large sintering window are advantageous [29].

In the case of cold wet comminution in stirred media mills, not only the polymer but also the grinding media and the grinding chamber are stressed, resulting in micro- and nano-fine abrasion of the grinding media and the grinding chamber material. In order to quantify the extent of abrasion and to investigate its influence on the intrinsic properties of the thermoplastic, such as its crystallization, comminution experiments were carried out with different grinding media materials [39]. For this purpose, PBT was comminuted with glass, chromium steel, cerium-stabilized zirconia, and yttrium-stabilized zirconia grinding media in ethanol for 15 h at a stress energy of 0.9 mJ. To prevent cross-contamination, separate grinding chambers were made of PTFE for each grinding media material and new virgin grinding media were used for each test. The analysis of the powders showed that there is no change in the crystalline structure or the molecular weights of the PBT due to the cold wet comminution [39]. However, the observed changes in the reflectance of the PBT powder and the dynamic and isothermal crystallization behavior could be attributed to the abrasion of the grinding media.

Of the materials tested, the SiO_2 and the ZrO_2 (Y) grinding media were found to be the most suitable for the production of powders for the PBF process in terms of the intrinsic properties of the powders. These grinding media showed little to no change in reflectance of the comminution product and the least altered sinter window and crystallization kinetics compared to the feed. Due to the higher density and the resulting faster comminution kinetics combined with the lower abrasion mass content (Fig. 2.5a), the ZrO_2 (Y) grinding media proved to be optimal compared to the untested grinding media.

Chromium steel grinding media are not suitable for the production of polymer powders to be used as feed material for making PBF feedstocks, as the comminution product is characterized by a remarkable discoloration, substantially increased crystallization kinetics, and a small sintering window (Fig. 2.5b).

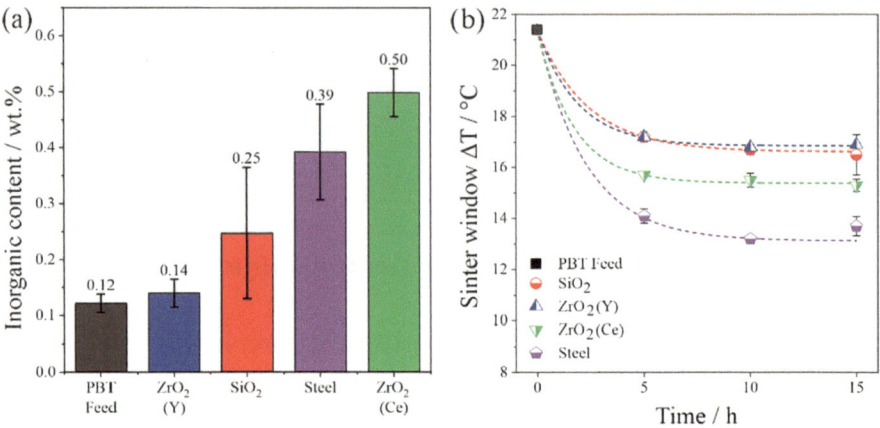

Fig. 2.5 **a** Grinding media abrasion mass content in a PBT powder after 15 h cold wet comminution in ethanol at a stress energy SE = 0.9 mJ; **b** Sinter window of the PBT feed material and cold wet comminuted PBT with different grinding media (reproduced from [5], CC-BY 4.0)

2.3 Thermally Induced Liquid-Liquid Phase Separation and Precipitation

Precipitation was studied as another process variant capable of generating polymer micro particles. The precipitation process is based on the phase behavior of polymers in so-called moderate solvents, i.e., liquid phases which only act as solvents for the polymer at elevated temperatures. As shown in Fig. 2.6a, and depending on the polymer concentration, a miscibility gap can develop upon cooling the polymer-solvent system. Liquid-liquid phase separation (LLPS) occurs, leading to the formation of droplets of high polymer concentration in a solvent-rich matrix [6].

Upon further cooling, the polymer crystallizes in these droplets, resulting in particle growth and precipitation of polymer particles. For our application, it is important to cool the system down into the metastable region below the binodal 'left' of the critical point to generate a viable particulate product. Interestingly, although thermal precipitation is applied for the production of commercial polyamide powder systems—which in addition to PBF-LB/P is also employed in coating applications—until recently the reports on this process for the production of powder systems were virtually limited to patent literature [2, 22, 23, 25].

Phase diagrams as shown in a simplified representation in Fig. 2.6a are not readily available and an optically accessible high-temperature, high-pressure cell was used to gain experimental insights into the phase behavior of a multitude of polymer/solvent systems investigated. In this device, the full dissolution of a polymer can be observed when the temperature is increased. Since the solubility temperature is frequently below the melting point, melt emulsification is not possible for these polymer-solvent systems. During cooling, the precipitation of polymer particles from the previously

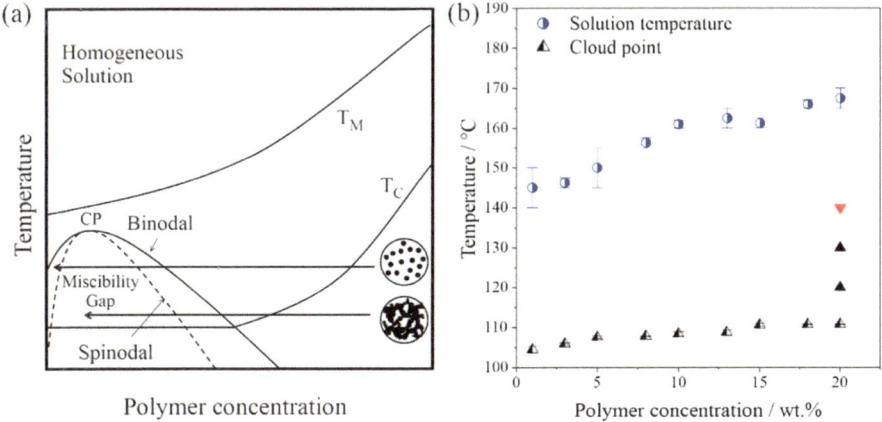

Fig. 2.6 a Simplified phase diagram for a binary system polymer-solvent (T_S: solution temperature, T_C: crystallization temperature, CP: critical point). Adapted from [26–29]. **b** Cloud point and solution temperature curves for the PA11-ethanol system. Cloud points (semi-filled triangles) were measured under continuous cooling (1.5 ... 2 K/min). Isothermal cloud points are given as filled triangle symbols. For the system containing 20 wt.-% polymer (upside-down triangle, T = 140 °C) no isothermal precipitation could be observed. Error bars depict the reproducibility (reproduced from [6])

homogeneous (and translucent) solution can be observed through the clouding of the polymer-solvent mixture. Corresponding experiments for a system PA11/ethanol are exemplarily depicted in Fig. 2.6b. The thermal precipitation was demonstrated as a viable approach for the production of micron-scale powder from engineering thermoplastics, for which melt emulsification proved to be very challenging.

To study the influence of the feed concentration on product particle size, experiments were performed in 200 ml laboratory autoclaves that were kept at 190 °C while being stirred with magnetic stirring bars (20 mm, ø 8 mm) at 600 rpm for 15 min to ensure full dissolution. Afterwards, the stirring speed was set to 100 rpm and the autoclaves were cooled at a rate of 0.7 K/min in the cloud point range (110 °C ± 10 °C, as shown in Fig. 2.6b). Figure 2.7a portrays the volume-weighted size distribution of the particles obtained after precipitation, filtration, and drying. The results indicate that mean diameters ($x_{50.3}$) of 30 µm, 83 µm, 75 µm, and 134 µm could be observed for polymer concentrations of 5 wt.-%, 10 wt.-%, 15 wt.-%, and 20 wt.-%, respectively. The span defined as ($x_{90.3} - x_{10.3}$) / $x_{50.3}$, was 1.86 (5 wt.-% polymer), 1.70 (10 wt.-%), 1.80 (15 wt.-%), and 1.34 (20 wt.-%), respectively. As shown by Matsuyama et al. [20] for dilute PP-diphenyl ether systems, larger particles are generated for higher polymer volume fractions due to the growth of the polymer-rich droplets formed during LLPS being dominated by coalescence.

Figure 2.7b shows the particle size distributions of PA11 particles precipitated in ethanol at different stirring conditions (100 rpm ... 1200 rpm), whereby it is evident that for increased stirring speeds the particle size distributions are shifted towards smaller particles. The exact mechanisms leading to the formation of particles of a

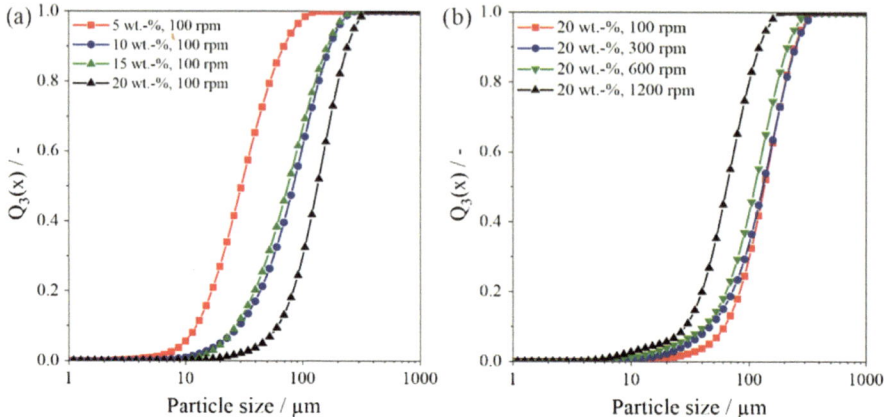

Fig. 2.7 Cumulative size distributions Q_3 of PA11 particles precipitated from ethanol under continuous cooling: **a** influence of initial feed polymer mass concentration, and **b** influence of stirring speed (reproduced from [6])

given size at varying stirrer speeds are complex. For instance, for poly(L-lactic acid) (PLLA) it is known that two opposing phenomena govern the phase separation and precipitation process [26]. On the one hand, droplet breakup is caused by increased stirring speeds due to higher shear stresses, hence, smaller particles are generated. This effect has previously been observed for phase inversion emulsification and thermally induced phase separation [10] where the droplet breakup is governed by the droplet size, and the shear stresses acting on it [35]. On the other hand, larger droplets, and, as a result larger particles, can be obtained due to coarsening effects such as coalescence or Ostwald ripening, which dependent on system characteristics such as mixing and droplet size distribution [8, 35]. For the precipitation of polypropylene particles, these coarsening effects were dominant [20].

In our experiments depicted in Fig. 2.7b, droplet breakup at increased stirrer speeds seems to be the dominant phenomenon. The particle size distribution shows a clear shift towards finer particles at 1200 rpm. However, for speeds between 100 and 600 rpm, only a minor effect of the mixing conditions can be observed. This could either allude to the competitive interplay between droplet breakup and coarsening mechanisms or indicate that a threshold for droplet breakup exists [6]. Furthermore, the torque provided by the experimental setup with magnetic stirring bars might be insufficient for systems with high volume fractions of polymers.

The different morphologies of particles precipitated under varying flow conditions in the reactor can also be seen in the SEM images shown in Fig. 2.8. Particles precipitated in smaller lab-scale autoclaves stirred with magnetic stirring bars tend to form aggregate structures with more and smaller primary particles while the well-defined flow conditions within the stirred tank autoclave lead to larger primary particles that are less aggregated. While the Reynolds number at the tip of the magnetic stirring bar during processing of the particles shown in the left image is roughly 8300, a value of

Fig. 2.8 SEM images of PA11 particles obtained by LLPS and precipitation in **a** small lab-scale autoclaves (200 ml, 20 wt.-% polymer, magnetic stirring bar at 600 rpm), and **b** a stirred autoclave (3000 ml, 20 wt.-% polymer, anchor stirrer at 350 rpm)

97700 can be reached for the flow agitated with an anchor stirrer within the stirred autoclave. Furthermore, due to the low torque transferred by the magnetic stirring bar, even lower effective Reynolds numbers might be realized for processing in small autoclaves. Hence, further investigations with the polymer/solvent system shown in this chapter were performed in a three-liter autoclave equipped with an anchor stirrer for more reliable and well-defined mixing conditions.

The molecular mass of a polymer used for PBF-LB/P influences the properties of additively manufactured parts in several ways. Mechanical part properties such as the Young's modulus, yield strength, and especially the elongation at break and impact strength can be impacted by a change in the material's molecular mass. Furthermore, it correlates with the polymer's sintering window and melt viscosity, which have a considerable influence on the processing properties. A low molecular weight and, hence, a low melt viscosity can lead to infiltration of the melt into the powder bed and poor shape retention in the build process, whereas a high molecular weight and melt viscosity can be detrimental due to poor coalescence of subsequent layers. This often leads to poor layer adhesion or the orange peel effect. The molecular mass distribution of PA11 powder precipitated from ethanol was quantified using gel permeation chromatography (GPC) and is shown in Fig. 2.9. In contrast to the feed material with a mean molecular weight (mass average, M_W) of (29.9 ± 0.2) kDa, the precipitated material exhibits a reduction of roughly 20 % to a value of (23.9 ± 0.1) kDa. This degradation, as well as a likely depletion of stabilizing agents during the precipitation process in ethanol, has to be accounted for during the production of a commercial-grade PBF-LB/P powder. The case of the commercial benchmark PA1101 material provided by EOS GmbH shows that a higher molecular weight of (37.5 ± 0.6) kDa can be achieved for a production via LLPS and precipitation, if necessary.

Additionally, the in-situ functionalization of PA11 with nano-sized additives during precipitation was investigated. Whereas mixing, blending, or coating of polymer powders with additives can result in inhomogeneous part properties due to demixing

Fig. 2.9 Molecular weight distributions (absolute values based on GPC light scattering) of Rilsan PA11 (Arkema) feed material, the precipitated powder, and PA1101 reference material (EOS GmbH). 1, 1, 1, 3, 3, 3-Hexafluoroisopropanol (HFIP) was used as an eluent in the GPC measurements (reproduced from [6])

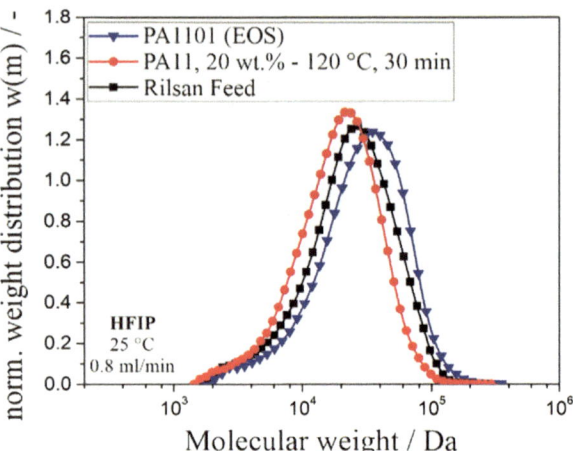

during feedstock handling or powder layer application, using a nanocomposite with additives embedded within a polymer matrix ensures homogeneity. Additives can be selected to tune specific properties, e.g., laser absorption, electrical or thermal conductivity [1], flame retardancy [24], hardness [27], friction and wear resistance [13], color, and many more. The influence of the presence of additives during LLPS and precipitation has been studied in detail for PA11-Al_2O_3/TiO_2/SiO_2 nanocomposite materials [38].

Figure 2.10a shows the particle size distributions of PA11 nanocomposites precipitated with varying mass fractions of Al_2O_3. The shift of the particle size distribution towards smaller values can be explained by the nucleating effect of the additives during precipitation. Due to the presence of heterogeneous crystallization nuclei, the crystallization of the thermoplast within the droplets formed by LLPS starts at a higher temperature. This gives the droplets less time to coalesce while the precipitation process is still ongoing, resulting in smaller particles. As shown in Fig. 2.10b, the total available nucleating surface area is decisive for the particle size of the precipitated powder. Data from specific surface area measurements (BET method) of the additives and TGA measurements of the additive weight content within the precipitated PA11 powder, combined with its mean particle size, reveal an exponential relation between the particle size and the amount of nucleating additive surface. When comparing different additives, their chemical composition seems to play a minor role in comparison to the influence of the surface area available for heterogeneous nucleation. However, a small and statistically significant material influence does exist [38].

Since melt crystallization is accompanied by a volumetric change and possibly warpage, slower crystallization kinetics are advantageous for good shape retention. Therefore, the crystallization kinetics of the additive-enhanced PA11 powders were analyzed by half crystallization times $t_{1/2}$ from isothermal DSC measurements according to the pattern shown in Fig. 2.11a. The half crystallization times for PA12

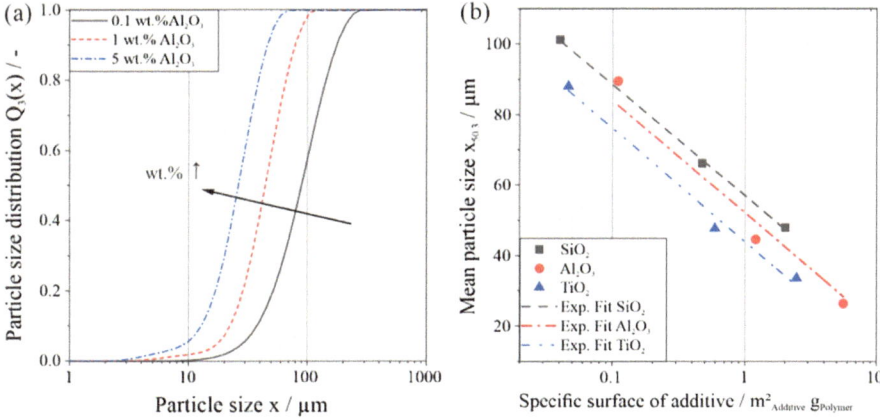

Fig. 2.10 Particle size distribution of PA11 composite powders with Al_2O_3, whereby **a** is the mean particle size $x_{50.3}$ plotted against the surface of the additive per gram polymer for PA11 composites with SiO_2, Al_2O_3 and TiO_2 (**b**) (reproduced from [38])

powders with varying mass fractions of SiO_2, Al_2O_3, and TiO_2 depicted in Fig. 2.11b indicate that the additives accelerate the crystallization by providing heterogeneous crystallization nuclei and reducing the nucleation energy barrier in the polymer melt. Accordingly, the half crystallization time curves are shifted towards lower times or increased temperatures and, hence, less undercooling, allowing for a crystallization at temperatures where pristine PA11 melt does not form sufficient homogeneous nuclei for solidification within the observed time frame. For Al_2O_3, this shift is pronounced, especially at higher additive concentrations. In contrast to nucleation during precipitation, a strong influence of additive material on the melt crystallization can be observed.

Single layers with additive-enhanced PA11 powders at varying laser energy densities and tensile rods made from pristine PA11 (see [6]) were produced. Binarized images of single layer samples obtained at an energy density of 0.4 mJ/mm^2 yielded the best results and are shown in Fig. 2.12a for reference. It is possible to build layers within a range of 0.3 to 0.67 mJ/mm^2, however, low energy densities did not allow for the generation of dense layers while the balling effect and shrinkage occurred at higher densities. As is to be expected from isothermal DSC data, warpage is most pronounced at high additive concentrations.

Furthermore, Fig. 2.12b shows a gyroid test specimen made from PLLA powder obtained from LLPS and precipitation in triacetin. The thermal precipitation of polymers from moderate solvents for the production of microparticles is applicable to various thermoplastics. Due to brevity, an in-depth presentation cannot be made in this chapter and for further information on the process, the reader is referred to [6, 7].

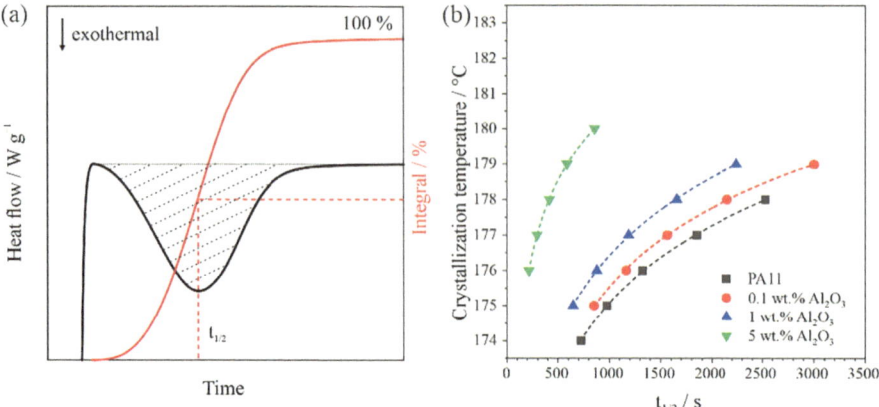

Fig. 2.11 Schematic determination of the half crystallization time of isothermal DSC measurements (**a**), and half crystallization time plotted vs. isothermal temperature TC for Al_2O_3 (**b**) composite powders (reproduced from [38])

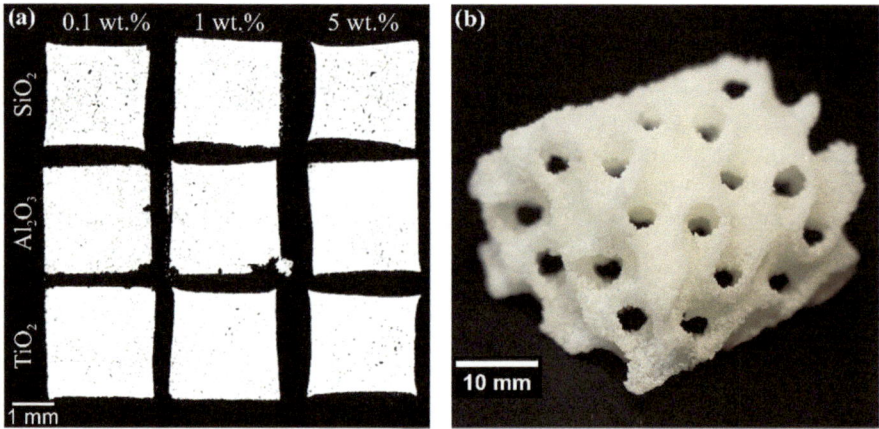

Fig. 2.12 **a** Single layers for PA11, samples, and different material systems for 0.4 mJ/mm^2, and **b** a detailed image of a PLLA gyroid test specimen (reproduced from [7, 38])

2.4 Melt Emulsification of Waxes and Polymers

The third top-down process investigated for the production of functional particle systems is melt emulsification [8, 9, 40]. The advantage of melt emulsification over other processes, such as comminution, lies in the production of highly spherical particles within a single process step, which can be advantageous for the PBF process in terms of powder flowability [5, 42]. In melt emulsification of waxes or polymers, two immiscible substances are used, usually water as the hydrophilic outer phase and the melt of a thermoplastic as the hydrophobic inner phase [36]. By increasing the temperature above the melting point of the polymer, the polymer liquefies and its melt is emulsified in the outer water phase through energy input. The energy input for droplet break-up can be realized by rotor stator devices as shown in our work, and also by high pressure nozzles or ultrasonic treatment [8, 21]. To stabilize the new micro and nanofine droplets against coalescence, the use of surfactants is mandatory [35]. Once the emulsion medium has cooled down, the polymer droplets solidify and the emulsion transforms into a dispersion of highly spherical polymer particles in water.

The melt emulsification of thermoplastics proves to be challenging [35] as the energy input must be high enough to emulsify the highly viscous melt while the surfactant must be potent enough to stabilize it. Especially the selection of suitable surfactants is challenging, as studies at room temperature do not allow conclusions to be drawn about their behavior at process temperatures. In general, polymers with a lower viscosity (low melting temperature) proved to be easier to emulsify than high-viscosity polymers. These challenges could be the reason why the available scientific literature on melt emulsification is very limited and reports on this process are mostly restricted to patents. In the following, our work on the melt emulsification of polyethylene wax, polyester waxes, and polycaprolactone (PCL) is discussed [8, 9, 34].

2.4.1 Melt Emulsification of Waxes in Aqueous Systems

The melt emulsification using water and paraffin wax, PE, or partially oxidized PE waxes stabilized with polyoxyethylene sorbitan trioleate (Tween 85), or polyester waxes stabilized with sodium dodecyl sulphate (SDS) was carried out in the melt emulsification setup shown schematically in Fig. 2.13. Water, polymer wax, and surfactant are introduced into tank T1, heated above the melting temperature of the polymer, and a pre-emulsion is generated by stirring. By opening the valves V1 and V2, the emulsion is formed by means of a rotor/stator mechanism. As soon as the desired process time is reached, the emulsion is transferred to the tank T2 by opening valve V3 and closing valve V2, where the emulsion cools down and the polymer solidifies. Using this setup, round particles can be produced, with the resulting particle size being controlled by the selected energy input and surfactant concentration.

The influence of the surfactant concentration and the energy input is shown in Figs. 2.14 and 2.15, respectively, for the melt emulsification of 10 wt.-% paraffin wax in water at a temperature of 70 °C and different surfactant concentrations, rotor stator speeds, and process times. Clearly visible in Fig. 2.14a is that with increasing surfactant concentration of 0.1−1 wt.-% the particle size distribution of the particles shifts to smaller sizes. Since the use of larger amounts of surfactant allows a larger interfacial area to be stabilized, smaller droplets, are stabilized. Here, the experimentally determined Sauter diameter $x_{1,2}$ of the particle is slightly above the size calculated by Eq. 2.8, which can be used to estimate the minimum achievable size

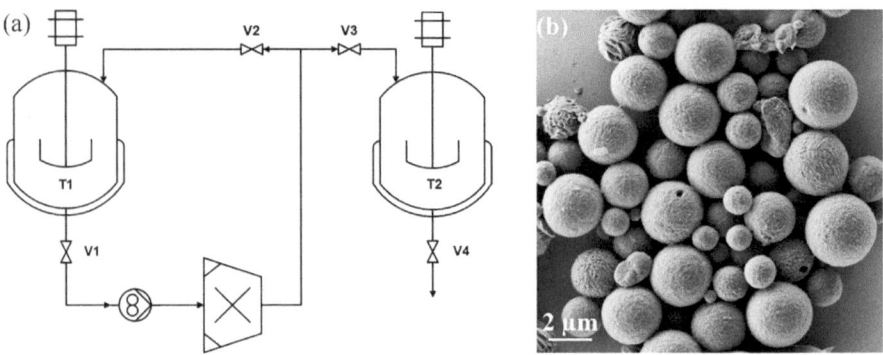

Fig. 2.13 Schematic layout of the melt emulsification setup (**a**), with a detailed image of the emulsified polyethylene particles (**b**) (reproduced from [9])

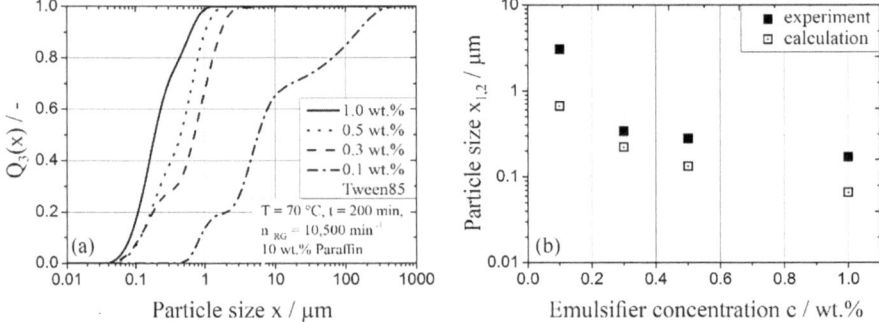

Fig. 2.14 Dependence of particle size distribution on surfactant concentration. Volume sum distribution Q₃ of emulsified paraffin particles at different surfactant concentrations (**a**); and Sauter diameter of emulsified paraffin particles at different surfactant concentrations (**b**) (reproduced from [9])

$x_{1,2\,min}$ in dependence of the emulsifier concentration c (g/L), the dispersed phase fraction ϕ, the area occupied by an emulsifier molecule at the droplet interface Ae (m^2), the Avogadro number NA, the molar mass of the emulsifier Me (g/mol), and the critical micelle concentration c_{cmc}. In this context, it is assumed that no coalescence takes place and that droplets are stable if they are covered by a monolayer surfactant.

$$x_{1,2\,min} = \left(\frac{6\,\phi}{1000\,A_e\,N_A} \right) \left(\frac{M_e}{\frac{c}{M_e} - c_{cmc}\,(1-\phi)} \right) \tag{2.8}$$

The energy input for the droplet break-up can be varied by changing the rotational speed of the rotor and the process time. In this case, the Sauter diameter $x_{1,2}$ can be described as a function of the energy input according to Eq. 2.9 [35, 37], whereby P is the power input, $\dot{V}$ the flow rate, and f and b are constants. Concerning the

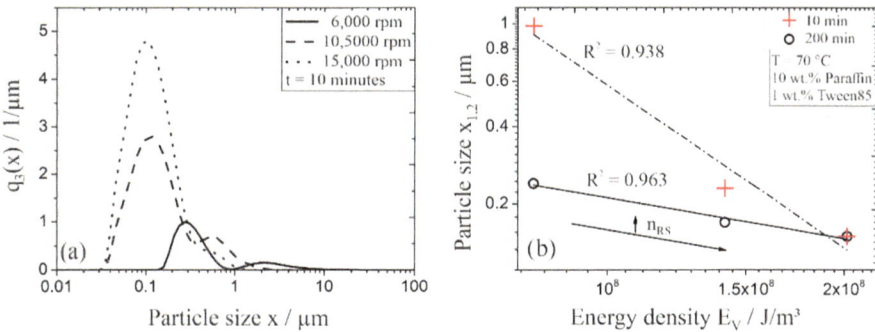

Fig. 2.15 Dependence of particle size distribution on energy input. Volume density distribution Q_3 of emulsified paraffin particles for different rotor speeds (**a**); and Sauter diameter of emulsified paraffin particles at energy densities E_V (**b**) (reproduced from [9])

fixed process time, it can be seen that by selecting higher speeds/energy inputs or a fixed rotor speed, the particle size decreases as the process time progresses until a plateau is reached. The reason for this is that the higher energy input produces smaller droplets, which results in a shift of the particle size distribution to smaller sizes until a plateau is reached at which the existing surfactant is no longer sufficient to stabilize the newly formed surface.

$$x_{1,2} = f E_V^{-b} = f \left(\frac{P}{\bar{V}} \right)^{-b} \tag{2.9}$$

Based on the data shown in Figs. 2.14 and 2.15, two possible routes for the production of particle systems for the PBF process emerge. Either fine particles are produced by high energy input, which are assembled into spherical microparticles in a bottom-up approach using spray drying of dispersions or agglomeration in fluidized beds (see Chap. 3). Alternatively, by using lower energy inputs and/or higher viscosity polymers, particles in a suitable range for the PBF process can be produced in a simple one-step process. The challenge here is to find a suitable polymer/surfactant combination and obtain a narrow particle size distribution despite low energy inputs. For the second process route, Fig. 2.16 shows the emulsification of higher viscosity PE and PCL. It is evident that with increasing viscosity of the polymer, and thus of the melt (at process temperature), the particle size distribution shifts to larger sizes compared to lower viscous waxes. Highly spherical particles in a suitable size range for PBF can be produced with this method for PP and PCL.

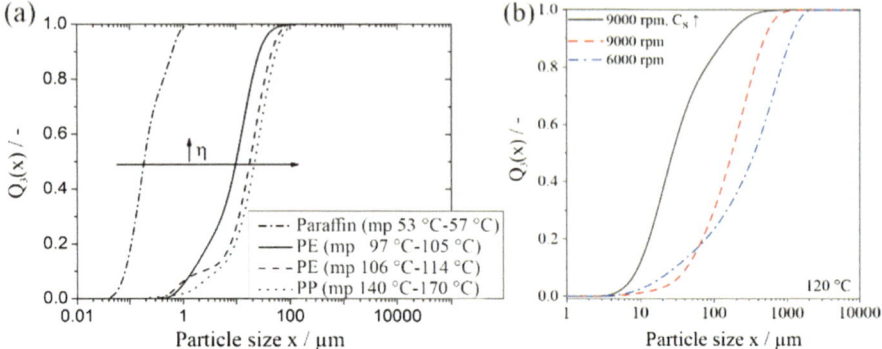

Fig. 2.16 Dependence of particle size distribution on polymer viscosity/melting temperature. Volume sum distribution Q_3 of emulsified systems with different viscosities/melting temperatures (**a**); and Volume sum distribution Q_3 of emulsified PCL at different process parameters (**b**) (reproduced from [9])

2.5 Conclusions

In this contribution, top-down approaches for producing polymer microparticles were discussed. Comminution of polymers in stirred media mills is a versatile method for the production of microparticles from a wide range of plastics, whereby the sole constraint is the crushability of the material. However, the shape and flowability of comminuted polymer powders is inappropriate for direct application of these materials in PBF-LB/P. The products typically are made up of irregular particles, which are characterized by cohesive material behavior, i.e., poor flowability and low packing fraction. The bulk solid properties of these comminution products can be improved, for example, by thermal rounding, which was successfully demonstrated for the production of spherical PBT or PBT/PC blend powders. Melt emulsification does not require any subsequent shaping process for the production of spherical particles, i.e., the product only needs to be separated and dried to obtain a powder made up of spherical particles. Moreover, this process—in addition to thermal precipitation—is an alternative for the size reduction of thermoplastics that are hard (or impossible) to comminute to the required size range due to their large elongation at break. However, the melt emulsification process has proven to be challenging due to the need for the stabilization of high-viscosity polymer melts in an appropriate continuous liquid phase. For low-melting polymers of low molecular weight, such as PCL or PE, spherical particles in an interesting size range for the PBF-LB/P process can be produced from aqueous systems, although for the melt emulsification of (most) engineering thermoplastics, the identification of proper solvent systems remains an open challenge for future research. Thermal precipitation of polymers was shown to be an excellent method for the production of good flowing polymer and polymer composite powders with a high crystallinity. The selection method developed for moderate solvents based on solubility parameters allows for a quick screening to identify promising polymer-solvent combinations. In summary, the three top-down processes discussed in this chapter were shown to be viable approaches to the development of novel thermoplastic micro powders, which at least showed initial processability in PBF-LB/P.

References

1. Athreya, S.R., Kalaitzidou, K., Das, S.: Processing and characterization of a carbon black-filled electrically conductive nylon-12 nanocomposite produced by selective laser sintering. Mater. Sci. Eng. A **527**(10–11), 2637–2642 (2010)
2. Baumann, F.E., Wilczok, N.: Preparation of precipitated polyamide powders with a narrow grain size distribution and low porosity (1998)
3. Becker, M., Kwade, A., Schwedes, J.: Stress intensity in stirred media mills and its effect on specific energy requirement. Int. J. Miner. Process. **61**(3), 189–208 (2001)
4. Bruyère, D., Simon, S., Haas, H., Conte, T., Menad, N.-E.: Cryogenic ball milling: a key for elemental analysis of plastic-rich automotive shedder residue. Powder Technol. **294**, 454–462 (2016)
5. Chatham, C.A., Long, T.E., Williams, C.B.: A review of the process physics and material screening methods for polymer powder bed fusion additive manufacturing. Prog. Polym. Sci. **93**, 68–95 (2019)
6. Dechet, M.A., et al.: Production of polyamide 11 microparticles for additive manufacturing by liquid-liquid phase separation and precipitation. Chem. Eng. Sci. **197**, 11–25 (2019)
7. Dechet, M.A., et al.: Development of poly(l-lactide) (PLLA) microspheres precipitated from triacetin for application in powder bed fusion of polymers. Addit. Manuf. **32**, 100966 (2020)
8. Fanselow, S., Emamjomeh, S.E., Wirth, K.-E., Schmidt, J., Peukert, W.: Production of spherical wax and polyolefin microparticles by melt emulsification for additive manufacturing. Chem. Eng. Sci. **141**, 282–292 (2016)
9. Fanselow, S., Schmidt, J., Wirth, K.-E., Peukert, W.: Production of micron-sized polymer particles for additive manufacturing by melt emulsification. In: Conference Proceedings, pp. 140007 (2016)
10. Feng, W., et al.: Synthesis and characterization of nanofibrous hollow microspheres with tunable size and morphology via thermally induced phase separation technique. RSC Adv. **5**(76), 61580–61585 (2015)
11. S.-u. E. Fraunhofer-Institut für Umwelt. Verdichtetes Kohlendioxid als Prozessadditiv zur Herstellung polymerer und mikronisierter Nanokomposite: Schlussbericht "nanocrosser"; Förderzeitraum: 01.01.2006–31.12.2008. Technical report (2009)
12. Hesse, N., Winzer, B., Peukert, W., Schmidt, J.: Towards a generally applicable methodology for the characterization of particle properties relevant to processing in powder bed fusion of polymers–from single particle to bulk solid behavior. Addit. Manuf. **41**, (2021)
13. Ji, Q.L., Zhang, M.Q., Rong, M.Z., Wetzel, B., Friedrich, K.: Friction and wear of epoxy composites containing surface modified SiC nanoparticles. Tribol. Lett. **20**(2), 115–123 (2005)
14. Kerkhof, F.: Anwendung der Bruchmechanik auf Hochpolymere. Kolloid-Z.u.Z.Polymere **251**(8), 545–553 (1973)
15. Knieke, C., Sommer, M., Peukert, W.: Identifying the apparent and true grinding limit. Powder Technol. **195**(1), 25–30 (2009)
16. Knieke, C., Steinborn, C., Romeis, S., Peukert, W., Breitung-Faes, S., Kwade, A.: Nanoparticle production with stirred-media mills: opportunities and limits. Chem. Eng. Technol. **33**(9), 1401–1411 (2010)

17. Kwade, A.: Determination of the most important grinding mechanism in stirred media mills by calculating stress intensity and stress number. Powder Technol. **105**(1–3), 382–388 (1999)
18. Kwade, A.: Wet comminution in stirred media mills–research and its practical application. Powder Technol. **105**(1–3), 14–20 (1999)
19. Liang, S.B., Hu, D.P., Zhu, C., Yu, A.B.: Production of fine polymer powder under cryogenic conditions. Chem. Eng. Technol. **25**(4), 401–405 (2002)
20. Matsuyama, H., Teramoto, M., Kuwana, M., Kitamura, Y.: Formation of polypropylene particles via thermally induced phase separation. Polymer **41**(24), 8673–8679 (2000)
21. Mehnert, W., Mäder, K.: Solid lipid nanoparticles: production, characterization and applications. Adv. Drug Deliv. Rev. **47**(2–3), 165–196 (2001)
22. K. R. D. I. . H. Meyer, K. H. Hornung, R. Feldmann, and H. J. D. I. Smigerski. Verfahren zur Herstellung von Pulverfoermigen Beschichtungsmitteln auf der Basis von Polyamiden mit mindestens 10 aliphatisch gebundenen Kohlenstoffatomen pro Carbonamidgruppe, Apr 17 1980
23. Monsheimer, S., et al.: Pulver enthaltend mit Polymer beschichtete Partikel (2013)
24. Morgan, A.B., Wilkie, C.A.: Flame Retardant Polymer Nanocomposites. Wiley-Interscience, Hoboken, N.J. (2010)
25. Mumcu, S., Winzer, H.: Process for the preparation of powdery polyamide coating compositions (1986)
26. Nie, T., He, M., Ge, M., Xu, J., Ma, H.: Fabrication and structural regulation of PLLA porous microspheres via phase inversion emulsion and thermally induced phase separation techniques. J. Appl. Polym. Sci. **134**(22), (2017)
27. Petrovi, Z.S., Javni, I., Waddon, A., Bnhegyi, G.: Structure and properties of polyurethane-silica nanocomposites. J. Appl. Polym. Sci. **76**(2), 133–151 (2000)
28. Ruggi, D., Barrès, C., Charmeau, J.-Y., Fulchiron, R., Barletta, D., Poletto, M.: A quantitative approach to assess high temperature flow properties of a PA 12 powder for laser sintering. Addit. Manuf. **33**, (2020)
29. Schmid, M., Amado, A., Wegener, K.: Materials perspective of polymers for additive manufacturing with selective laser sintering. J. Mater. Res. **29**(17), 1824–1832 (2014)
30. Schmidt, J., et al.: Optimized polybutylene terephthalate powders for selective laser beam melting. Chem. Eng. Sci. **156**, 1–10 (2016)
31. Schmidt, J., Peukert, W.: Dry powder coating in additive manufacturing. Front. Chem. Eng. **4**, (2022)
32. Schmidt, J., Plata, M., Tröger, S., Peukert, W.: Production of polymer particles below 5 μm by wet grinding. Powder Technol. **228**, 84–90 (2012)
33. Schmidt, J., Romeis, S., Peukert, W.: Production of PBT/PC particle systems by wet grinding. In: Proceedings of the International Conference, pp. 50003 (2017)
34. Schmidt, J., Sachs, M., Fanselow, S., Wirth, K.-E., Peukert, W.: New approaches towards production of polymer powders for selective laser beam melting of polymers. In: Conference Proceedings, pp. 190008 (2017)
35. Schubert, H.: Emulgiertechnik: Grundlagen, 2nd edn. Verfahren und Anwendungen. Behr's Verlag, Hamburg (2010)
36. Schuchmann, H.P., Danner, T.: Emulgieren: Mehr als nur Zerkleinern. Chemie Ingenieur Technik **76**(4), 364–375 (2004)
37. Stang, M., Schuchmann, H., Schubert, H.: Emulsification in high-pressure homogenizers. Eng. Life Sci. **1**(4), 151 (2001)
38. Tischer, F., Cholewa, S., Düsenberg, B., Drummer, D., Peukert, W., Schmidt, J.: Polyamide 11 nanocomposite feedstocks for powder bed fusion via liquid-liquid phase separation and crystallization. Powder Technol. 118563 (2023)
39. Tischer, F., Düsenberg, B., Gräser, T., Kaschta, J., Schmidt, J., Peukert, W.: Abrasion-induced acceleration of melt crystallisation of wet comminuted polybutylene terephthalate (PBT). Polymers **14**(4), 810 (2022)
40. Tischer, F., Düsenberg, B., Peukert, W., Schmidt, J.: Production of spherical ester wax powders by melt emulsification for additive manufacturing. Procedia CIRP **111**, 33–36 (2022)

41. Wilczek, M., Bertling, J., Hintemann, D.: Optimised technologies for cryogenic grinding. Int. J. Miner. Process. **74**, S425–S434 (2004)
42. Zou, R.P., Yu, A.B.: Evaluation of the packing characteristics of mono-sized non-spherical particles. Powder Technol. **88**(1), 71–79 (1996)

Wolfgang Peukert has been a full professor at the Friedrich-Alexander-Universität Erlangen-Nürnberg at the Institute of Particle Technology. After spending 7 years in industry, he was a full professor at the Technical University of Munich at the Institute of Particle Technology from 1998 to 2003. His research interests are widespread in Particle Technology with an emphasis on nanoparticle technology, product engineering and formulation technology, interface science, and engineering.

Jochen Schmidt is the leader of the additive manufacturing group at the Institute of Particle Technology at Friedrich-Alexander-Universität Erlangen-Nürnberg. He obtained his doctoral degree in Physical Chemistry from University of Jena in 2008. His research focuses on developing processes for the production and functionalization of materials for Additive Manufacturing, emphasizing novel particle systems for laser-based powder bed fusion of polymers and the characterization of Additive Manufacturing materials.

Chapter 3
Gas Phase Functionalization of Polymer and Metallic Materials for Powder-Based Additive Manufacturing

Björn Düsenberg, Juan Sebastian Gomez Bonilla, Marius Christ, Jochen Schmidt, Karl-Ernst Wirth, Wolfgang Peukert, and Andreas Bück

3.1 Introduction

The majority of commercially available particulate materials in additive manufacturing are single-material systems. Multi-material systems are typically generated by bulk mixing as heterogeneous systems that are difficult to reproduce due to segregation effects during mixing and later deposition in the manufacturing process. This limits the properties of the additively manufactured components. For further application areas of additively manufactured components, it is therefore imperative to produce process-optimized single-material particles in addition to multi-material systems, whereby the production of the material systems and their handling require new process routes. The base materials for powder-based additive manufacturing must meet various requirements for processing, and the particles should be between 70 and 120 μm in size and have good flowability. These targeted properties also apply to multi-material systems.

There are different approaches to modify powders in order to achieve targeted properties such as good flowability. In the physical process route, the particles used for the laser-based powder bed fusion process are modified by using mechanical forces. In a chemical process route, the modification of the particles is achieved by chemical modification of the particle surface.

This chapter presents several new and extended processes to adjust, modify, and structure the particles and powders for powder-based additive manufacturing. The chapter commences with an overview of shape adjustment by thermal rounding, followed by the surface modification by nanoparticles due to dry coating or

B. Düsenberg · J. S. Gomez Bonilla · M. Christ · K.-E. Wirth · W. Peukert · A. Bück (✉)
Friedrich-Alexander-Universität Erlangen-Nürnberg, Institute of Particle Technology, Cauerstr. 4, 91058 Erlangen, Germany
e-mail: andreas.bueck@fau.de

D. Drummer and M. Schmidt (eds.), *Progress in Powder Based Additive Manufacturing*, Springer Tracts in Additive Manufacturing,
https://doi.org/10.1007/978-3-031-78350-0_3

plasma-enhanced chemical vapor deposition, before closing with particle structuring using fluidized bed spray agglomeration. In this context, the advantages and limitations of all process options are discussed.

3.2 Shape Adjustment by Thermal Rounding

The uniform in-process deposition of a particle layer requires good flowability, i.e., a spherical shape of the individual particles. However, depending on the production process, e.g., wet comminution [5–23], particles exhibit not only different sizes but also non-sphericity and shape anisotropy. The main approach of thermal rounding is to apply heat to the non-spherical particles, partially melting them and transforming the particles into a spherical shape by interfacial forces. Especially polymer particles are modifiable by this approach [4]. Shape transformation occurs in thermal downstream reactors (Downer) (Fig. 3.1), in which the polymer starting material is fed into a N_2 gas stream by means of a dispersing unit while the aerosol is fed into the heated section of the downer by a head unit [22, 23]. The head unit allows a further, secondary gas stream to be fed, which encloses the aerosol and thereby minimizes contact between the heated reactor wall and the molten polymer particles. Furthermore, the secondary gas stream can be used to control the overall residence time of the particles. The reactor can be heated directly or indirectly, with the direct method involving heating of the secondary gas stream and indirect heating via heating elements outside the reactor (Fig. 3.1) [13].

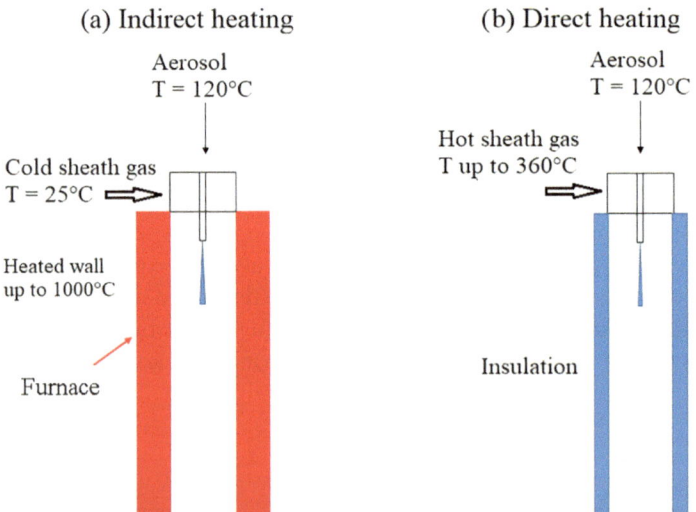

Fig. 3.1 Schematic representation of **a** indirect heating and **b** direct heating of a downer reactor from [13]

The maximum temperature set-point results from the respective melt temperatures of the polymers used, while the residence time of the particles in the molten state required for rounding can be estimated with the aid of an established sintering model [19]. The required residence time for rounding can be estimated using the dimensionless time $t_{rounding}$:

$$t_{rounding} = \frac{t\sigma}{\eta x_f} \tag{3.1}$$

with the particle size x_f, the surface energy σ, and the melt viscosity η. The structure of the non-spherical starting material is not relevant for the rounding step [19]. The residence time required for complete rounding of polystyrene particles, using literature values for σ and η [23], is obtained from Eq. (3.1) to be 2.6 s. Experiments under variation of the residence times of the particles in the downer show good agreement with the estimate. This was demonstrated in Dechet and Gomez et al. [4], where PBT-PC particles were thermally rounded at average residence times of 4.52 s and 2.61 s, respectively. Gomez et al. [17] further studied the influence of different particle sizes and the process variables of temperature and volume flows of the gas and sheet gas on the residence time distributions, thereby establishing the ranges of operation and throughput of the downer reactor.

An in-depth study was performed for high density polyethylene (HD-PE). Starting with the raw material displayed in Fig. 3.2a, the polymer shape resulting from the indirect heating approach [13] is displayed in Fig. 3.2b, showing a significant improvement in the sphericity of the individual particles.

A comparison of the specific results from direct and indirect heating is performed by using the shape factors of circularity χ and roundness R [25]:

$$\chi = \frac{4 \cdot \pi \cdot A}{P^2} \tag{3.2}$$

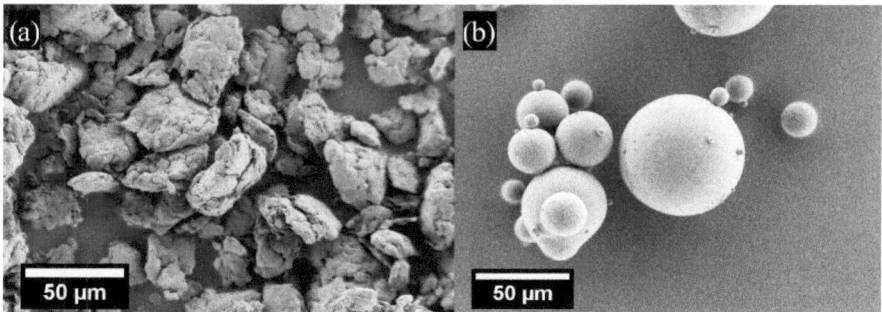

Fig. 3.2 Polymer particles (here HD-PE) **a** before rounding and **b** after the rounding process (image from [4])

$$R = \frac{4 \cdot A}{\pi \cdot x_{\max}^2} \tag{3.3}$$

with A being the projected area of the particle, P the parameter, and $x_{\max}$ the maximum measurable diameter. The circularity combines the surface roughness with the aspect ratio of a particle and shows how spherical ($C = 1$) or star-like ($C = 0$) a single particle is. The roundness depends solely on the aspect ratio and determines whether the particle is a sphere or elongated [25]. The results (Fig. 3.3) show a strong influence of the heating method in both shape factors, namely circularity (a) and rounding (b). Firstly, the shape of the particles becomes much more spherical with both heating methods in comparison with the raw material, as already shown in the SEM image (Fig. 3.2). Secondly, the indirect heating performs worse in comparison to the direct heating approach. This results from a strong coalescence of the partly molten polymer particles in the downer reactor due to turbulent flow patterns instead of a directional, rather laminar flow [17, 27]. This leads to agglomeration, which results in a larger particle size or a non-spherical shape. As a consequence, the powder yield resulting from the use of the indirect heated downer reactor is lower than with the directly heated reactor, which indicates that the gas flow pattern is a crucial process variable.

The results of the rounded powders' flowabilities are displayed in Fig. 3.4a. Since mainly the fine fraction is discharged by the use of the indirectly heated reactor, the powder flowability is poorer, despite better shape factors. This results from the interparticle adhesive forces which have a stronger effect on fine particles than on coarser ones, as the ratio between the adhesive force and gravitational force of each particle decreases. This shows that the shape factors alone are not decisive for the flowability of the powders, but that it is a complex interaction of several factors. In Fig. 3.4b, the corresponding layers of powder display the deposition behavior of

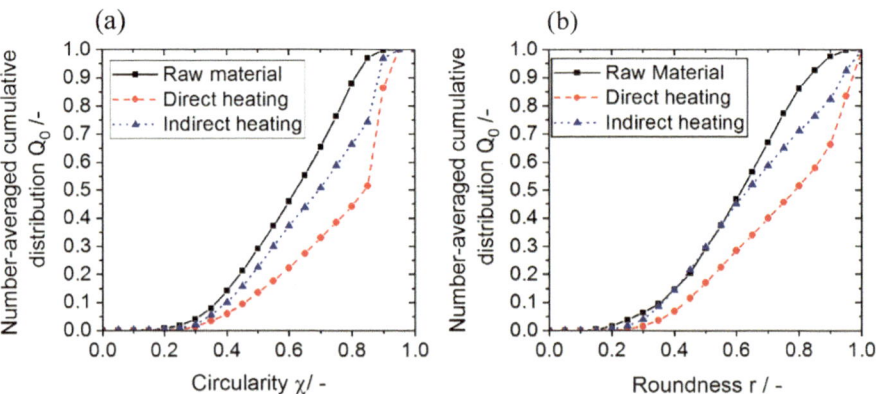

Fig. 3.3 Comparison of circularity (**a**) and roundness (**b**) distributions between direct and indirect heating, according to [13]

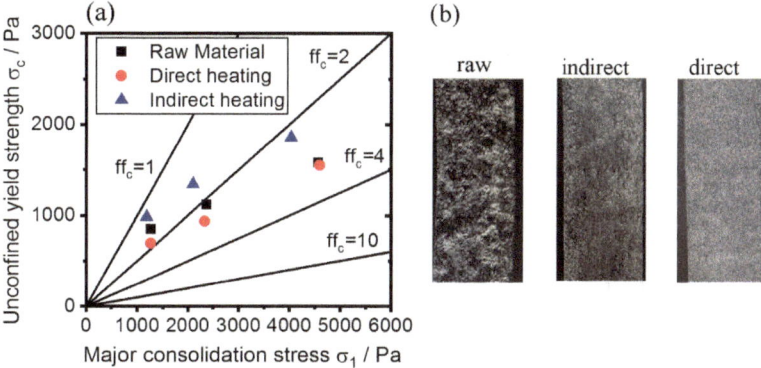

Fig. 3.4 Comparison of rounded particles with the raw material, **a** flow function according to Jenike [12], at different consolidation stresses and **b** corresponding deposited layers [13]

the particles after thermal rounding in comparison with the raw material. This also confirms that the direct heating method is preferable.

In summary, the thermal rounding of polymers was developed and exemplified by both direct and indirect heating of the gas stream for various thermoplastics, e.g., polypropylene, polyethylene, polybutylene terephthalate, and polystyrene [13, 22, 24], including high-temperature thermoplastics (e.g., PEKK). Composite particles made from polybutylene terephthalate and polycarbonate [4] by co-milling were also successfully rounded as displayed in Fig. 3.5—opening up new avenues for the shape modification of complex polymer composite materials.

The thermal rounding of polymer powders is a method for improving the flow properties without the addition of further additives, which may have a negative impact on component quality. For example, some additives worsen the sintering window or the crystallinity of the polymers [26]. Furthermore, rounding offers the possibility of merging multi-materials with each other to already mix several materials at the particle level (cf. Fig. 3.5).

3.3 Surface Modification

The surface of the materials can be modified by a wide variety of physical and chemical processes, whereby plasma enhanced chemical vapor deposition (PE-CVD) and dry particle coating are especially important in this context. The methods were used to create novel functionalities, such as electrical conductivity [8], thereby enabling electrophotographic powder deposition [7, 20] and magnetizable polymers [6], or enhancing existing properties such as the powder flowability [3, 16].

Fig. 3.5 Polybutylene
terephtalate—polycarbonate
composite particle, thermally
rounded (image taken from
[4])

3.3.1 Dry Coating

The principle of dry coating is displayed in Fig. 3.6. It is applicable for the surface
modification of polymers and metals [18]. The nanoparticles, so-called guest particles
(GP), are added to the host particles (HP) and mixing aids (often glass beads) into
a mixer. Usually around 0.1 to 0.5 wt.-% of GP are used to modify the HP. During
the mixing process, the mixing aids deagglomerate the nanoparticles, which are
distributed due to the motion within the mixer on the HP surface. Adhesive forces
(electrostatic and van der Waals) acting between the host and guest particles lead to a
strong bonding between the particles. The final product (Fig. 3.6e) can have different
qualities, depending on the process properties

To estimate the amount of guest particles (in wt.-%) needed to cover the host
particle surface with one single layer, the following equation [29] can be used:

$$m_{(g,\text{wt\%})} = \left(\frac{N \cdot d_{GP}^3 \cdot \rho_G}{d_{HP}^3 \cdot \rho_{HP} + N \cdot d_{GP}^3 \cdot \rho_{GP}} \right) \cdot 100 \tag{3.4}$$

with the host particle diameter d_{HP}, the GP diameter d_{GP}, as well as the mass
densities ρ_{HP} and ρ_{GP} of the host and guest particles, respectively. The number of
guest particles N is obtained as:

$$N = \frac{4(d_{HP} + d_{GP})^2}{d_{GP}^2} \tag{3.5}$$

Various amounts (in wt.-%) of nanoparticles produce different degrees of coverage
on the host particle surface, whereby the amount used is crucial for the final product,
as a low amount may not achieve the intended purpose. For example, in Fig. 3.7,
charge control substances are deposited on polypropylene particles to control the
polarity and the amount of charge accumulating and thereby enable photoelectric

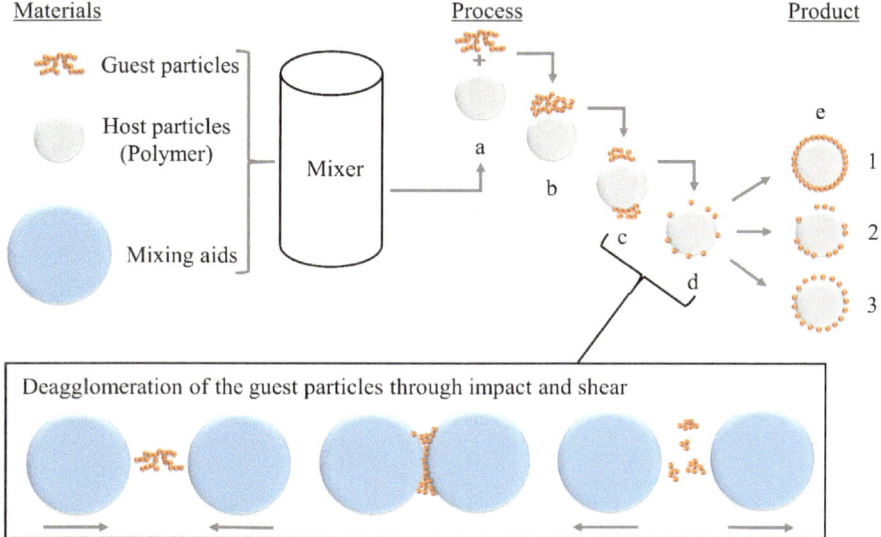

Fig. 3.6 Dry coating process according to Alonso et al. [1, 2] and obtained from Düsenberg et al. [6]; **a** addition of GP to HP and **b** first contact between HP and GP; **c** deagglomeration and **d** dispersion of the guest particles; and **e** the different product qualities that can be achieved

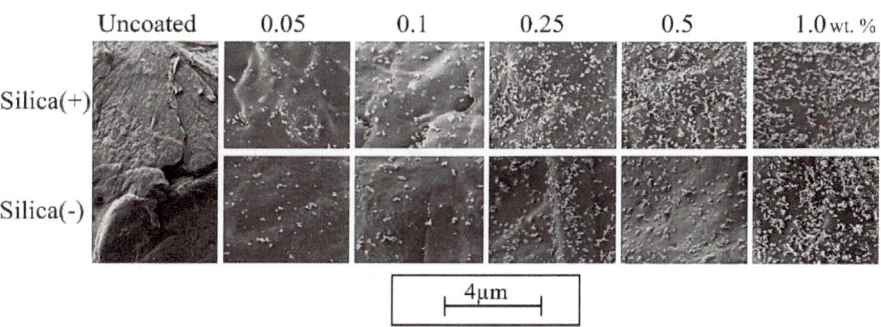

Fig. 3.7 Exemplary development of the degree of coverage on the HP surface. Here, charge control substances were deposited on polypropylene [7]. The uncoated polymer surface (without GP) on the left is displayed to showcase the difference to the coated surfaces

powder deposition in additive manufacturing, as was shown in cooperation with the Bavarian Laser Center (Prof. Dr. Michael Schmidt) in [7, 20].

The degree of coverage on the HP is dependent on the process properties such as mixing intensity, ratios between HP and GP and, of course, the amount, size, and density of the mixing aids, whereby the temperature of the dry coating process is also especially important [10]. The temperature influences the triboelectric charge of the powder and leads to an increased degree of coverage, which significantly increases the flowability of the powders [10] but also decreases the yield. The temperature has

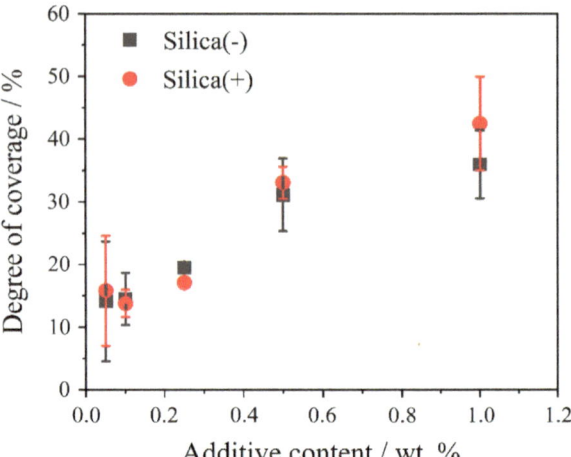

Fig. 3.8 Degree of coverage (nanoparticles on host particle surface) at different additive contents from [7]. The datapoints correspond to Fig. 3.7

to be chosen carefully to define the optimum temperature for a specific dry coating process. Figure 3.7 displays the degree of coverage corresponding to the SEM images obtained. As suggested by the SEM images, the degree of coverage increases by the increase of the guest particle content and, with the same process conditions but different additives, similar coating results are obtained. The process of dry coating with the use of guest particles can thus be generalized, for example, to coating with iron oxide nanoparticles [6]. Karg et al. [18] and Gärtner et al. [15] also showed the coating of aluminum and copper host particles with silica nanoparticles to enhance the flowability and reduce the segregation of mixtures of both.

To demonstrate the effectiveness of nanoparticle coating in powder-based additive manufacturing, Fig. 3.9a shows a powder bed created with uncoated polypropylene. Here, many furrows and grooves in the powder bed are observable, which make it impossible to produce components. Figure 3.9b displays two produced tensile strength specimens, as they can only be produced by an ideally smooth powder bed, i.e., after dry coating of the polymer particles (Fig. 3.8).

Dry coating offers an easy-to-use approach to modify powders at different scales. It is usable in laboratories during the development phase, as well as in industrial production scales. As the process is completely solvent-free, its energy consumption is low in comparison with liquid-phase processes, as no post-processing by drying is required.

3.3.2 Plasma Functionalization

The second method for modification of the particle surface investigated is via a fluidized bed with a plasma unit (Fig. 3.10a). The use of a fluidized bed has several advantages. In this system, the powder can be plasma functionalized [14] or coated by

(b) Tensile test specimens producible
with 0.1 wt. % nano silica

(a) Powder bed *without* nanoparticles
to enhance the flowability

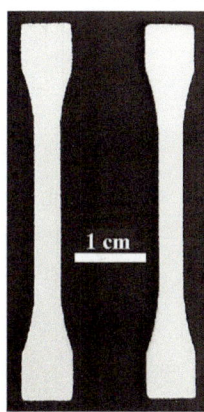

Fig. 3.9 **a** Powder bed from polypropylene in an industrial laser powder bed fusion machine with many furrows and grooves. **b** Tensile test specimens produced with 0.1 wt.-% of nanosilica

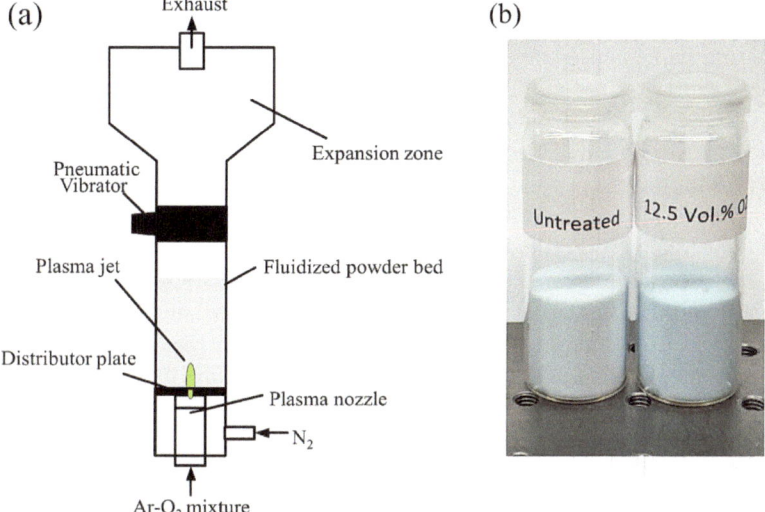

Fig. 3.10 Schematic representation of the fluidized bed PECVD plant (image from [14])

PECVD [16]. The functionalization, for example, opens up the possibility of direct coloring of the polymer powder (Fig. 3.10b), whereby the mixing of the primary particles within the reactor is more or almost ideal, so that they are almost uniformly functionalized.

As an example, the in situ production of silica (SiOx) nanoparticles by means of PE-CVD is presented. A constant flow of hexamethyldisiloxane (HMDSO) into

Ar   2.5 Vol. % O$_2$ 5 Vol. % O$_2$

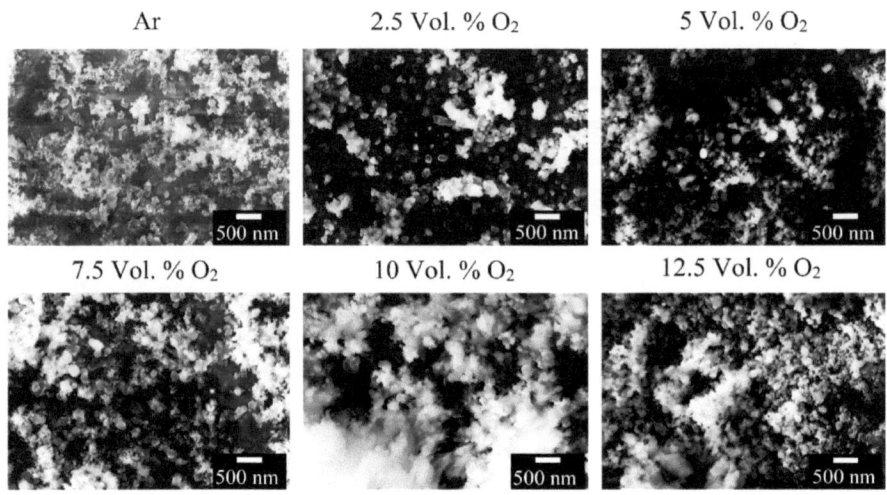

7.5 Vol. % O$_2$ 10 Vol. % O$_2$ 12.5 Vol. % O$_2$

Fig. 3.11 Nanoparticles, generated with a PECVD particle generation unit, deposited on an Si-Wafer with different argon-oxygen ratios taken from [16]

the plasma flame causes the formation of silica nanoparticles, which adhere to the fluidized particles (for more details, see [16]). The silica nanoparticles generated (Fig. 3.11) depend on the plasma flame, which in turn can be manipulated by increasing the oxygen content that controls the combustion. The specific surface areas of the nanoparticles differ between $23\,m^2g^{-1}$ to $25.9\,m^2g^{-1}$ leading to Sauter mean diameters between 39 and 45.5 nm. The observable structures and interparticle porosity are similar to those described by Walliman et al. [28].

The generation of silica nanoparticles from HMDSO on polymer powder (polypropylene in Fig. 3.12) shows a homogeneous distribution of individual silica particles up to silica aggregates, with sizes ranging from 18 nm to 160 nm (obtained by image analysis). ICP-OES measurements confirmed that these are silica particles and also determined the mass (additional details in [16]).

Comparing the modification via dry coating and PE-CVD in terms of flowability, the value of the flow function ffc (Fig. 3.13) of the dry coated powders is slightly below of those modified by PE-CVD. This results from a much better distribution of nanoparticles on the host particle surface in PE-CVD, as the fluidized bed technology is predestined for this task. This is especially noticeable since less than 0.1 wt.-% additive was used, leading to the conclusion that an advanced distribution and deposition method is able to enhance the flowability of powders as much as larger quantities of nanoparticles deposited by dry coating.

Tensile strength specimens were prepared from the PE-CVD modified powders by Gomez et al. [16] and subsequently analyzed for their mechanical properties. Compared to the raw powders, it was shown (see Fig. 3.14) that the Young's modulus and stress at break could be increased for all specimens and the elongation at break

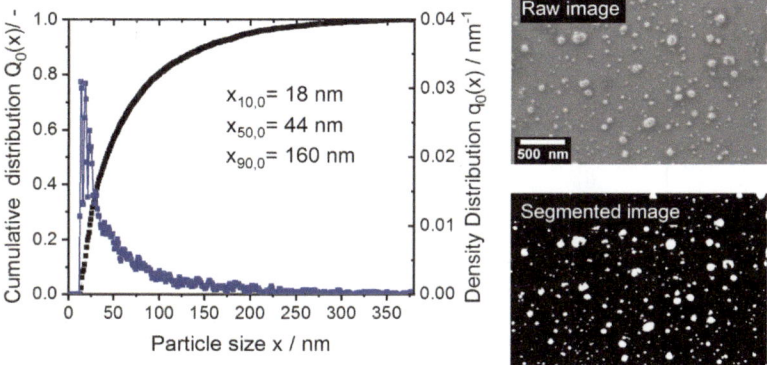

Fig. 3.12 Particle size distributions (number averaged), of the deposited SiOx particles from SEM images and taken from [16]. The SEM images are binarized (example in segmented image) to extract the data. In this image with 7.5 vol.-% O_2

Fig. 3.13 Development of the powder flowability at different consolidation stresses and several quantities of silica nanoparticles in comparison with a standard dry coating formulation from [16]

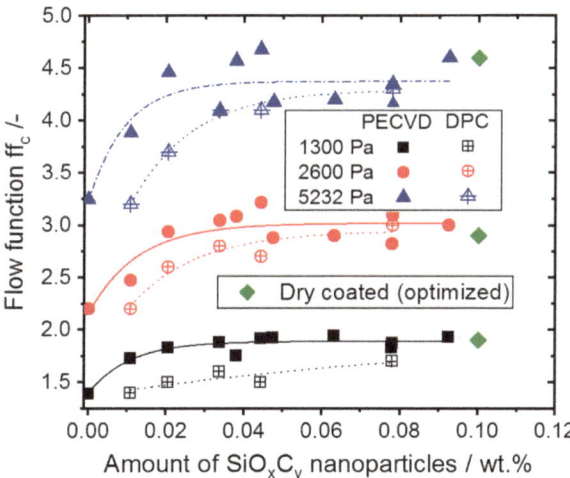

remained more or less constant. This makes PE-CVD an expensive but very effective method of powder functionalization for additive manufacturing.

Both methods are suitable to modify powders, metals, and polymers for the use in powder-based additive manufacturing processes. If the focus is on using as little additive as possible, and the cost of a process is not a decisive factor, the fluidized bed with a PECVD unit is a high-quality method for modifying particles. However, if simplicity, speed, and low investment costs are important, dry coating is preferred.

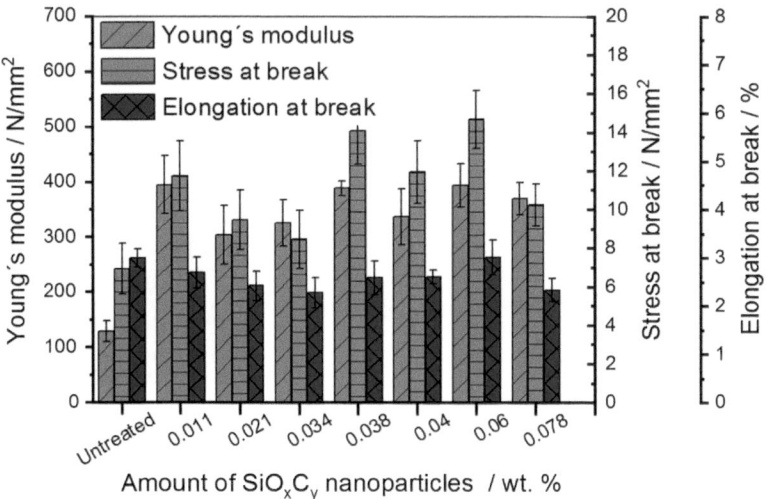

Fig. 3.14 Development of the mechanical properties from additive manufactured tensile strength specimens at different amounts of silica nanoparticles

3.3.3 Structuring of Polymer Particles: Towards Multi-materials

Particle structuring, especially for building secondary particles (agglomerates) for powder-based additive manufacturing can be performed by fluidized bed spray agglomeration. The structure of the agglomerates is crucial for the flowability of the powder and the evaluation of the structuring behavior of small primary particles in fluidized bed spray agglomeration processes is a crucial prerequisite to produce multimaterial agglomerates [9].

In fluidized bed spray agglomeration, the powder is fluidized by a gas flow in a fluidized bed reactor. At the top of the chamber, a nozzle is mounted and, from this nozzle, the binder liquid is atomized to form fine droplets (cf. Fig. 3.15). The droplets collide with the particle surface (1) to form a sticky film (2). After particle-particle contact of the primary particles, they adhere to each other based on liquid bridges between them. Subsequent drying (3) leads to the formation of solid bridges on the liquid bridges, which create a stable agglomerate (4).

The first challenge in structuring particles for use in powder-based additive manufacturing is the underlying primary particle size and (for polymers) low density, which result in the Geldart group C (cohesive) [12]. Group C is characterized by the fact that the interparticle adhesive forces are large due to its small size and density, which results in strong cohesion of the particles and difficult fluidization behavior. To overcome this, the surface of the primary particles must already be modified and the dry coating method is suitable for this purpose. As an example, polystyrene primary particles ($x_{90,3} < 100\,\mu$m and $\rho = 1.06\,$g/cm^3) were coated with 0.1 wt.-% silica

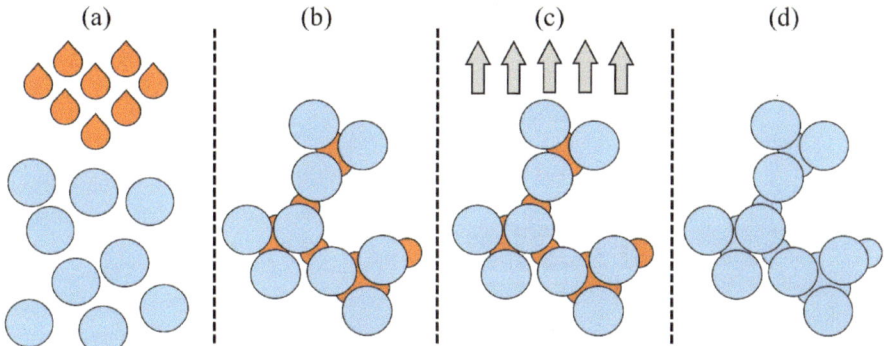

Fig. 3.15 Representation of the steps of particle formation during agglomeration. **a** Wetting of primary particles with binder liquid, **b** particle-particle contacts lead to liquid bridges between the primary particles, **c** drying and solid bridge formation **d** (sketch from [9])

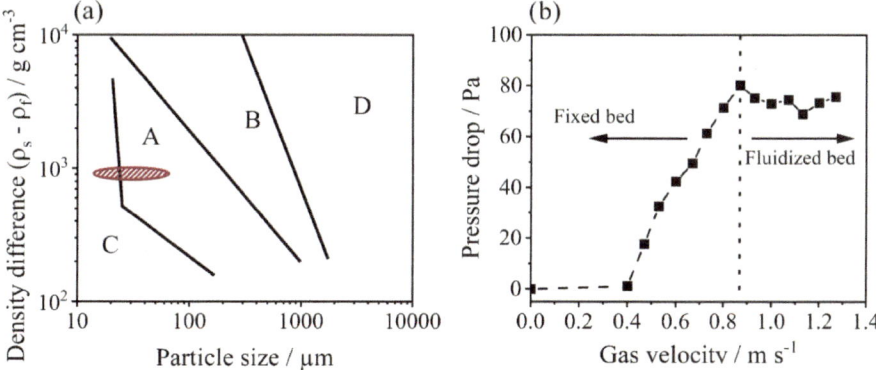

Fig. 3.16 **a** Geldart classification of the primary particles (here polystyrene as a model polymer) and **b** measured pressure-drop in regard to the gas velocity indication the created fluidized bed after dry coating

nanoparticles. This does not change the original Geldart classification as shown in Fig. 3.16a, but the fluidization behavior moves in the direction of group A (aeratable), enabling fluidization and thus agglomeration (Fig. 3.16b).

The agglomeration leads to agglomerates of different sizes, which are shown for example in Fig. 3.17. Thus, the primary particles agglomerate to form medium-sized (approx. 15 primary particles) to large (more than 30 primary particles) agglomerates. This finding is consistent with the Stokes criterion [11] and the calculations of Rieck et al. [21] which describe that small particles are more likely to agglomerate than large ones. This is especially true at lower densities.

Over the investigated range of volume flow ratios between the atomizer gas volume flow and the volume flow of the binder liquid, two process ranges are formed

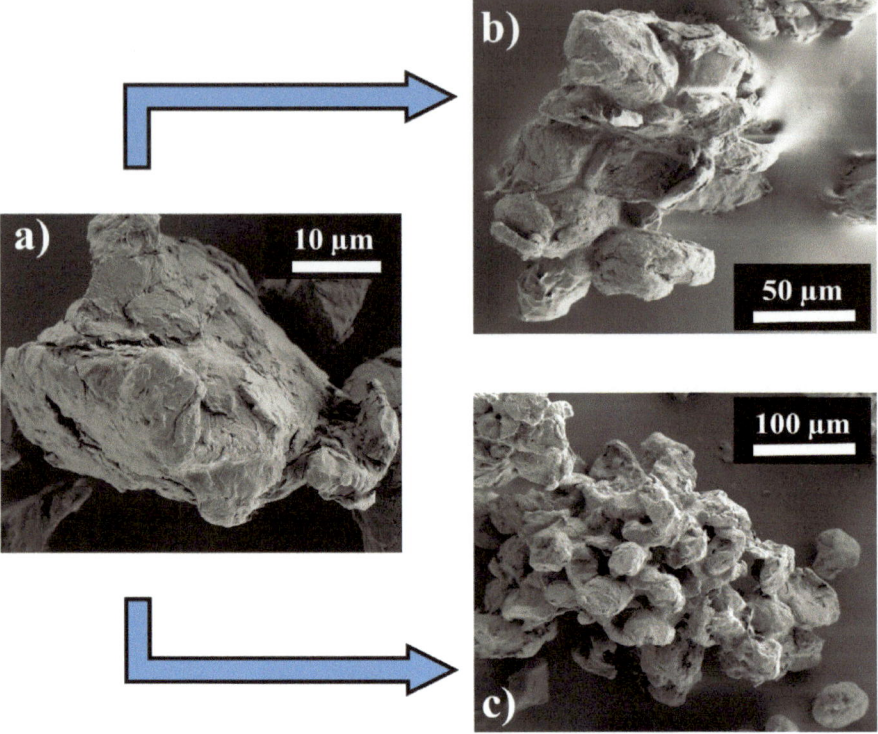

Fig. 3.17 SEM images of the **a** primary particle, **b** a medium size agglomerate, and **c** a large agglomerate (sketch from [9])

(Fig. 3.18), one of which is stable while the other is unstable. The unstable process range is up to a volume flow ratio of 3000 (corresponding to larger droplets). In these, clustering and sticking of large amounts of polymer occurs, resulting in chunks of 5 cm in size. The remaining particles and agglomerates seem to have a suitable size for the laser-powder bed fusion process, but as the agglomeration is unstable, it is not reproducible, the yield is not economical, and the cleaning effort is disproportionately high. From a volume flow ratio of 3300, reproducible agglomeration then occurs without major chunks affecting the fluid bed. The agglomerate size varies between 100 and 600 μm. These agglomeration experiments demonstrated the ability to agglomerate fine Geldart C/A primary particles.

The evaluation of the shape factors circularity and roundness and example results are shown in Fig. 3.19. In this figure, the graph marked (a) represents a volume flow ratio of 3300 (start of the stable process regime) and the graph marked (b) represents a volume flow ratio of 6700 (maximum value in the stable process regime). The two volume flow ratios do not differ significantly in their values. Compared to the circularity and roundness of the primary particles (mean circularity equal to 0.32; mean roundness equal to 0.5), a pronounced improvement was achieved by

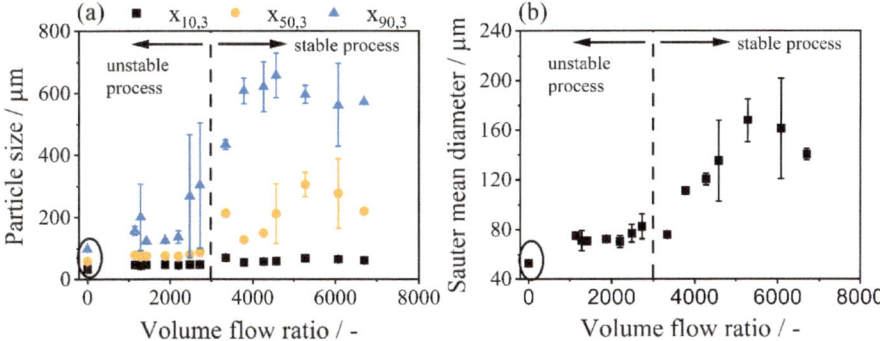

Fig. 3.18 Particle size distributions of the produced agglomerates at different volume flow ratios at the nozzle. **a** Key values $x_{10,3}$, $x_{50,3}$ and $x_{90,3}$; **b** with Sauter mean diameter (data from [9])

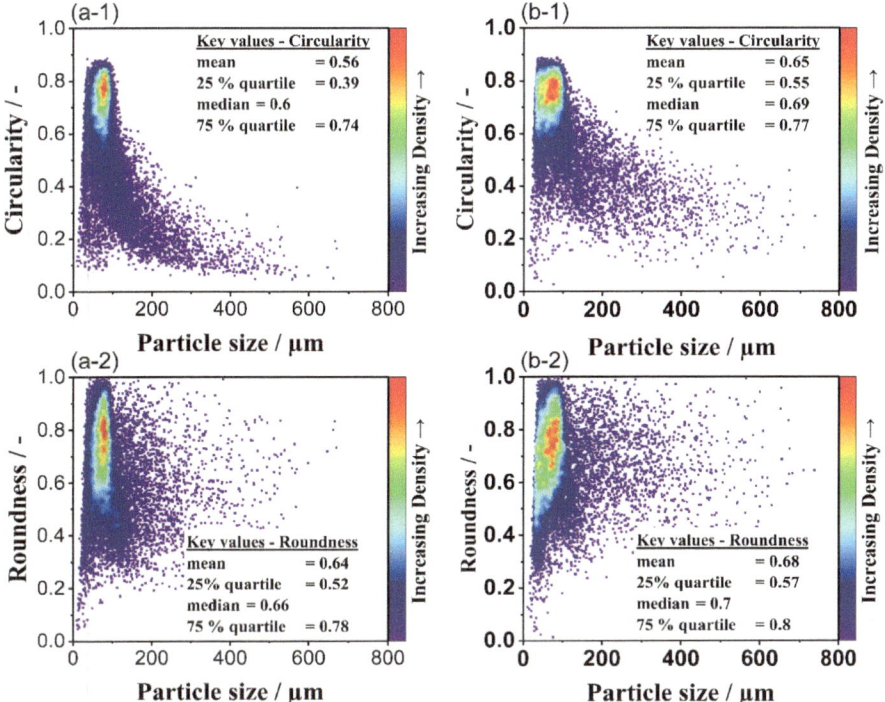

Fig. 3.19 Circularity according to Eq. 3.2 and roundness according to Eq. 3.3 of the exemplary experiments. **a** Volume flow ratio 3300 and **b** volume flow ratio 6700

agglomeration. Since the purpose of any agglomeration is to produce particles with a high flowability, this can be confirmed by the significant improvement in these two parameters [13].

In summary, concerning the structuring it can be concluded that a previous modification of the surface is essential to make the fine particles fluidizable. Based on this, the fluidized bed spray agglomeration offers a wide range of options for further integrating the finest particles that would otherwise be fed into thermal recycling and production cycles. Furthermore, this work provides a basis for creating multi-material agglomerates from different primary particles to generate synergies on the particle level, which later have a positive effect on the component. Combined with thermal rounding, multi-material aggregates can be transformed into spherical particles with specific material ratios on the particle level.

3.4 Summary and Outlook

This chapter presented a summary of the results concerning shape adjustment, surface modification, and structuring to produce powders for powder-based additive manufacturing. The improvements were carried out on various materials such as polyamide, polypropylene, polystyrene, polyethylene, polybutylene terephthalate, and high temperature thermoplastics, leading to the conclusion that the processes are generally applicable for the functionalization of powders for additive manufacturing. The materials were used in the laser powder bed fusion process [8, 14, 16, 24], with a special focus on electrophotographic powder deposition methods [7, 20]. For the surface adjustment, thermal rounding in a downstream reactor is utilized. This reactor uses heat energy to melt the particles, thereby yielding spherical droplets that retain their shape after cooling down. The spherical shape significantly increases the powder flowability. Two different approaches were developed and investigated: direct and indirect heating, whereby the direct heating results in a higher yield but slightly lesser rounding than the indirect heating. This is not concerning, as even partially rounded particles perform much better in terms of flowability. Subsequent to the thermal rounding, or as a single process, the surface modification via dry coating or plasma enhancing CVD can be performed, whereby especially the dry coating is a fast, solvent-free process that consumes low amounts of energy to modify the powders, regardless of whether polymers or metals are to be modified. PECVD is, from a safety point of view, as well as from its energy consumption, a more advanced process and worth using for high-priced products. When using PECVD to modify metal powders, it is crucial to create an inert environment, as metals in combination with the hot plasma flame are able to burn. The great benefit of PECVD, in comparison to dry coating, is the enhanced distribution of nanoparticles on the host particle surfaces, leading to the same results as dry coating with only a fraction of the additive used. The structuring of the particles is accomplished by using spray fluidized bed technology and the small primary particles are agglomerated to roughly spherical shapes. This knowledge is crucial for the development of multi-material particles as combinations of two polymers, two metals, or a mixture of these. Subsequently, thermal rounding can be used to make the agglomerates more spherical to enhance the processability.

In the future, the presented methods could be used to create—either through stand-alone use or in combination—customized materials with unique properties for powder-based additive manufacturing. They will accelerate the development of novel materials and bringing them to market by producing them in bulk. Bulk production by continuous operation is achievable by extending the underlying unit operations to continuous operation. New opportunities and challenges will arise from the interplay (e.g., recycles) and adjustment of individual times scales of the process steps.

References

1. Alonso, M., Alguacil, F.: Dry mixing and coating of powders. Revista de metalurgia **35**, 315–328 (1999)
2. Alonso, M., Satoh, M., Miyanami, K.: Powder coating in a rotary mixer with rocking motion. Powder Technol. **56**, 135–141 (1988)
3. Blümel, C., Sachs, M., Laumer, T., Winzer, B., Schmidt, J., Schmidt, M., Peukert, W., Wirth, K.-E.: Increasing flowability and bulk density of PE-HD powders by a dry particle coating process and impact on LBM processes. Rapid Prototyp. J. (2015)
4. Dechet, M., Gömez Bonilla, J., Lanzl, L., Drummer, D., Bück, A., Schmidt, J., Peukert, W.: Spherical polybutylene terephthalate (PBT)—polycarbonate (PC) blend particles by mechanical alloying and thermal rounding. Polymers **10**, 1373 (2018)
5. Düsenberg, B., Esper, J., Maußner, F., Mayerhofer, M., Schmidt, J., Peukert, W., Bück, A.: Control of crystallization of PBT-PC blends by anisotropic SiO2 and GeO2 glass flakes. Polymers **14**, 4555 (2022)
6. Düsenberg, B., Groppe, P., Müssig, S., Schmidt, J., Bück, A.: Magnetizing polymer particles with a solvent-free single stage process using superparamagnetic iron oxide nanoparticles (SPION) s. Polymers **14**, 4178 (2022)
7. Düsenberg, B., Kopp, S.-P., Tischer, F., Schrüfer, S., Roth, S., Schmidt, J., Schmidt, M., Schubert, D., Peukert, W., Bück, A.: Enhancing photoelectric powder deposition of polymers by charge control substances. Polymers **14**, 1332 (2022)
8. Düsenberg, B., Seidel, A., Kopp, S.-P., Schmidt, J., Roth, S., Peukert, W., Bück, A.: Production and analysis of electrically conductive polymer-carbon-black composites for powder based additive manufacturing. Procedia CIRP **111**, 18–22 (2022)
9. Düsenberg, B., Singh, A., Schmidt, J., Bück, A.: Spray agglomeration of polymer particles: influence of spray parameters on shape factors. Powder Technol. **422**, 118491 (2023)
10. Düsenberg, B., Tischer, F., Valayne, E., Schmidt, J., Peukert, W., Bück, A.: Temperature influence on the triboelectric powder charging during dry coating of polypropylene with nanosilica particles. Powder Technol. **399**, 117224 (2022)
11. Ennis, B., Tardos, G., Pfeffer, R.: A microlevel-based characterization of granulation phenomena. Powder Technol. **65**, 257–272 (1991)
12. Geldart, D.: Types of gas fluidization. Powder Technol. **7**, 285–292 (1973)
13. Gomez Bonilla, J., Dechet, M., Schmidt, J., Peukert, W., Bueck, A.: Thermal rounding of micron-sized polymer particles in a downer reactor: Direct vs indirect heating. Rapid Prototyp. J. **26**, 1637–1646 (2020)

14. Gomez Bonilla, J., Szymczak, T., Zhou, X., Schrüfer, S., Dechet, M., Schmuki, P., Schubert, D., Schmidt, J., Peukert, W., Bück, A.: Improving the coloring of polypropylene materials for powder bed fusion by plasma surface functionalization. Addit. Manuf. **34**, 101373 (2020)
15. Gärtner, E., Jung, H., Peter, N., Dehm, G., Jägle, E., Uhlenwinkel, V., Mädler, L.: Reducing cohesion of metal powders for additive manufacturing by nanoparticle dry-coating. Powder Technol. **379**, 585–595 (2021)
16. Gomez Bonilla, J., Düsenberg, B., Lanyi, F., Schmuki, P., Schubert, D., Schmidt, J., Peukert, W., Bück, A.: Improvement of polymer properties for powder bed fusion by combining in situ PECVD nanoparticle synthesis and dry coating. Plasma Process. Polym. **18**, 2000247 (2021)
17. Gomez Bonilla, J., Unger, L., Schmidt, J., Peukert, W., Bück, A.: Particle Lagrangian CFD simulation and experimental characterization of the rounding of polymer particles in a downer reactor with direct heating. Processes **9**, 916 (2021)
18. Karg, M., Rasch, M., Schmidt, K., Spitzer, S., Karsten, T., Schlaug, D., Biaciu, C.-R., Gorunov, A., Schmidt, M.: Laser alloying advantages by dry coating metallic powder mixtures with SiOx nanoparticles. Nanomaterials (Basel, Switzerland) **8**, (2018)
19. Kirchhof, M., Schmid, H.-J., Peukert, W.: Three-dimensional simulation of viscous-flow agglomerate sintering. Phys. Rev. E **80**, 26319 (2009)
20. Kopp, S.-P., Düsenberg, B., Eshun, P., Schmidt, J., Bück, A., Roth, S., Schmidt, M.: Enabling triboelectric charging as a powder charging method for electrophotographic powder application in laser-based powder bed fusion of polymers by triboelectric charge control. Addit. Manuf. 103531 (2023)
21. Rieck, C., Bück, A., Tsotsas, E.: Estimation of the dominant size enlargement mechanism in spray fluidized bed processes. AIChE J. **66**, e16920 (2020)
22. Sachs, M., Friedle, M., Schmidt, J., Peukert, W., Wirth, K.-E.: Characterization of a downer reactor for particle rounding. Powder Technol. **316**, 357–366 (2017)
23. Schmidt, J., Sachs, M., Blümel, C., Winzer, B., Toni, F., Wirth, K.-E., Peukert, W.: A novel process route for the production of spherical LBM polymer powders with small size and good flowability. Powder Technol. **261**, 78–86 (2014)
24. Schmidt, J., Sachs, M., Fanselow, S., Zhao, M., Romeis, S., Drummer, D., Wirth, K.-E., Peukert, W.: Optimized polybutylene terephthalate powders for selective laser beam melting. Chem. Eng. Sci. **156**, 1–10 (2016)
25. Takashimizu, Y., Iiyoshi, M.: New parameter of roundness R: circularity corrected by aspect ratio. Prog. Earth Planet. Sci. **3**, 1–16 (2016)
26. Tischer, F., Düsenberg, B., Gräser, T., Kaschta, J., Schmidt, J., Peukert, W.: Abrasion-induced acceleration of melt crystallisation of wet comminuted polybutylene terephthalate (PBT). Polymers **14**, 810 (2022)
27. Unger, L., Gömez Bonilla, J., dos Santos, D., Bück, A.: Particle residence time distribution in a concurrent multiphase flow reactor: Experiments and Euler-Lagrange simulations. Processes **10**, 996 (2022)
28. Wallimann, R.: Nanoparticle formation in atmospheric pressure plasma for the enhancement of powder flowability (2018)
29. Yang, J., Sliva, A., Banerjee, A., Dave, R., Pfeffer, R.: Dry particle coating for improving the flowability of cohesive powders. Powder Technol. **158**, 21–33 (2005)

Jochen Schmidt is the leader of the additive manufacturing group at the Institute of Particle Technology at Friedrich-Alexander-Universität Erlangen-Nürnberg. He obtained his doctoral degree in Physical Chemistry in 2008. His research focuses on developing processes for the production and functionalization of materials for Additive Manufacturing, emphasizing novel particle systems for laser-based powder bed fusion of polymers and the characterization of Additive Manufacturing materials.

Wolfgang Peukert has been a full professor at the Friedrich-Alexander-Universität Erlangen-Nürnberg at the Institute of Particle until retirement in 2024. After spending 7 years in industry, he was a full professor at the Technical University of Munich at the Institute of Particle Technology from 1998 to 2003. His research interests are widespread in Particle Technology with an emphasis on nanoparticle technology, product engineering and formula-tion technology, interface science, and engineering.

Andreas Bück is professor for Particle Technology since at Friedrich-Alexander-Universität Erlangen-Nürnberg since 2017. He holds a Ph.D. in chemical engineering (graduated: 2012) from Otto von Guericke University Magdeburg, Germany. His research interests focus on particle formulation processes involving drying, e.g., spray-drying, coating, and agglomeration, combining experimental and numerical methods towards tailored processes and products.

Chapter 4
Process Adapted Multiscale Material Characterization

Simon Cholewa, Maximilian Marschall, Michael Schmidt,
and Dietmar Drummer

Abstract Since the laser based powder bed fusion of polymers process involves previously unknown boundary conditions, process-adapted powder characterization at the particle and powder layer levels is essential. In this chapter, new measurement methods are presented, and the results are discussed within the context of process-adapted boundary conditions for powder flow behavior and optical and thermal properties.

4.1 Introduction

Laser based powder bed fusion of polymers (PBF-LB/P) is an additive manufacturing technology that overcomes the geometry and flexibility limitations of conventional technologies such as injection molding through sequential layer-by-layer fabrication. While the advantages of PBF-LB/P have already been sufficiently discussed, to achieve and exceed the current advantages with new materials, the fundamentals of the process must be known. This involves the interaction between different states of material and laser during processing to a final component. Simulating the process for future monitoring and prediction of mechanical properties, as well as tailoring material and powder, requires information about why the commonly used materials perform well compared to others not used so far, and how to influence this processability. Since the process has previously unknown boundary conditions,

S. Cholewa · D. Drummer (✉)
Friedrich-Alexander-Universität Erlangen-Nürnberg, Institute of Polymer Technology, Am Weichselgarten 10, 91058 Erlangen-Tennenlohe, Germany
e-mail: dietmar.drummer@fau.de

M. Marschall
Bayerisches Laserzentrum GmbH, Konrad-Zuse-Straße 2-6, 91052 Erlangen, Germany

M. Schmidt
Friedrich-Alexander-Universität Erlangen-Nürnberg, Institute of Photonic Technologies, Konrad-Zuse-Straße 3/5, 91052 Erlangen, Germany

© The Author(s) 2025   57
D. Drummer and M. Schmidt (eds.), *Progress in Powder Based Additive Manufacturing*, Springer Tracts in Additive Manufacturing,
https://doi.org/10.1007/978-3-031-78350-0_4

a process-adapted powder characterization on the particle and powder layer level is essential.

4.2 Investigations of Powder Bulk Properties and Their Relevance for PBF

During powder bed fusion, the flowability of the bulk is a critical factor for uniform powder distribution along the building plate and the manufacture of dense parts with good mechanical properties. Polymer powder materials for PBF-LB/P have a narrow particle size and shape distribution, which is conditioned by the production process, composition, and additivation. To investigate the influence of particle size on particle-particle interactions, the flowability represented as powder tensile strength between glass sphere particles can be measured with atomic force microscopy (AFM). The smaller the particles, the more cohesive they are, resulting in a reduced bulk density and high powder tensile strength. In this context, powder tensile strength refers to the amount of normal force needed to separate two layers of powdery particles on its favored sliding plane. High bulk density is associated with narrow sliding planes, resulting in a low amount of force needed to separate them, as shown in Fig. 4.1.

Most experiments have to be conducted with an approximation or behavior of a controlled bulk, whose behavior observations will then be downsized to the surface level. As an example, the dry coating additivation of a hydrophobic polymer with a hydrophilic fumed silica is not successful, whereas hydrophobic polymer and hydrophobic fumed silica show a good surface agglomeration of powder and additive [3, 25]. These hydrophilic/hydrophobic interactions can be explained but hardly investigated directly [2].

Additivation by dry coating resulted in a decrease of cohesive forces between the particles, followed by better flowability and therefore related to higher bulk density.

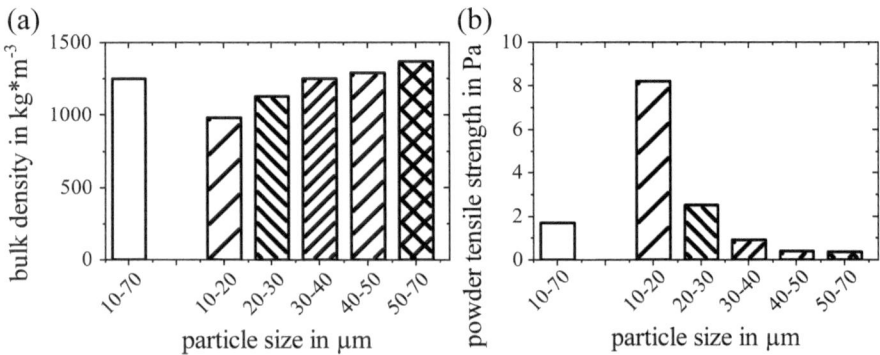

Fig. 4.1 a Bulk density and **b** powder tensile strength of layers of glass microspheres with varying particle size distribution

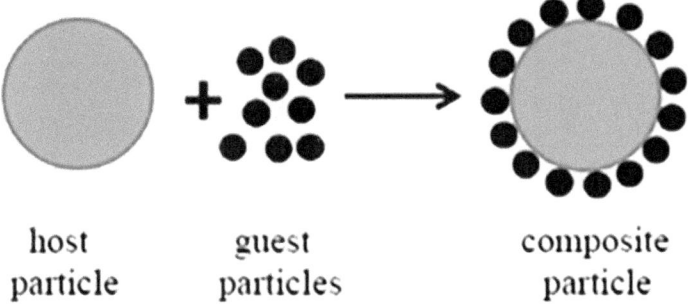

host   guest   composite
particle   particles   particle

Fig. 4.2 Schematic illustration of a dry coating process with additives as guest particles [3]

This effect was shown for various polymer and metal powders, as long as the additives are chosen correspondingly to matching polymer powder materials [2, 28]. As also indicated in other chapters, enhanced bulk density correlates with a smaller tensile stress and less porosity of the built part (Fig. 4.2).

Hence, the additive particles are coated to the surface, whereby a long coating process duration or mechanical stress and impacts on the surface cause the guest particles to be buried in the surface, leading to a decreased flowability of the powder. Flowability correlates with the adhesive friction on a single particle level inside of the particle collective. Those interparticle forces are manipulated by dry coating additives as flow agents to the bulk material. The additivated polymer particles develop remarkably lower cohesive forces compared to the unmodified material. This was investigated by AFM while pulling two particles apart, whereby it was shown that the host polymer particles deform during the process. They have multiple particle contacts and the additives, which only adhere to the surface, can transfer or agglomerate. Thus, they are no longer uniformly distributed and the inter-particle adhesion force increases in comparison to that of the virgin powder [14].

Geometrical changes such as material transfer, deformation, contact electrification, and electrostatic charge occur in dependency on normal forces, contact time, quantity, and contact geometry between polymer particles and substrate. Polymers behave differently to these external induced influences. For example, polystyrene (PS) particles gain electrostatic charges even before contact, although the charge gradually decreases with each contact to a stable level. Polymethylmethacrylate (PMMA) needs a lot more contact time to get charged up but remains stable for a long time. Polyamide (PA) and polypropylene (PP) charge up very fast and accumulate over a long time. On PA, this electrostatic charge effect was even notable a month later during scanning electron microscopy and exceeded the measurability by AFM. Changes in the local topography on the particle surface can be observed, as well as domain formation after the contacts. These domains can either originate from charges or recrystallization, which results in locally differing contact adhesive forces over the particle surface. This charge accumulation on PA particles increases with each contact and slowly relaxes within hours. Particles that are coated with fumed

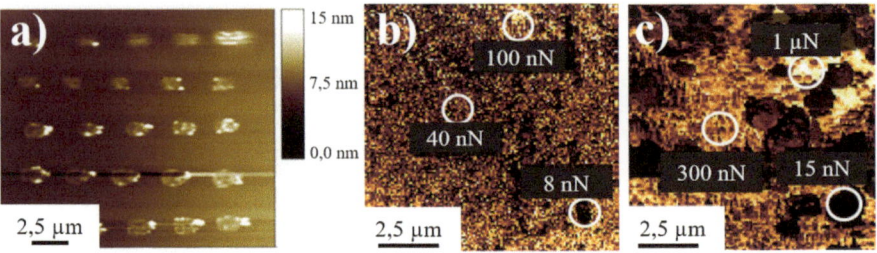

Fig. 4.3 **a** Material transfer at AFM after repeated contact between a PS particle and an Si wafer (normal force 9500 nN, contact area 1.68 μm^2); **b/c** change in cohesive force of a PA substrate **b** before and **c** after friction measurement

silica do not charge up that fast and the accumulation degrades more rapidly. These triboelectric charges on polymer particles without additives are unipolar, extensive, and not controllable in and for a heterogeneous bulk material. Hence process-relevant bulk properties, such as flowability and powder bed or layer density are strongly correlated to the tribo-electrostatic charging behavior of the material [13]. However, since discharge from a bulk of insulating material is a very slow process, charging up has to be prevented (Fig. 4.3).

Most polymer powders possess an almost symmetrical charge distribution related to a particle diameter with low quantities of strong charges. These collectives are mostly bipolar with a net charge distribution that lies close to zero [13]. Depending on the charge slope of additives such as fumed silica, the charge distribution after tribo-electric charging can be tailored to be unipolar by dry coating the particles [10].

Bulk properties in general, like flowability, are strongly dependent on the particle size and particle size distribution, and narrowly distributed powders are often used. The bulk density is a function of particle size and smaller particle sizes refer to less dense packing due to the larger cohesiveness of the material. Narrowly distributed powders of close to spherical shape were shown to obey a random close packing (RCP). RCP is thus the highest density to be achieved for a bulk of similar particles without external influences besides gravity, although it is not the closest packing density. This is the case for the distributed powder layers for PBF, as a notable influence of the particle shape on the packing was only observed for sphericity values below 0.6.

While doctor blading powder layers with a rake, particle segregation becomes more pronounced with wider particle size distributions and slower blading velocity [11]. Simulations can prove this experimentally observed behavior [22].

Taking into account that PBF-LB/P is a heat-affected process at a few degrees below the melting temperature of the material, bulk properties are also dependent on the thermal conditions. Flowability correlates with the normal stress and the shear stress that is necessary to separate the powder layers. The shear stress increases dramatically between room temperature and the preheating temperature of the process. Hence, this effect is also dependent on the thermal properties of the polymer powder.

The change in flowability is represented by the progression of the angle of internal friction. Used powder possesses a higher angle of internal friction after the thermal load and hence a lower flowability for the next built job. This correlates to poorer part properties compared to virgin powder [14].

Polymer powder materials are not only additivated by flow agents. The demand for tailored polymer blends is increasing. To qualify the success of the blending, consisting of co-communition and a reforming process, the powder particles have to be investigated at the single particle level and, with Raman spectroscopy, the polymer-specific band peaks of a single particle can be obtained. This distribution can be visualized by diffusive coloring in which particles are impinged with diffusive ruthenium tetroxide gas to achieve a contrast by different diffusivity in the various polymers. The gas enriches more in the amorphous polymer compared to higher crystalline polymers and can be investigated using electron microscopy [8].

4.3 Laser–Material Interaction in Powder Bed Fusion of Polymers

The thermal history during laser exposure is one of the key factors that determines the part properties in PBF/LB-P. So far, however, it is not possible to investigate the interaction between the laser beam and the particle on a micro level. Therefore, a new measurement setup using flash-DSC equipped with a CO_2 laser and optical elements for focusing and beam separation was developed. The approach allows the process to be simulated on the smallest possible scale, as individual or small numbers of particles are heated directly by the CO_2 laser [15]. With this methodology, a deeper understanding of heating rates in the process, process parameter investigations, and degradation effects was gained for the first time [34].

Figure 4.4 shows process investigations for Polyamide11 (PA11) with differing exposure parameters, such as exposure time and laser power, at a preheated temperature that mimics the building temperature. Figure 4.4a illustrates the maximum temperature observed with temperature-time curves for different combinations of impact times and laser powers. As the impact times and laser powers are progressively increased, the maximum temperature attained during laser exposure demonstrates an almost linear rise. In addition to the maximum temperatures and the heating rates, a statement can be made about the material-laser interaction after cooling during the subsequent heating process. The steps are the same as those for PA12's Degree of Particle Melt, but in this case, a flash DSC is used instead of a DSC. This approach takes advantage of the properties of the PA systems with markedly different positions of the melting peak at the first and second heating. If the energy provided by the laser is insufficient, the unmelted particle will have the same melting pattern as the first heating rate would have. This is observed in Fig. 4.4b by a single melting peak at an elevated temperature. If the energies gradually increase, a partial melting

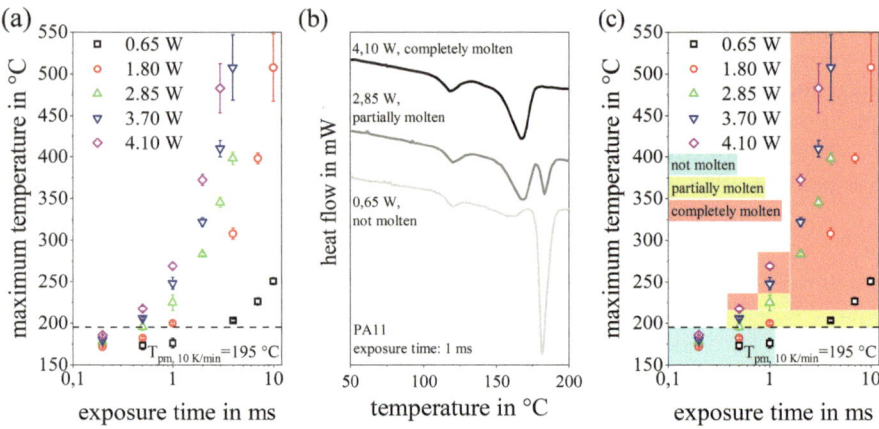

Fig. 4.4 Maximum temperature reached during exposure depending on parameters (**a**), second heating after exposure (**b**), and the process matrix for the evaluation of suitable process parameters (**c**) for PA11

can be observed displaying a double peak, whereas the melting peak at low temperatures is assigned to the share of the melted particle, and the other to the share that has not been melted. Consequently, in the case of complete melting with increasing energy, only a single peak is observed. If this evaluation is carried out for a variety of parameters, it results in a process matrix that can be used for assessment. With this method, interactions may be represented and statements about process parameters can be made even with a small amount of material input.

The influence of laser power and impact time on the resulting maximum temperatures and heating rates during laser exposure are also investigated for PA12. With increasing laser power and impact time, the maximum temperature rises to approximately 450 °C for the chosen settings. Heating rates up to 2×10^6 K/min for impact times of up to 3 ms are shown in Fig. 4.5a. A key question that arises in the process with the high maximum temperatures is when decomposition will take place. This measurement setup also allows statements to be made about the thermal stability of the polymer. At the maximum temperatures reached, degradation would already be detected in the classical approach using TGA, as shown in Fig. 4.5b. The unspecific boundary condition with a constant and "slow heating rate" is thus the issue, since the PBF-LB/P has extremely low exposure times and thus low interactions with the high energies. The thermal decomposition can be evaluated in different ways, either using the melting point T_m or by crystallization T_c. These measurements are carried out after laser irradiation using the flash DSC without laser. Figure 4.5c shows the crystallization curves for different exposures, ranging from 210 °C to 400 °C (for short to long exposure times). If thermal decomposition were to take place, crystallization would already occur at higher temperatures, since decomposition leads to increased mobility of the chains [34].

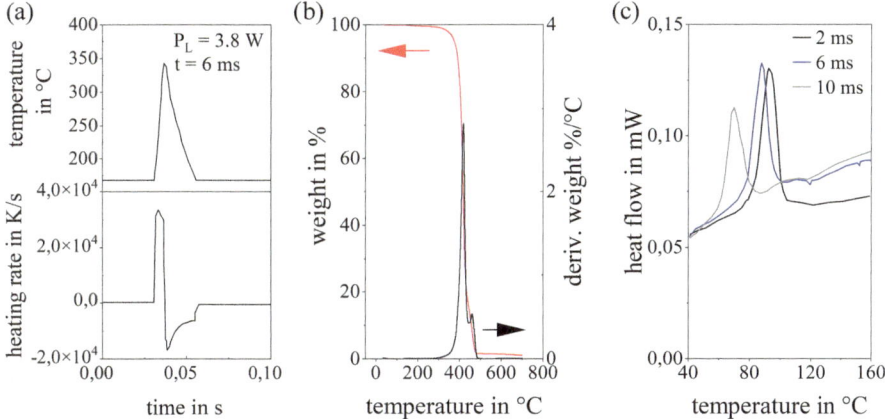

Fig. 4.5 Temperature profile during exposure and corresponding heating rates for PA12, **b** TGA analysis for PA12, crystallization curve after laser exposure for different impact times for a laser power of 3.8 W for degradation statements [34]

4.4   Crystallization in Powder Bed Fusion of Polymers

The prevailing concept of quasi-isothermal laser sintering postulates that the polymer melt and powder exist together throughout the building phase [27]. However, as per the Hoffmann-Lauritzen theory, the applicability of the simplified quasi-isothermal model is restricted to a specific timeframe [33]. This limitation arises due to the occurrence of crystallization, where a transition from a liquid state to a solid state takes place at temperatures below the equilibrium melting temperature (T_{m0}). While it follows that crystallization occurs at any temperature below T_m, whether or not this time is attained depends on the process.

Figure 4.6 shows that for relevant temperatures, such as those in the procedure, crystallization begins within the first minutes although the complete crystallization can take up to several hours. The impact of temperature on material recoating in PA12 (PA2200) was investigated in simulations conducted by Amado et al. Their findings suggest that when the building temperature reaches 172°C, approximately 30% of the relative crystallization conversion is achieved within 5 min during the building process. Drummer et al. [9], in a separate study, substantiated the simulation results of Amado et al. through experimental validation. They employed an in-house developed drop test for polyamide 12. The experimental results revealed that crystallization and subsequent solidification of the material were already observed in the uppermost layers.

In Sect. 4.5, the answer to the vital question of how the crystallization process is related to the material's solidification is provided. This topic is crucial because internal stresses caused by crystallization and thermal shrinkage must be considered.

Existing simulative approaches, such as that of Amado et al. [1], have the limitation that they only take into account the cooling of the freshly applied powder and not the

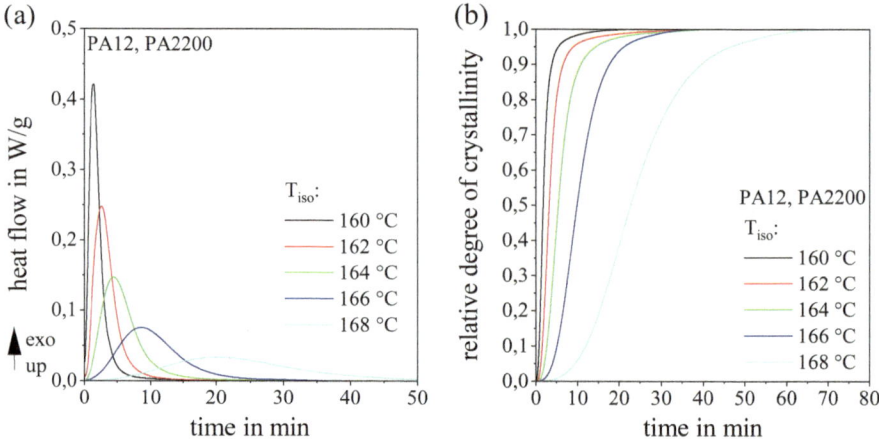

Fig. 4.6 Heat flow for varied isothermal crystallization (**a**), and progress of degree of crystallinity (**b**) for PA12 [35]

heating caused by the re-irradiation of the laser of the overlying layers. Experiments with classical DSC are not possible because the heating and cooling rates are too slow. Therefore, flash DSC is used in current investigations to consider these effects (cooling and reheating) [4]. Since direct detection of the crystallization progress during the holding phase at temperatures above 140 °C is not possible—primarily because of the limited measurement signal due to the low sample weight—a new approach was chosen. In this technique, the crystallization progress was recorded indirectly by measuring the melting rate. Here, cold crystallization is still used even for high cooling rates such as 2000 K/s and heating rates of 1000 K/s, see Fig. 4.7a. When measured isothermally at 166 °C for 300 s, two peaks appear, whereby the one at the higher temperature is attributed to crystallization at the isothermal holding temperature of 166°C. However, since there is insufficient time for complete crystallization, cold crystallization occurs during heating and is attributed to the first peak. Only one peak is attributed to cold crystallization after the crystals in procedure 3 were melted briefly by simulating the laser beam. At this point, it can be already stated that there is a significant impact of reheating by upper layers.

To investigate the effects of undercooling resulting from powder application on crystallization, as well as the temperature increase caused by re-exposure, two separate methodologies were employed. The temperature and time profiles were obtained using three in situ thermocouple measurements [9]. In the first procedure, the particle was melted, and the sample was rapidly cooled to 139 °C at a rate of 2000 K/s, held for 0.5 s, and then heated to the holding temperature of 166 °C to simulate subcooling. This allowed for the examination of the impact of undercooling (procedure 2). To investigate the exposure of the top layer, the sample was heated to 180 °C following a holding step at 166 °C, then cooled to the building temperature of 166 °C, and subsequently melted at 1000 K/s to determine the progress of crystallization (procedure

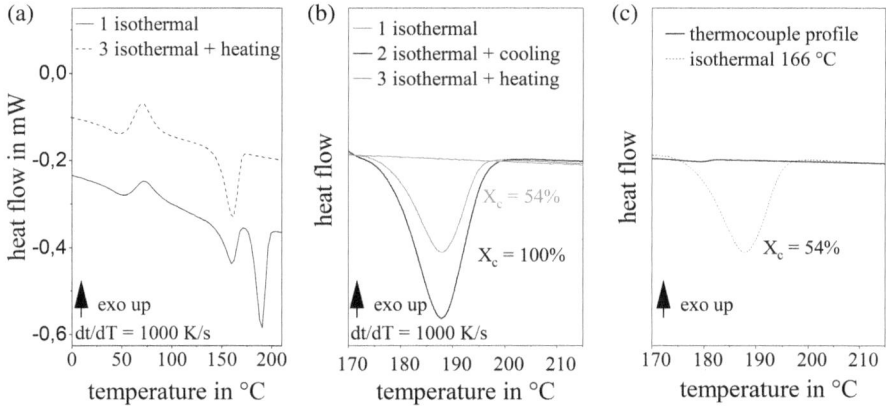

Fig. 4.7 Heat flow measured during the second heating after quenching (**a**), crystallization progress comparison as a function of subcooling and reheating (**b**) and a comparison between thermocouple measurement and isothermal holding stage at 166 °C (**c**) [4]

3). As a reference point, isothermal crystallization experiments were conducted at 166 °C, with holding times of 300 s, 600 s, 900 s, and 1200 s (procedure 1). These experiments were necessary to indirectly determine the time required for complete crystallization, based on the subsequent melting process. In this way, it is possible to describe the relative crystallization process by considering the completion of the crystallization process after 1200 s at an isothermal temperature of 166 °C, as observed in measurement process 1.

This information can then be expressed in the form of a ratio. Finally, the temperature profile for the first 10 layers, until a temperature continuum was achieved, was adapted from the in-situ measurements conducted by Drummer et al. [9] and compared to the isothermal measurements performed for the same duration. When an initial subcooling is introduced before the isothermal holding stage, such as through powder application, the sample is deemed fully crystallized Fig. 4.7b (2). This is because the heightened subcooling of the melt triggers the formation of crystallization nuclei, which in turn accelerates the progress of crystallization at the specified isothermal temperature.

Upon reheating, it becomes apparent that the heat has completely melted the crystalline phases, as no endothermic peak was detectable Fig. 4.7b (3). Figure 4.7c shows an isothermal measurement at 166 °C versus the thermocouple-specific measurement program for the first 300 s after exposure. Following the isothermal holding stage, a crystallization progression of 54% was observed. However, when using the process-adjusting measuring program involving multiple subcooling through powder application and the associated temperature increase during powder exposure, no detectable crystal formation was measured beyond that point.

These results, along with the temperature increase observed in the sublayer, align with previous measurements and support the notion that reheating plays a dominant role in the PBF-LB/P process. The methodology allowed for experimental crystalliza-tion investigations under process conditions, and revealed the impact of the freshly applied powder, especially for the thinner parts that are typical in PBF-LP/P. Soldner et al. [31] and Chap. 14 "Macroscopic Modeling, Simulation and Optimization" have demonstrated a similar effect where the crystallization time is prolonged due to the presence of a process induction time, resulting in an increased crystallization time for thicker-walled components. Furthermore, this measurement method enables the analysis of new process strategies [26]. It is demonstrated that this knowledge of crystallization is a key building block for the development of materials within this complex process [7].

4.5 Structure Development—Hardening

The impact of melt hardening under low melt undercooling and atmospheric pressure creates boundary conditions that have yet to be extensively studied, as conventional techniques do not typically require such information. However, in the case of PBF-LB/P, the transition from the melt state to an elastically dominant melt after exposure becomes crucial [12]. This transition occurs during the building phase and involves crystallization under conditions that induce stresses due to volume shrinkage. Con-sequently, a tailored evaluation is necessary to determine the duration for which the molten polymer remains viscously dominant and the point at which stresses are stored in the melt. In this study, the crystallization of a semi-crystalline melt is investigated by combining rheological data with FTIR (Fourier-transform infrared spectroscopy) microscopy. A modified measurement setup involving an ATR (atten-uated total reflectance) crystal on the rheometer enables simultaneous characteriza-tion of crystallization through FTIR spectroscopy and measurement of the material's rheological behavior.

Therefore, the crystallization of the FTIR measurements were validated and, for this purpose, the bands resulting from the crystallization were set with a reference band. The results indicated that the measurement range is extended, since the mea-surement is absolute and too low, and the measurement signals cannot be lost in the noise compared to DSC [5]. The FITR approach is very well suited for evaluating the progress of crystallization. Looking at the parallel solidification curves in Fig. 4.8a, after a short initiation time, a strongly decreasing loss factor can be noted. This obser-vation is due to the crystallization. If we now observe the crystallization progress necessary for the transition from liquid to solid at 45°, we find that for PP (Fig. 4.8b) and for PA12 this already occurs at low levels, i.e. at values significantly lower than 10% [12]. Another important factor in the process is the point at which stresses can be absorbed into the material. For this purpose, isothermal relaxation tests were carried out from the melt and these clearly show that no stresses are absorbed in the material until the crossover point is reached.

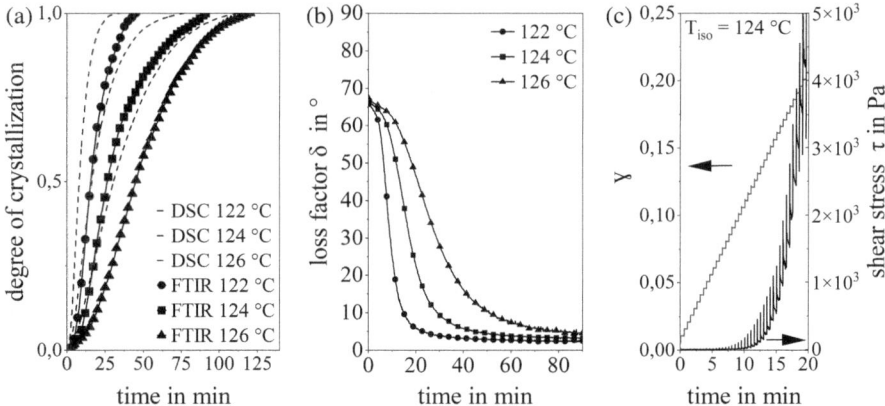

Fig. 4.8 Comparison of the relative degree of crystallization from DSC measurements with FTIR (**a**), crystallization hardening for different isothermal holding temperatures (**b**), and relaxation experiments of the melt (**c**) for Polypropylene [6]

4.6 Structure Development—Shrinkage

Shrinkage varies in extent depending on the type of plastic. However, it cannot be completely avoided. While it is less pronounced in amorphous thermoplastics, greater shrinkage is to be expected in semi-crystalline thermoplastics. This is due to the higher density caused by the crystalline domains that are formed. When the melt or component shrinks inhomogeneously due to different temperatures, which also leads to different crystallization shrinkage locally, residual stresses accumulate that can lead to warpage and curling. Warpage is one of the main challenges in PBF/LB-P and is defined as the unwanted deformation of the processed components that occurs due to shrinkage caused by an uneven crystallization process. Inhomogeneous temperature distribution can also contribute to this effect. Shrinkage and warpage can also result in another effect known as curling in which the edges of a component lift as a result of shrinkage differences and the resulting residual stresses between the upper and lower surfaces. Curling usually occurs on the surface of the part while the first layers are being processed.

Thermomechanical analysis is a suitable method for investigating the thermal expansion of a specimen, in which the specimen is investigated as a function of temperature under a defined furnace atmosphere. For this purpose, the specimen is placed in a holder and loaded with a static or dynamic force, and then subjected to a specific temperature profile. In this way, the TMA can be used to measure the linear coefficient of expansion of solid bodies until softening. However, there are limitations in the temperatures at the phase transition, and therefore an investigation in the molten state is not possible. This hurdle can be overcome with a so-called Pressure-Volume-Temperature (PVT) measurement since both solid and molten samples can be investigated. The problem that emerges, however, is the high pressures at which the measurement must be carried out, and only an extrapolation to normal pressure is possible [19].

A further approach that has been used in these studies is the gap change due to isothermal crystallization on a rheometer at the normal pressure used in the process. After melting the sample, it is cooled to the isothermal holding temperature and, once this temperature is reached, the measurement of isothermal crystallization shrinkage begins by detecting the change in the measuring gap over time, whereupon the plate moves in response to shrinkage and thus the measuring gap becomes smaller. In this setup, it is important that no pressure is applied to the plate from above, as this could change the gap and thus influence the result of the measurement. For this reason, a small force of $-0.01\,N$ is applied as a normal force. In addition, a thermo-gap is used during the test to compensate for the thermal shrinkage of the two plates of the rheometer, so that this effect also does not influence the measurement.

An important result is the linear growth of shrinkage with crystallization as already at the beginning at low turnovers the shrinkage increases disproportionately, which confirms—similar to the solidification studies—that even low turnovers are of great importance and a typical reference value such as $t_{1/2}$ is only conditionally suitable for the process. In addition to the constant normal force and the observation of the gap change, the inverse measurement is also possible, in which the gap is kept constant and the normal force generated by the crystallization is measured. This shows a parallel plot in which even small conversions lead to a large increase in the normal force (Fig. 4.9).

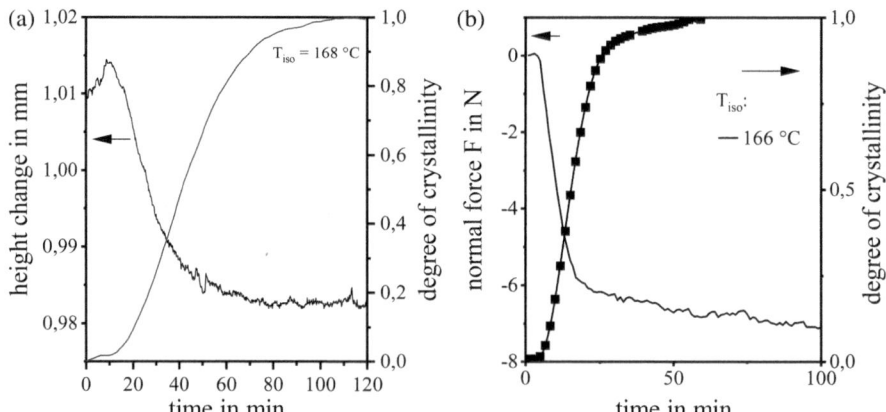

Fig. 4.9 Isothermal gap change of PA12 at 168 °C (**a**), and isothermal normal force development due to isothermal crystallization at the constant gap at 168 °C (**b**)

4.7 Investigations Concerning Optical Properties

When developing multi material approaches for PBF-LB/P, deviations in sintering behavior for the various materials due to different absorptance and heat distribution played a significant role in the early stages of the investigations presented in Chap. 6 concerning multi material additive manufacturing. Additionally, the optical properties of the powder layer were shown to include significant deviations from the optical properties of the solid material from the same polymer [16–18]. For this reason, a measurement setup was employed to gain access to the characterization of optical properties in process-near conditions of a single polymer powder layer. A preliminary version of this technique was developed to analyze the preheating capabilities and absorptance resulting in varying optical energy inputs for polymers to achieve a multi material powder bed fusion process route [17, 18]. More details about this multi material approach are provided in Chap. 6. These investigations opened up a whole scientific field in which the development, improvement, and extension of the method took place. In this context, it should be noted that these investigations are not only relevant for multi material PBF. To obtain a setup to perform fundamental research, this early version was reworked. It contains a narrow heating chamber with a silver solid heater and transmission hole. On this heater, a powder sample is prepared on top of BaF2 glass windows to allow high transmittance for the CO_2-laser at 10.6 μm wavelength. The chamber is closed with the same BaF2 glasses to achieve temperature stability inside. This heated sample chamber is placed between two integrating spheres (also known as Ulbricht spheres) with open ports to access total reflectance, directed reflectance, total transmittance, and directed transmittance and to irradiate the sample with the laser with an angle of 8°. Additionally, the setup was equipped with an FTIR modular unit with $\lambda = 1.3$–25 μm to analyze spectroscopically accessible structural changes in the material, such as isothermal crystallization. Further details concerning this measurement setup are provided in the publications of Schuffenhauer et al. [30] and Marschall et al. [20].

4.7.1 Optical Properties Without Thermal Influence

Before commencing with further investigations, the influences of the powder on the optical behavior at room temperature must be known. Optical properties are strongly conditioned by the microstructure of the mostly semi-crystalline thermoplastic polymers in PBF-LB/P. While this behavior is already known for the raw material, it has to be reconsidered for the dominant influences in a high entropic powder layer structure. To demonstrate the different morphologies, a quenched and a slowly cooled PA12 sample are compared. Thin films in transmission microscopy (Fig. 4.10) show different spherulite sizes of the crystalline phase within the amorphous matrix. The slow cooling exhibits spherulites of approximately 40 μm, whereas the quenched ones only reached fractions of this size, ranging from about 5 μm. The quenched

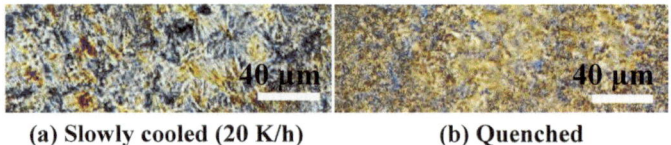

(a) Slowly cooled (20 K/h)   (b) Quenched

Fig. 4.10 Polarization microscopy images of PA12 thin sections with crystallization morphology differences related to cooling rate [29]

samples contained far more of the small nucleating spherulites compared to the slow cooled ones. This high count of spherulites results in numerous scattering interfaces within the polymer, thereby measurably reducing the transmittance of the material. Controlled slow cooling results in bigger spherulites impinging each other, thus reducing the number of scattering interfaces. Therefore, the transmittance is higher on the slowly cooled samples [29].

This implies that the total number of scattering events within the powder layer is critical on a scale thoroughly below polymer particle size. Polymer powder materials for PBF are modified with a range of additives to control powder properties such as flowability, humidity, color, sometimes absorbers, and more. The most frequently used additives are flow aid additives such as fumed silica. To evaluate the influence of these additives on the optical properties of the powder, powder was additivated with additional amounts of fumed silica, titanium dioxide, alumina, and functional additives such as copper and iron oxide nanoparticles. As illustrated in the graph below, additional flow aids on an already processable material will still enhance the absorptance of the layer and, within range of the measurement uncertainty, they show the same trending and can be presumed as comparable. At the same time, however, the optical properties of a defined layer of pure additives can differ markedly, although the additivated PA12 powder layers show similar trending when varying these additives. This indicates that it is not the optical material properties that mainly determine the optical behavior of the powder layer in terms of flow aid additives. The quantity of scattering events exerts the main influence, which is enlarged by additional additives on the surface, but also by a slightly enhanced powder bed density conditioned by powder flowability—even though better flowability does not directly presume enhanced powder bed density. Therefore, choosing an additive to manipulate powder properties should always consider the change in optical behavior and, therefore, the holistic powder layer properties [21]. The optical penetration depth can be derived from these measurements via the inverse adding doubling mechanism [23, 24]. When the optical penetration depth diminishes below the powder layer height, the layer interconnection will be affected. Additionally, the energy input in the already molten layer beneath is important to prevent isothermal crystallization, which will occur significantly earlier than previously assumed. This effect will be addressed later on in this chapter.

Having Rayleigh scatterers like nano-scale SiO_2 broadens the energy input area and reduces the penetration depth. While this effect is surpassed by the strong Mie scattering events in the powder layer, it plays a significant role in evaluating the optical behavior of these thin powder layers. As PBF-LB/P material particle sizes are mostly between 50–100 μm, a layer of 200 μm height (applied on a flat surface) is assumed to have not more than two or three particles above each other, but contain a multitude of nano additive particles on the host particle surface and volume. Related to the investigations in bulk properties, the flowability increases while adding nanoparticles to the powder. Enhanced flowability alludes to a powder bed density that is also dependent on the blading velocity and thermal expansion properties. Increasing the powder bed or layer density by reducing the interparticle friction and charging—and thus changing the Mie scattering behavior and influencing the Rayleigh scattering events with additives—results in the trends shown above for enhancing the resulting absorptance. The PA11 powder systems shown with different additive contents in Fig. 4.11 are processed as single layers with the same energy density ED of 0.4 mJ/mm^2. In all systems, the weight of single layers is lower compared to the reference material, which is the pure PA11 system. Moreover, as the additive content increases, there is a noticeable decrease in weight across all systems. When considering the thickness of each layer (as shown in Fig. 4.12b), it becomes apparent that the weight decreases despite the higher density of the additive. These characteristics of the part are attributed to changes in the interaction between the beam and the powder. On one hand, the addition of additives alters the flow behavior, resulting in a higher packing density of the powder bed. This, in turn, leads to a decrease in the optical penetration depth due to multiple reflections. At the same time, the additives act as scattering centers, further reducing the optical penetration depth. Consequently, the depth of the melt pool is reduced, causing a decrease in component thickness and, subsequently, reducing the weight of the specimen.

Likewise, this set of influences is relevant for choosing a suitable particle size distribution (PSD) as fine fractions result in higher absorptance, which is related

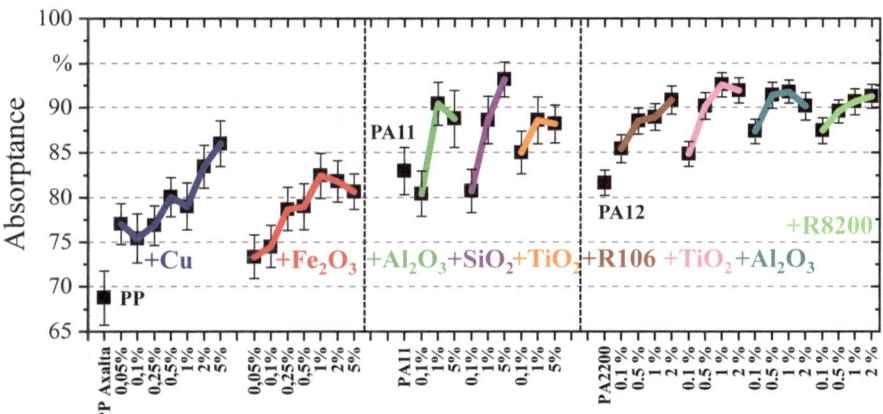

Fig. 4.11 Absorptance of powder layers (PP: 400 μm; PA11/12: 200 μm) with different content of nanoparticle additives (dry coated: Cu, Fe$_2$O$_3$, SiO$_2$ (R106, R8200), TiO$_2$, Al$_2$O$_3$; inside PP-particle: Al$_2$O$_3$, SiO$_2$, TiO$_2$); beam source CO$_2$ laser λ =10.6 μm

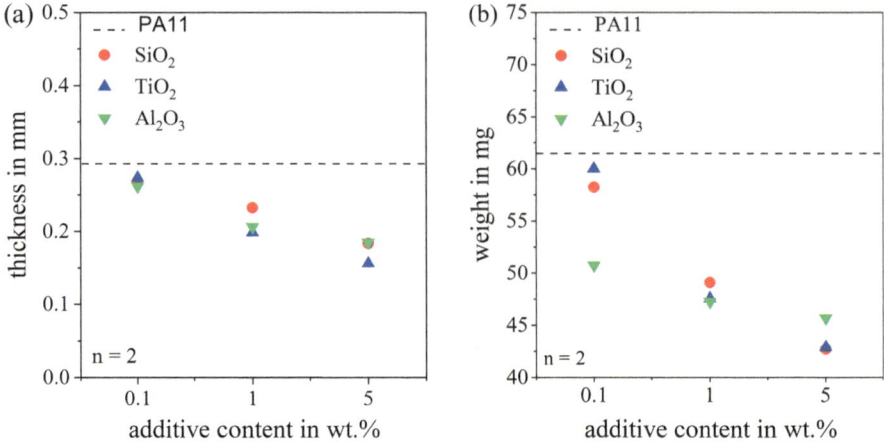

Fig. 4.12 Single layer at an energy density of $0.4 \, mJ/mm^2$, comparison of sample thickness (**a**) and sample weight (**b**) across different material systems with PA11 serving as the reference (represented by the dashed black line) [32]

to more particles per defined layer height. The opposite applies in the case of the coarse fractions having only one or two partly overlapping particles in the beam path, whereby the absorptance decreases. This effect can be measured for particle fractions. However, when limiting PSD by only sieving out coarse particles, the absorptance will not significantly decrease until the mesh sizes are below the D_{50} value of the powder material. This is validated for layer heights of 100–250 µm and commercially available PA12 PBF-LB/P powders at room temperature.

4.7.2 Optical Properties at Process Temperatures

Process temperatures during PBF-LB/P include a process window for preheating between the polymer powder melting onset and the crystallization temperature. This preheating temperature is to be chosen as close to the melting temperature as possible to reduce solidification processes such as curling and warpage, and thereby to ensure that the process does not abort while facilitating good melting with a minimal required laser energy input. Polymers expand in a pronounced fashion under thermal influence before the crystalline phase starts to melt. This changes the powder layer height but does not impact the absorptance of the layer, as the optical density did not change. Therefore, Schuffenhauer et al. also investigate the influence of fillers on the thermal dependent optical properties [30].

With these detailed measurements, the Inverse Adding-Doubling technique can be applied to access the material constants for the optical properties of the powder [23, 24]. This technique is more accurate than estimations using the Lambert-Beers law

Fig. 4.13 Optical material coefficients for a PA12 powder material for PBF-LB/P

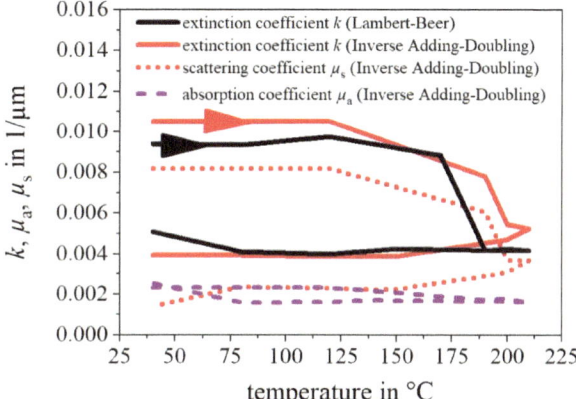

for this application. Material constants depending on the temperature and phase of the powder material can be set up and used as universally valid for the development of process parameters and simulations (Fig. 4.13).

Additionally, the melting behavior of the particles is strongly dependent on their size as small particles need less thermal energy to be melted. Furthermore, the phase of the material determines its optical behavior. Correlating all these effects, a dependency within the degree of particle melt (DPM), the PSD, temperature, and absorptance of the powder results and Fig. 4.14a illustrates this dependency. Small PSD by the same layer height gives a higher absorptance as the layer is optically denser. However, when starting to melt, this turns around until the melting process is finished. In the molten material, only the layer height conditions the absorptance, and

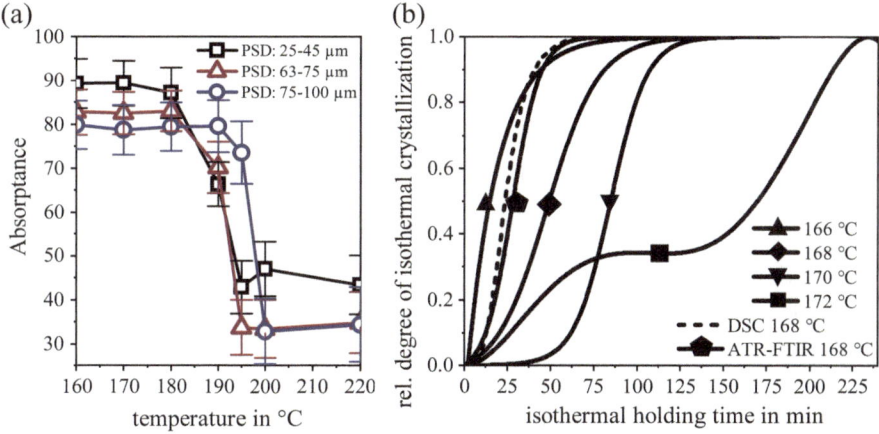

Fig. 4.14 a Correlation between PSD and DPM while particle melting endures, affecting Absorptance of the powder layer; **b** isothermal crystallization of PA12 in γ-phase (FTIR 945 cm^{-1}) at different temperatures concerning DSC measurement [20]

thus the denser packed powder layer of small PSD again results in higher absorptance. As we know, DPM in PBF-LB/P is not 100% and hence the most interesting information lies between the onset of melting and the completed melting process. In this area, the PSD-absorptance correlation turns around, as the bigger particles need more thermal energy input to gain the same proportion of molten particle shells to unmolten particle cores.

4.7.3   Long Layer Downtime and Process Upscaling Regarding Isothermal Crystallization

PBF-LB/P is meant to be an isothermal process between the crystallization temperature of the polymer and the melting onset. In between the building chamber, the preheating temperature is chosen. The temperature in the powder bed depth decreases slowly, leading to solidification processes within longer built jobs, although the main crystallization and solidification should only appear while slowly cooling afterward. In the context of process upscaling and increasing chamber dimensions, temperature homogenization across the whole volume is a challenge to address. Even though the temperature is stable, isothermal crystallization occurs depending on the temperature variation from the melting temperature and the layer downtime until new thermal energy is distributed on the same spot. These processes are believed to take place in the powder bed depth volume. However, it can be shown that small temperature variations from optimized preheating already lead to isothermal crystallization within layers that are still involved in the active building process. This leads to different optical energy inputs due to changes in optical properties. Additionally, it changes the rheology of the viscous phase which can result in deviations from the desired geometry. As visualized in Fig. 4.14b, small temperature variations already lead to crystallization within a few minutes. More information concerning this topic is provided in the publication [19].

4.7.4   Model of Absorptance in Polymer Powder Layer Systems

For a most suitable simulation base for a polymer bulk material in the case of optical behavior, the bulk has to be described somewhere between a volume consisting of fractals and a thin layer system of partly overlapping spherical particles. These have to be described with a static deviation formfactor towards the potato-like real particle shape, depending on the powder forming process route. The given layer model includes all optically relevant interactions between the beam and powder to describe and evaluate a particle layer system of diffuse scattering and partly transmitting shares. Therefore, they do not relate to applicable Van-der-Hulst single particle light scattering phenomena. Based on this fundamental research, process parameters could be predicted to ensure a homogeneous optical energy input throughout the building chamber. This is the key to ensuring process quality in up-scaling to enlarged building volumes.

As shown below, the beam material interaction for a polymer powder layer is not a trivial process. Due to the rough powder bed surface, there is only a limited amount of unscattered reflection, mostly diffuse reflection. The first interaction is the scattering event at the polymer particle surface. Assuming that the particle consists of a homogeneous solid material, the intensity decreases according to the Lambert Beers law. However, inside the polymer are additives besides the spherulites of the semi-crystalline thermoplastic and each of these leads to different scattering phenomena. Spherulites and polymer particles as a whole act as Mie scatterers, as they are close to the wavelength of the incident beam, distributing a forward-oriented scattering distribution. The particles that are much smaller than the wavelength interact as Rayleigh scatterers, without a special direction-dependent distribution. However, as every surface or inhomogeneity acts as an optical resistance, the Rayleigh scattering phenomena distributes the remaining intensity in the direction of the less optically dense regions. Rayleigh is much less dominant compared to Mie scattering but together they create multi scattering/multi reflection channels leading deeper into the polymer layer. This generates a strong dependency of the layer height and density on the absorptance of the layer, whereby the classical Lambert Beers law only applies to a smaller portion of the layer when describing the optical behavior of a homogeneous particle. This correlates to the deviations to inverse adding doubling when investigating the absorption coefficient as already shown. When the measurement with the double integrating spheres is performed and the optical coefficients are determined, they summarize the influence of all the occurring optical interactions to a value to determine optimized layer height and exposure parameters (Fig. 4.15).

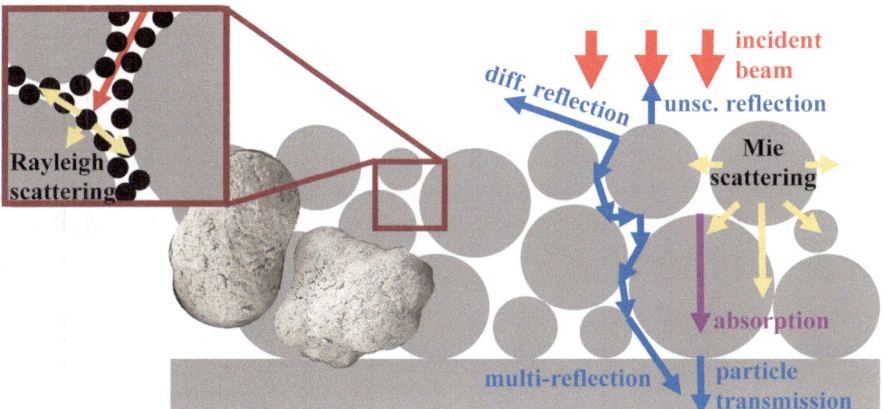

Fig. 4.15 Schematic illustration of beam matter interactions during exposure of a polymer powder layer to a far infrared laser beam; with examples of real PA12 particles

4.8 Summary

Considering all influences on the model of absorptance in a powder layer, the knowledge gained and measurement technique can be used for the powder design process with fast feedback circles in the early material development stage. Additionally, powder refreshing effects can be redirected to measurable optical properties of a polymer powder layer. With a set of referenced values to compare to, this is a rapid method to ensure that the build quality with old or refreshed powder remains unchanged. Depending on the polymer, prepared old powder will be added to the new powder, and in the case of PA12, 50/50 is commonly used although it is more a practicable best guess than a scientific decision, as long as the process remains stable. Additionally, the FTIR can then provide further information about the crystallization state but limited information about oxidation and powder aging. Problems that occurred might be solved by using Raman spectroscopy instead, for distinct aging information, but requires a surrounding and material that allows for the evaluation of Raman signals without fluorescent ingredients. As a process observation and controlling device, the elaborated setup has to be used in reflection mode to analyze the optical properties in process of PBF-LB/P. However, reflectance was no value to give significant information about the powder layers state in this experimental setup. The information to be obtained in-situ can only relate to reflectance change due to phase borders in correlation to optical penetration depth with relation to the beam engaging angle. The fundamental scientific outcome of this research enables reverse engineering of process parameters from the side of a polymer powder layer. Its resulting absorptance and optical penetration depth relates to suitable process parameters for sufficient interlayer connection while keeping the energy input on a minimal level to reduce negative effects such as curling or thermal damage. Especially the investigations concerning isothermal crystallization provide a promising tool for further use, and upscaling and bigger building chambers are key topics in the AM community. The layer time differences and temperature stability of a process mostly suffer when the thermodynamic mass increases. Thus, the information about isothermal crystallization in the upper layers will have an even greater influence. The handling of optical energy input to the polymer will benefit from an implementation of the investigated information about the crystallization kinetics and the triggered change in absorptance and rheology. Material-specific crystallization kinetics can be correlated to the layer time depending on the built part geometry and the thermal conditions of the building chamber. This can be used to adjust the process parameters with an algorithm to ensure higher process quality, even within upscaled processes.

References

1. Amado, A., Schmid, M., Levy, G.: Characterization and modeling of non-isothermal crystallization of polyamide 12 and co-polypropylene during the sls process. In: Proceedings of the 5th International Polymers & Moulds Innovations Conference (2012)
2. Blümel, C., Sachs, M., Laumer, T., Winzer, B., Schmidt, J., Schmidt, M., Peukert, W., Wirth, K.-E.: Increasing flowability and bulk density of pe-hd powders by a dry particle coating process and impact on lbm processes. Rapid Prototyp. J. **21**(6), 697–704 (2015)
3. Blümel, C., Schmidt, J., Dielesen, A., Sachs, M., Winzer, B., Peukert, W., Wirth, K.-E.: Dry particle coating of polymer particles for tailor-made product properties. In: The 29th International Conference of the Polymer Processing Society (2014)
4. Cholewa, S., Drummer, D.: Crystallization behavior under process conditions in powder bed fusion of polymers. In: Proceedings of the 12th CIRP Conference on Photonic Technologies LANE 2022, Procedia CIRP (2022)
5. Cholewa, S., Drummer, D.: Prozessnahe simultane bestimmung des kristallisations- und fließverhaltens von pa 12 im lasersintern. In: Proceedings of the 18th Rapid. Tech 3D Conference (2022)
6. Cholewa, S., Jaksch, A., Drummer, D.: Structure development of semi-crystalline polymers in laser based powder bed fusion. In: Proceedings of the Antec (2023)
7. Cholewa, S., Stieglitz, L., Jaksch, A., Rieger, B., Drummer, D.: Tailored syndiotactic polypropylene feedstock material for laser-based powder bed fusion of polymers: Material development and processability. ACS Appl. Polym. Mater. **5**(4), 2430–2439 (2023)
8. Dechet, M., Gómez Bonilla, J., Lanzl, L., Drummer, D.,, Bück, A., Schmidt, A., Peukert, W.: Spherical polybutylene terephthalate (pbt)—polycarbonate (pc) blend particles by mechanical alloying and thermal rounding. Polymers **10**(12), 1373 (2018)
9. Drummer, D., Greiner, S., Zhao, M., Wudy, K.: A novel approach for understanding laser sintering of polymers. Additive Manufact. **27**, 379–388 (2019)
10. Düsenberg, B., Tischer, F., Valayne, E., Schmidt, J., Peukert, W., Bück, A.: Temperature influence on the triboelectric powder charging during dry coating of polypropylene with nanosilica particles. Powder Technol. **399**, 117224 (2022)
11. Fanselow, S., Emamjomeh, S., Wirth, K., Schmidt, J., Peukert, W.: Production of spherical wax and polyolefin microparticles by melt emulsification for additive manufacturing. Chem. Eng. Sci. **141**, 282–292 (2016)
12. Greiner, S., Jaksch, A., Cholewa, S., Drummer, D.: Development of material-adapted processing strategies for laser sintering of polyamide 12. Adv. Ind. Eng. Polym. Res. **4**(4), 251–263 (2021)
13. Hesse, N., Dechet, M., Bonilla, J., Lübbert, C., Roth, S., Bück, A., Schmidt, J., Peukert, W.: Analysis of tribo-charging during powder spreading in selective laser sintering: assessment of polyamide 12 powder ageing effects on charging behavior. Polymers **11**(4), 609 (2019)
14. Hesse, N., Winzer, B., Peukert, W., Schmidt, J.: Towards a generally applicable methodology for the characterization of particle properties relevant to processing in powder bed fusion of polymers – from single particle to bulk solid behavior. Additive Manufact. **41** (2021)
15. Lanzl, L., Wudy, K., Drexler, M., Drummer, D.: Laser-high-speed-dsc: process-oriented thermal analysis of pa 12 in selective laser sintering. In: Proceedings of the 10th CIRP Conference on Photonic Technologies LANE 2018, Physics Procedia (2018)
16. Laumer, T., Stichel, T., Bock, T., Amend, P., Schmidt, M.: Characterization of temperature-dependent optical material properties of polymer powders. In: Proceedings of the Polymer Processing Society 30th Annual Meeting (2014)
17. Laumer, T., Stichel, T., Nagulin, K., Schmidt, M.: Optical analysis of polymer powder materials for selective laser sintering. Polym. Test. **56**, 207–213 (2016)
18. Laumer, T., Wudy, K., Drexler, M., Amend, P., Roth, S., Drummer, D., Schmidt, M.: Fundamental investigation of laser beam melting of polymers for additive manufacture. J. Laser Appl. **26**(4) (2014)

19. Launhardt, M., Drummer, D.: Determination of the fundamental dimension development in building direction for laser-sintered parts. J. Polym. Eng. **39**(2), 197–206 (2019)
20. Marschall, M., Cholewa, S., Kopp, S.-P., Drummer, D., Schmidt, M.: Holistic characterization of pbf-lb/p powder regarding isothermal crystallization, rheology and optical properties under process conditions. In: ICAT 2023 - 8th International Conference on Additive Manufacturing (2023)
21. Marschall, M., Heintges, C., Schmidt, M.: Influence of flow aid additives on optical properties of polyamide for laser-based powder bed fusion. In: Proceedings of the 12th CIRP Conference on Photonic Technologies LANE 2022, Procedia CIRP (2022)
22. Parteli, E., Pöschel, T.: Particle-based simulations of powder coating in additive manufacturing suggest increase in powder bed roughness with coating speed. In: EPJ Web of Conferences. EDP Sciences (2017)
23. Prahl, S.: Everything i think you should know about inverse adding-doubling. Date not provided
24. Prahl, S., van Gemert, M., Welch, A.: Determining the optical properties of turbid media by using the adding-doubling method. Appl. Opt. **32**(4), 559–568 (1993)
25. Sachs, M., Blümel, C., Winzer, B., Schmidt, J., Peukert, W., Wirth, K.-E.: Dry particle coating of hydrophobic and hydrophilic polymer particles for improvement of flowability and bulk density. In: Partec Proceedings (2013)
26. Schlicht, S., Cholewa, S., Drummer, D.: Process-structure-property interdependencies in non-isothermal powder bed fusion of polyamide 12. J. Manufact. Mater. Process. **7**(1), 33 (2023)
27. Schmachtenberg, E., Seul, T.: Model of isothermic laser-sintering. In: 60th Annual Technical Conference of the Society of Plastic Engineers (ANTEC) (2002)
28. Schmidt, J., Peukert, W.: Dry powder coating in additive manufacturing. Front. Chem. Eng. **4** (2022)
29. Schuffenhauer, T., Stichel, T., Schmidt, M.: Experimental determination of scattering processes in the interaction of laser radiation with polyamide 12 powder. In: Proceedings of the 11th CIRP Conference on Photonic Technologies LANE 2020, Procedia CIRP (2020)
30. Schuffenhauer, T., Stichel, T., Schmidt, M.: Employment of an extended double-integrating-sphere system to investigate thermo-optical material properties for powder bed fusion. J. Mater. Eng. Perform. **30**, 5013–5019 (2021)
31. Soldner, D., Greiner, S., Burkhardt, C., Drummer, D., Steinmann, P., Mergheim, J.: Numerical and experimental investigation of the isothermal assumption in selective laser sintering of pa12. Additive Manufact. **37** (2021)
32. Tischer, F., Cholewa, S., Düsenberg, B., Drummer, D., Peukert, W., Schmidt, J.: Polyamide 11 nanocomposite feedstocks for powder bed fusion via liquid-liquid phase separation and crystallization. Powder Technol. **424** (2023)
33. Vyazovkin, S., Sbirrazzuoli, N.: Isoconversional approach to evaluating the Hoffman-Lauritzen parameters (u* and kg) from the overall rates of nonisothermal crystallization. Macromolecular Rapid Commun. **25**(6), 733–738 (2004)
34. Wudy, K., Drexler, M., Lanzl, L., Drummer, D.: Analysis of time dependent thermal properties for high rates in selective laser sintering. Rapid Prototyp. J. **24**(5), 894–900 (2018)
35. Zhao, M., Wudy, K., Drummer, D.: Crystallization kinetics of polyamide 12 during selective laser sintering. Polymers **10**(2), 168 (2018)

Dietmar Drummer succeeded Prof. Ernst Schmachtenberg as Head of the Institute of Polymer Technology at Friedrich-Alexander-Universität Erlangen-Nürnberg on May 1, 2009. He was the spokesperson for the Collaborative Research Center 814 Additive Manufacturing, which focused on process understanding, development, and process-adapted material characterization. Furthermore, he heads the Polymer Group of Neue Materialien Fürth GmbH and 2 Keylabs at the Bavarian Polymer Institute.

Michael Schmidt   has headed the Institute of Photonic Technologies since its founding in 2009 at Friedrich-Alexander-Universität Erlangen-Nürnberg. His research interests include laser application from micro- to macroscopic scales within industrial manufacturing, additive manufacturing, and medical engineering. He was the vice spokesperson of the Collaborative Research Center 814 Additive Manufacturing.

Part II
Processes and Exposure Strategies

Chapter 5
New Process Strategies for Laser Powder Bed Fusion of Polymers

Sandra Greiner, Samuel Schlicht, and Dietmar Drummer

Abstract This chapter investigates the effects of thermal and temporal influences on the resulting part properties, particularly the temperature profile in the powder bed fusion process with a laser beam for polymers (PBF-LB/P). Through detailed process analysis, new strategies have been developed to increase the robustness of the process, control thermal history, and improve part quality while promoting sustainability and productivity. Key insights are provided into the heating and powder coating processes, which are critical constraints for achieving uniform part properties with minimal material variation. Furthermore, a non-isothermal PBF approach is presented for processing materials that are not traditionally suitable for PBF.

5.1 Introduction

Laser powder bed fusion of polymers (PBF-LB/P) is an additive manufacturing technology that fuses powder particles layer-by-layer with a laser to generate complex-shaped polymer parts from digital data. For the predictable fabrication of parts with desired properties, the material (see Chap. 4) and the process have to be understood and must fulfill certain requirements. Therefore, specific thermal boundary conditions, as described by the model of isothermal laser sintering [32], have to be met. According to this model, the processing window for semi-crystalline materials is defined by the onset of melting and crystallization (Fig. 5.1). The larger this temperature range, the higher the process robustness.

S. Greiner · S. Schlicht · D. Drummer (✉)
Friedrich-Alexander-Universität Erlangen-Nürnberg, Institute of Polymer Technology,
Am Weichselgarten 10, 91058 Erlangen-Tennenlohe, Germany
e-mail: dietmar.drummer@fau.de

© The Author(s) 2025
D. Drummer and M. Schmidt (eds.), *Progress in Powder Based Additive Manufacturing*, Springer Tracts in Additive Manufacturing,
https://doi.org/10.1007/978-3-031-78350-0_5

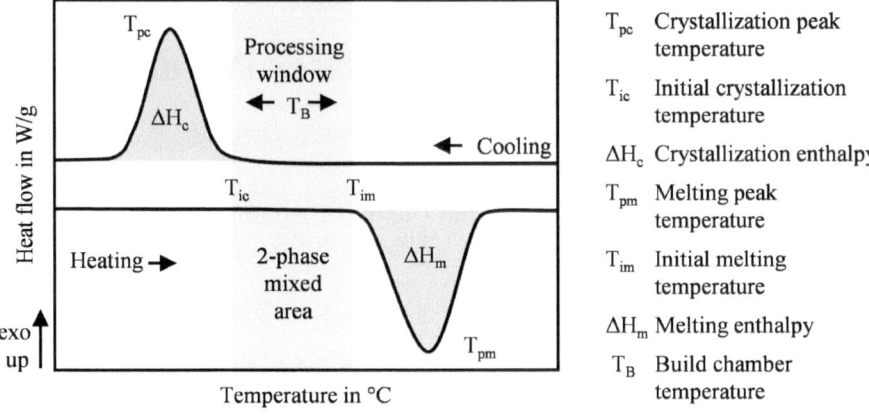

T_{pc} Crystallization peak
      temperature

T_{ic} Initial crystallization
      temperature

ΔH_c Crystallization enthalpy

T_{pm} Melting peak
      temperature

T_{im} Initial melting
      temperature

ΔH_m Melting enthalpy

T_B Build chamber
      temperature

Fig. 5.1 Schematic depiction of the model of isothermal PBF

The three process phases of preheating, processing, and cooling (Fig. 5.2a), and their specific interactions and emerging influences on the part properties have to be understood, particularly during the processing phase in which the three sub-processes of powder coating, exposure, and consolidation (Fig. 5.2b) are repeated for each layer. A deep understanding of the complex thermal interactions is essential to derive new process strategies for a highly productive PBF process.

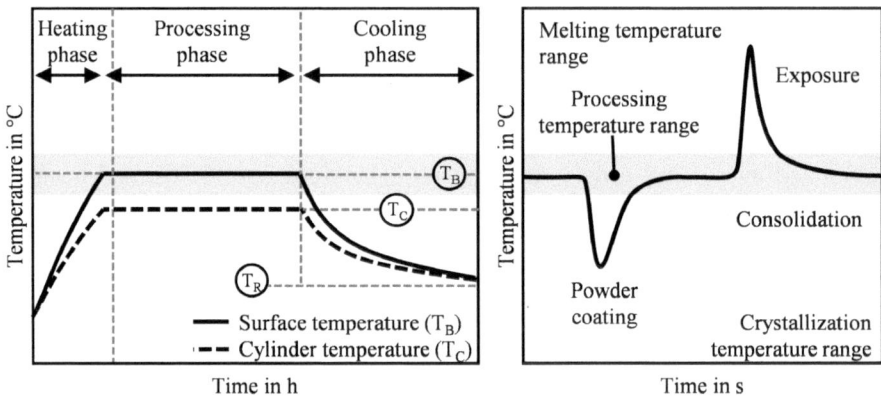

Fig. 5.2 Temperature history during **a** the three superordinate process phases and **b** the thermal impact of the three sub-process steps of the PBF-LB/P process

In the state of the art, these manifold and uncontrolled thermal interactions lead to a process with limited robustness and reproducibility of the resulting part properties, such as dimensional accuracy, shape, and position deviations for part densities or mechanical properties. The locally significantly varied temperature histories affect the part properties and induce a comparatively low reproducibility of part properties. Furthermore, the variability of the thermal boundary conditions sets high requirements for the qualification and application of new material systems to the PBF process, especially with regard to an industrial context [1]. In the following sections, these relevant process phases and the sub-process steps are analyzed under various boundary conditions with regard to state-of-the-art processing. In particular, generating an understanding of time- and location-resolved thermal effects during the iterative build process is of great importance for the further development of the process. In addition, based on this fundamental knowledge, new process strategies are derived that allow for an increased process stability, process acceleration, or the homogenization of part properties throughout the build chamber.

5.2 Process Phases, Boundary Conditions, and Influences

The selective laser sintering process is divided into three distinct phases that are governed by various process- and material-related influences and boundary conditions. These will be discussed with regard to their existing constraints, relevant boundary conditions, process-induced influences, and process-based enhancements.

5.2.1 Preheating Phase

During the preheating phase, the PBF system and the contained material are continuously heated to the build chamber temperature, which represents a material-specific temperature level on the powder bed surface (T_B). In addition, the build cylinder is heated to a specific temperature (T_C) level to temper the material within the deeper layers in the build direction (z-height). This temperature level depends on the size and shape of the build chamber and is usually several degrees below T_B. According to the model of isothermal PBF, an indication of the valid build chamber temperatures can be derived from DSC measurements (Fig. 5.1). The model describes the presence of melt next to the powder bed within the complete processing phase. According to this theory, high T_C values are beneficial to induce uniform shrinkage due to very low cooling rates and slow material crystallization and consolidation during the cooling step. Nevertheless, in [26, 37], isothermal crystallization was detected for polyamide 12 (PA12), which can be correlated to material solidification in the processing phase. In current research [9], an experimental validation of this theory was obtained, enhancing the model representation with significant aspects regarding material consolidation. Based on this extended process model, new processing strategies emerge that involve early material solidification as a distinguishing aspect.

5.2.2 Processing Phase

In PBF, the three constantly repeating sub-process steps of powder coating, laser exposure, and material consolidation characterize the processing phase. Apart from analyzing specific interrelations, the impact of these sub-processes has to be understood in isolation to allow for targeted process adaptions.

5.2.2.1 Powder Coating

The process step of powder coating can be regarded as the initial step of the three continuously repeating sub-process steps. After lowering the build platform, new powder from the respective feed area is supplied. Therefore, a powder coating mechanism, such as a doctor blade or a counter-rotating roller, applies new powder by a translatory application movement at a defined coating speed. Counter-rotating rollers allow for higher process robustness compared to that obtained with doctor blades. Through powder fluidization, the probability of a layer tearing out at the slightest sign of curling is significantly reduced. In addition to the provision of new powder, the powder coating process is responsible for the generation of homogeneous boundary conditions in terms of powder spreading, powder bed density, temperature distribution, and thus process stability. Hence, the powder coating sets the foundation for the subsequent exposure process.

Powder spreading

For homogeneously spreading the powder on the surface, sufficient powder has to be provided. In [6], it was shown that the temperature distribution on the powder bed surface was significantly affected by the amount of powder provided. However, only a minor impact on part properties, such as part density or morphology, was found. A lower amount of powder is correlated to faster heating and a slight tendency to higher part densities exhibiting an interaction with the exposure process.

Powder bed density

The powder bed density is a process variable that directly results from the coating strategy in conjunction with the material used and its powder properties. Due to limited accessibility within the build chamber, this parameter is difficult to measure directly. However, the powder bed density can be determined indirectly via the fabrication of large powder-filled hollow containers or through the direct characterization of the powder bed surface using optical methods [19]. Fringe projection measurements allow for a direct measurement of the powder bed surface roughness and its correlation to the powder bed density. Apart from the shape of the recoating mechanism, the coating speed also represents a parameter that directly influences the surface roughness of the powder bed. The acceleration of the recoating speed, as well as an increase of the powder temperature, is correlated with higher surface roughness and higher standard deviations. Regarding the part properties produced from a high flowability PA12 powder, the powder spreading characteristics do not show a direct correlation with the emerging part surface roughness or part density. In contrast, a significant influence is observed when considering the processing of powders with aspherical particle geometries—which are associated with a significantly reduced flowability—impeding emerging part properties. Therefore, a compromise has to be found regarding the process robustness, process speed, and part quality. In current research, it was found that the powder bed density affects the homogeneity of the local particle distribution and therefore the variability of the initial boundary condition obtained for each layer. In addition, the powder bed density represents a significant influence on the energy input, the beam propagation within the powder bed, and the heat transport within the powder system [3], necessitating the consideration of the interaction of the powder system and corresponding powder spreading parameters for optimizing emerging powder bed characteristics.

Thermal impact of powder coating

Controlling the impact of powder coating places particular significance on thermal aspects. Coating the previously molten cross-sections with new and cooler powders can affect the process stability and induce curling, especially when the temperature of the newly applied powder is below the crystallization temperature range of the polymer used. In Fig. 5.3a, spherulites whose growth is initiated by the presence of new powder particles can be detected, whereby the respective setup of the PBF system is decisive for the temperature level of the newly applied powder. In general, higher stability towards thermal inconsistencies can be achieved by employing a reduced amount of powder or by preheating the powder [4]. In spite of the constraints mentioned, applying new (and eventually cooler) powder to every layer may present a reproducible boundary condition that is accompanied by a lower aging impact on the part cake powder, as well as on the powder that is deposited into the overflow containers. In the current research, it was found that the powder bed density and therefore the surface roughness obtained during powder coating is dependent on the material temperature (see Fig. 5.3b). With increasing temperature, the surface roughness of the resulting powder bed is increased through the facilitated stickiness and the corresponding agglomeration of particles, leading to poorer slipping and unrolling of the powder particles. Accordingly, a lower powder coating temperature possesses the potential to contribute to the powder bed density through improved powder flowability. The powder flowability exhibits a dependency on the material temperature, thereby impairing the significance and transferability of standardized flowability tests that are commonly performed at room temperature.

Preheating of the powder coating mechanism and the heat transport to the powder can induce higher powder temperatures [4]. Especially for low coating speeds, a high temperature of the coating system induces heating of the newly applied powder, and thus a slightly higher mixing temperature is generated between the powder and melt. Therefore, targeted modifications and enhancements of the applied machinery and the corresponding process parameterization embed the potential for enhancing the homogeneity of the temperature distribution and the process stability. Powder

(a)   (b)

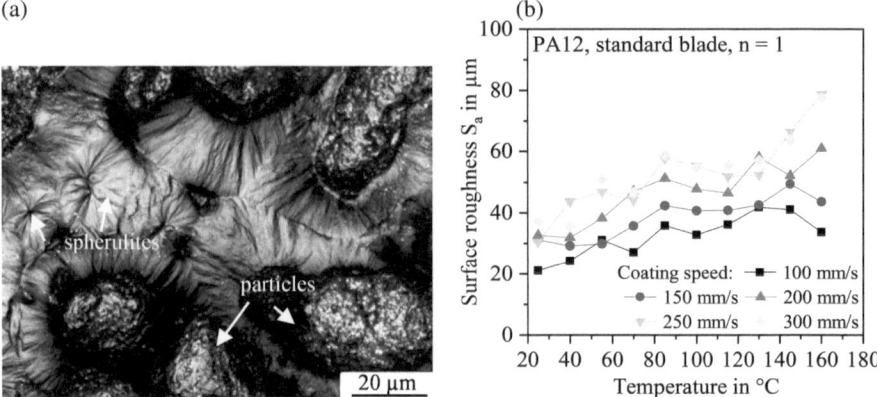

Fig. 5.3 a Laser scanning microscopy of a part surface indicating particle and thermally-induced crystallization [5], and **b** surface roughness of the powder bed under variation of the coating speed and the build chamber temperature

densification

Apart from thermal stabilization and decreasing coating speed, compaction tests indicated a high potential for the generation of powder beds with increased densities [18]. In theory, fluctuations can be reduced by the compaction of the powder bed in relation to the bulk density as statistically-distributed powder beds are avoided. Due to a more homogeneous and smoother surface, a locally more uniform energy input can be achieved, leading to a stabilization of the layer thickness and lower curling probability [18]. In contrast, laser propagation is limited within powder beds of higher density [3]. When applying densification mechanisms to the PBF process, the shape of the recoating mechanism, in particular, plays a significant role. Since standard coating mechanisms are based on the free-flowing application of powder, a random particle distribution with high porosity and local uncertainties is obtained. Through compaction of the powder bed system, this uncertainty can be reduced, allowing for reproducible boundary conditions and corresponding thermal and optical powder bed properties for the following exposure step.

Coating geometries

In general, commercially available laser-based PBF systems are equipped with doctor blades or counter-rotating roller systems. The objective of the available machinery focuses on the homogeneous application of a new layer of powder under a high degree of process stability. The findings in [8] highlight that blade systems lead to slightly higher powder bed densities than counter-rotating rollers. However, in the current research, the impact of a variation of applied forces through the variation of the coating speed and the coating mechanism cannot entirely be confirmed on multi-layer part mechanics [5]. Despite these limitations, slight compaction can be achieved by increasing the rotational speed of the roller [23] or by varying the blade geometry or combining multiple recoating mechanisms [24]. Niino [24] proposed the

(a)

(b)

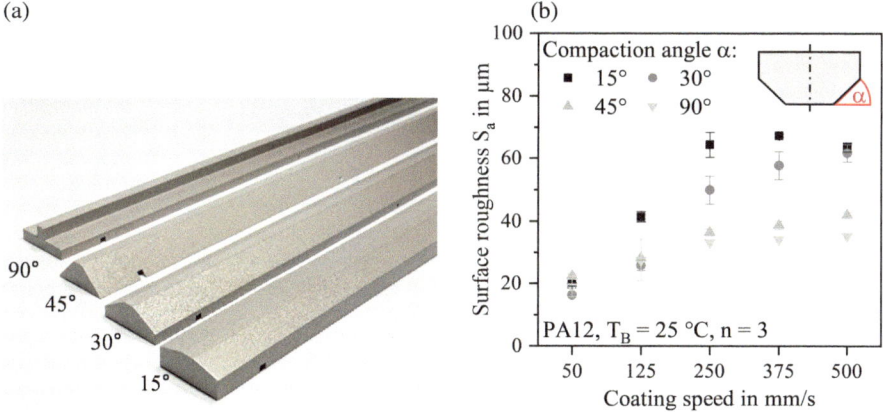

Fig. 5.4 a Temperature controllable blade systems with systematically varied compaction angles and **b** surface roughness depending on the coating speed and compaction angle

combination of two coating mechanisms that can be used to unlink powder spreading from powder compaction. In order to transfer compaction from compressing tests to the PBF process, in current research, doctor blades with different opening angles (see Fig. 5.4a), which result in a varied force effect on the powder bed, were applied, showing a positive correlation of the surface roughness and the underlying recoating speed (see Fig. 5.4b). In addition, a negative correlation of the compaction angle and the surface roughness contradicts previous assumptions and thereby reflects the complex interaction of powder rheological properties and process-induced boundary conditions. Furthermore, through the application of pressure to the powder bed and on the melt, the risk for process abortion due to implicitly applied shear loads on slightly curled monolayers is significantly higher. Therefore, a combination of a counter-rotating roller for powder spreading with a concurrent roller for powder compaction is considered the most promising approach.

The effects shown indicate that the complex interaction between the coating geometry, the applied coating speed, and the resulting powder bed quality necessitates material-specific machinery optimizations for obtaining a stable process and for the generation of additive parts with reproducible part properties. However, despite representing a significant prerequisite for the overall stability and process robustness, only minor impacts can be derived from the sub-process of powder coating compared to the sub-process of laser exposure.

5.2.2.2 Exposure

Subsequent to powder coating, the newly coated powder is heated to its target value. For a robust PBF process, a homogeneous surface temperature distribution within the process window is essential, representing the initial and most crucial thermal

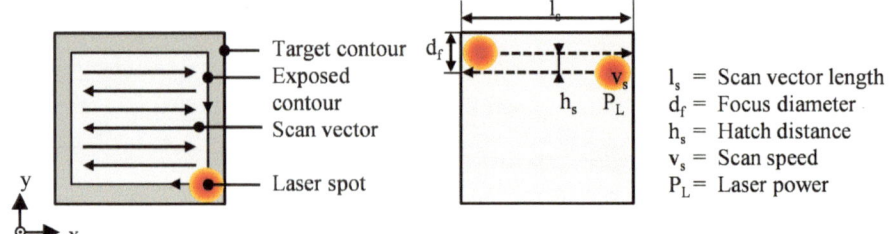

Fig. 5.5 Schematic depiction of the meander exposure strategy and relevant exposure parameters for the exposure of a quadratic cross-section

boundary condition for part generation. For stable processing, low thermal gradients are tolerable, with excessive temperatures leading to uncontrollable melting and temperatures that are too low inducing layer curling and interlinked process abortions. After the target build chamber temperature is reached, the laser exposure of the cross-sectional geometries to be manufactured is triggered.

The predominant laser-based PBF process is characterized by the laser irradiation of respective cross-sections, which is usually applied line by line. Considering the material-beam interaction and the interlinked heating of the material, the melting of the polymer is induced. In consequence, the resulting part properties are influenced by isothermal crystallization-induced consolidation and part shrinkage that is evoked by slow material cooling in the height direction (z-direction) [13, 16]. The exposure process itself is characterized by many influencing parameters such as the laser source, the laser focus d_f, the laser power P_L, the scan speed v_s, or the hatch distance h_s (see Fig. 5.5).

Apart from scalar process parameters, the applied scan pattern has a significant effect on the emerging temperature fields and interlinked part properties, which is associated with the thermal superposition of distinct vectors [12]. As a result, a holistic understanding of the temperature fields is essential for characterizing the material behavior during the process and its impact on the resulting part properties. Furthermore, understanding thermal interactions during the process can promote the development and enhancement of process-adapted material characterization [35] (see Chap. 4), process simulation [33] (see Chap. 14), or the processing of multi-material compounds in PBF (see Chap. 9). In addition to influences on the melting and flow behavior, material aging (e.g., post-condensation of polyamide 12 (PA12)) and degradation are initiated. The exposure process can be used to tailor the melting and crystallization behavior associated with the melt temperature, the respective viscosity, and flow behavior [7]. The exposure process thus plays a key role in the development of part properties such as part porosity, morphology, and mechanical properties [7]. As previously discussed, the processing of a powdery good with excellent flowability inherently limits the impact of powder coating on the exposure process (see Fig. 5.6a). At identical and optimum energy density levels, and under variation of the scan speed and the coating speed, a velocity-dependent influence of the exposure parameters is visible as a significant and decisive process impact.

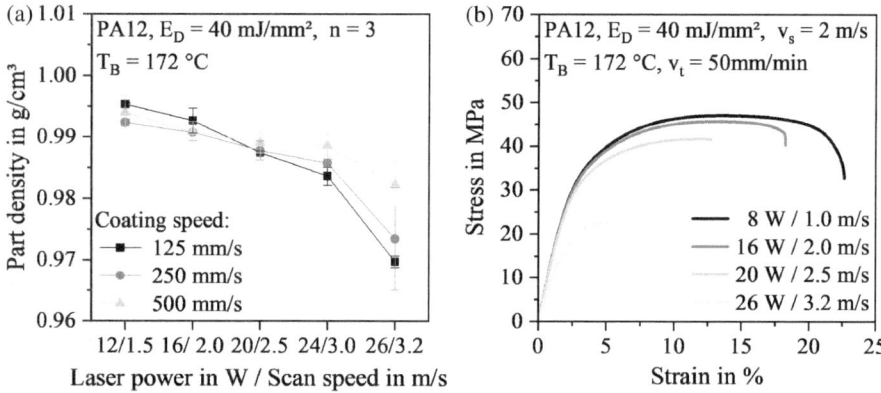

Fig. 5.6 **a** Part density as a function of the coating speed and exposure parameters [19], and **b** a stress-strain curve of tensile bars fabricated under variation of the scan speed at identical energy density [11]

Regarding the variation of the powder coating speed within a large speed range, no significant influences are observed. The strong velocity-dependence of the exposure process can be traced back to the variation of the impact time and the heating rate dependency of the thermal penetration depth, which is well described in [2]. Hence, a lack of fusion of the respective layers may result, leading to comparatively low part densities. According to these findings, a negative correlation between the melt pool depth and the applied exposure speed is obtained. The time-dependency of thermal conduction represents an inherent limitation to process accelerations, potentially inducing structural changes, e.g., the presence of unmolten particles or insufficient layer bonding [7, 22]. This impact can result in compromised or negatively affected part mechanics at identical energy density levels (see Fig. 5.6b). Hence, for process acceleration, a disproportionate increase in laser power is purposeful to achieve equivalent part properties.

Geometry-dependency of the exposure process

Beyond the previously discussed influences on the exposure process and formed transient temperature fields, recent research described the impact of the macroscopic part geometry on the interaction of the energy level and velocity-dependent influences on the emerging thermal history of a particular part and the interlinked part properties [21]. Through the thermal analysis of the exposure process, using an infrared camera, correlations between process parameters, part geometry, the resulting temperature fields, and the respective part properties can be derived. Parallel scan vectors characterize the meander exposure strategy that is commonly applied for the PBF process and the processing of polymer powders. The current research shows that during the subsequent exposure of scan vectors, a significant temperature build-up is present (see Fig. 5.7). This build-up is a result of the low thermal conductivity of the powder bed and the thermal superposition of previously exposed regions. The

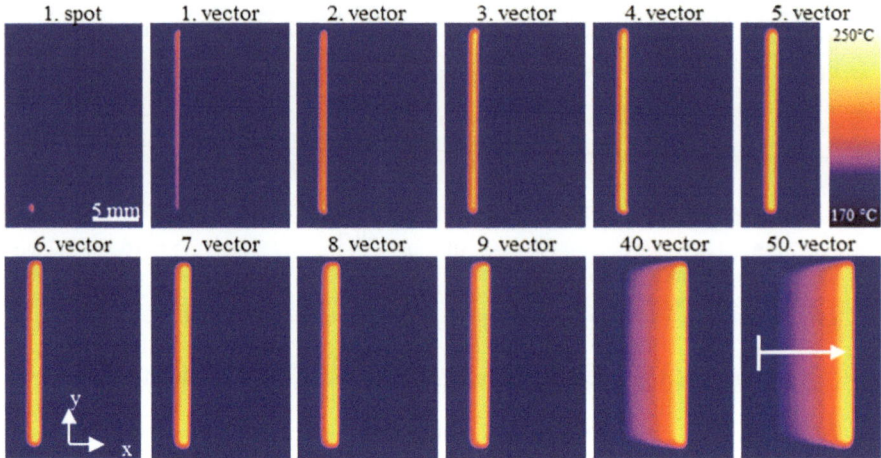

Fig. 5.7 Temperature build-up during the exposure of a rectangular cross-section using the meander exposure strategy, according to [11]

direction of the temperature build-up is perpendicular to the direction of the scan vector for the meander exposure strategy, indicating the influence of the sequence of consecutively exposed vectors on the evolving thermal gradient.

Regarding the exposure of various cross-sectional areas that are correlated with the variation of the scan vector length, a strong dependency on the measured maximum temperature profile is observed (see Fig. 5.8a). It is evident that the scan vector length directly correlates to the laser return time, leading to significant initial heat accumulations within small part cross-sections that are not present in larger cross-sections due to intermediate cooling. After a certain number of scan vectors, a geometry-dependent stable temperature level—corresponding to a thermal equilibrium—is reached and the number of parallel scan vectors becomes rather subordinate. Although a significant temperature increase occurs within the first scan vectors, the impact on monolayer properties is rather low, induced by the homogenization of the surface temperature after a few seconds [14]. For the prediction of emerging part properties, especially the thermal interactions between consecutive layers have to be taken into consideration (see Fig. 5.8b).

Especially regarding the resulting part properties, significant morphological alterations can be observed for parts with different cross-sectional areas when applying similar exposure parameters (see Fig. 5.9). As mentioned earlier, geometric variations are linked to the variation of the maximum temperature profiles.

However, in the current research, it was found that these changes cannot directly be matched to the present maximum temperatures, with the thermal interactions between the subsequently exposed layers having been shown to be more relevant for the development of part properties. According to [13, 16], thermal inter-layer interactions can be characterized by detecting parameters such as the decay time, the temperature increase above the previously exposed cross-sections (see Fig. 5.8b),

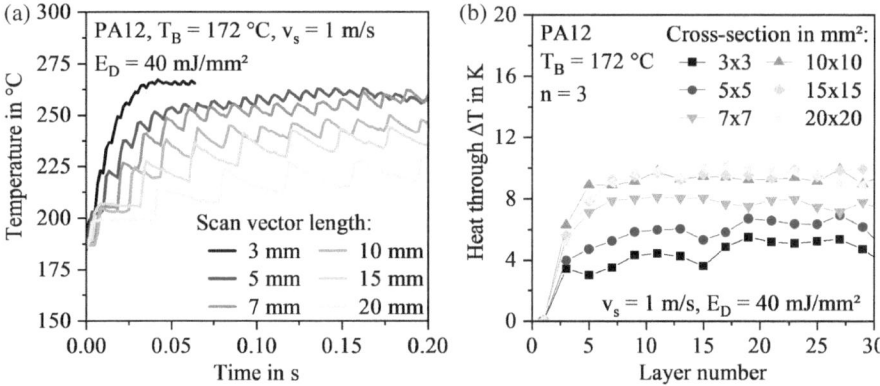

Fig. 5.8 **a** Heat build-up in response to the scan vector length and **b** geometry-dependent heat induced by previously exposed part cross-sections relative to the initial layer, according to [11]

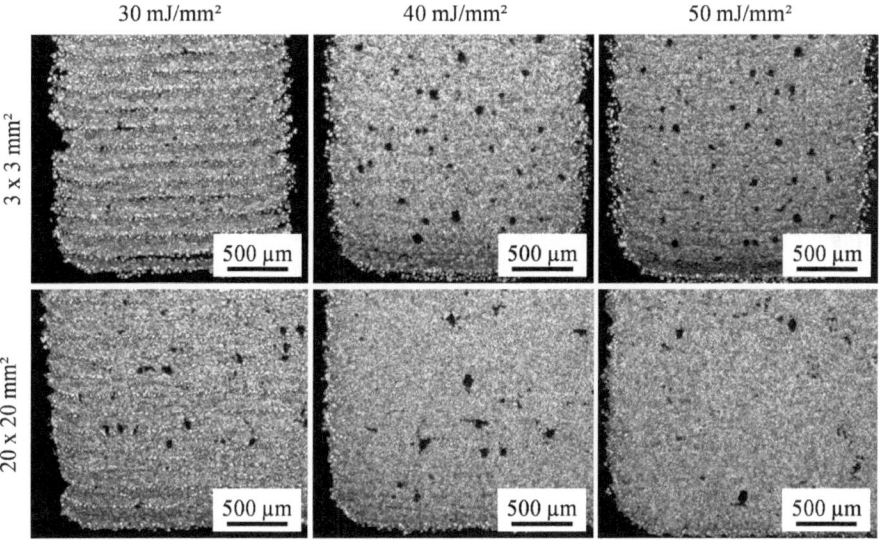

Fig. 5.9 Morphological alterations in parts with different cross-sectional areas under similar exposure parameters, illustrating the impact of geometry on temperature profiles

and inherently by evaluating the cooling rates. It was found that the laser return time exhibits a positive correlation with the cross-sectional areas, leading to higher mean melting temperature levels alongside an increased timespan of thermal exposure. Owing to the significant influence of geometric effects, the exposure-induced increase in the mean temperature level does not show a correlation with the initial maximum temperature. Arising from the increasing timespan and the elevated mean temperature level, an extended period of time promotes the material coalescence, leading to improved part properties.

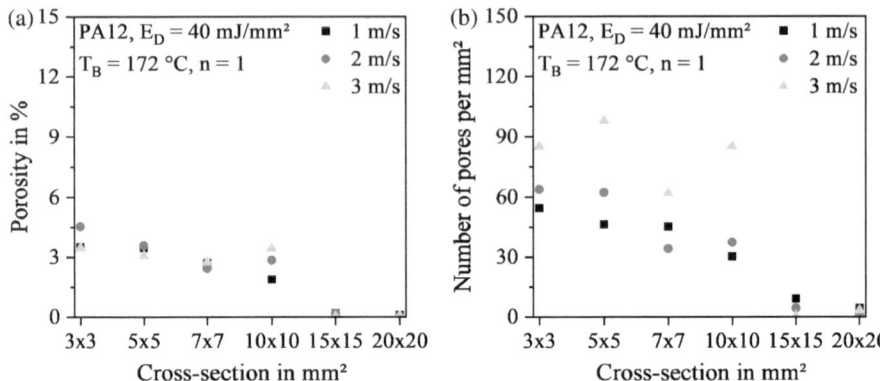

Fig. 5.10 **a** Part porosity and **b** number of pores under variation of the part cross-section and the scan speed at an identical optimum energy density level, according to [11]

This phenomenon can be traced to geometrical aspects, such as the surface to volume ratio, and to material-specific properties such as the thermal conductivity or the heat capacity. The latter material characteristics significantly influence the heat transport from the part center to the cooler powder bed surface, implicitly influencing the thermal characteristics and interlinked material coalescence. Regarding the part porosity acquired by computed tomographic measurements, a strong velocity dependency is visible at the optimum energy density level (see Fig. 5.10). For parts with small cross sections of 3×3 mm^2, comparatively high porosities of more than 5% are visible, whereas the largest cross-sections are almost non-porous.

However, not only the porosity but also the pore size distribution and the sphericity of pores are affected by geometrical boundary conditions (see Fig. 5.11). For parts with large exposed areas, the predominant formation of small pores is derived, whereas small cross-sections are characterized by the emergence of pores with diameters in the range of one layer (approximately 100 µm). In addition, the particle shape becomes rounder with a higher cross-sectional area. This is visible through a higher proportion of pores that are connected to a sphericity value of "1". Monolayer experiments on geometrically more complex shapes, such as triangles, ovals, or hollow structures illustrate a general transferability of the finding shown with regard to identical surface areas [15]. The results of this process analysis indicate that new material-adapted process strategies are necessary to fabricate parts with predictable part properties, thereby avoiding geometry-dependent influences on the resulting part properties.

This prerequisite can be fulfilled by the adaption of the exposure process and the manipulation of the temperature fields and the thermal history with regard to the scan pattern or the exposure parameters. In general, two methods are available for the development of new exposure strategies, as temperature fields can either be simulated prior to the process or experimentally induced. In the current research, temperature fields that were acquired during the process were used to set up an FEM simulation

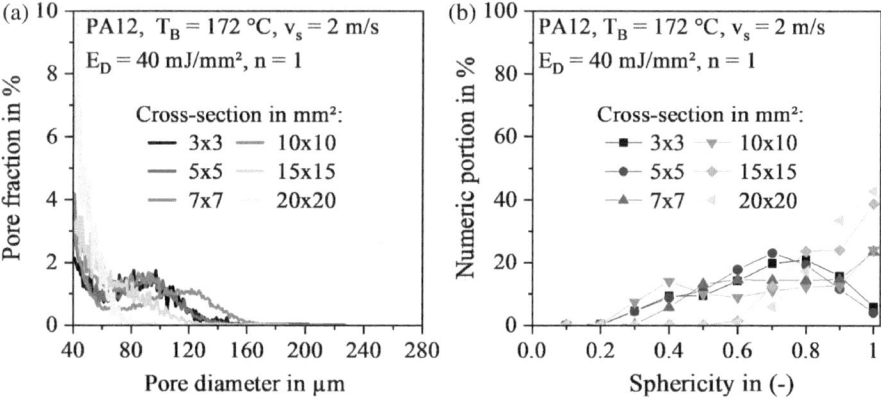

Fig. 5.11 a Pore size distribution and **b** sphericity distribution under variation of the part cross-section at the optimum energy density level, according to [11]

(see Chap. 14) that is capable of predicting the maximum temperatures and layer-wise recrystallization [33], indicating the significance of novel processing strategies for homogenizing emerging temperatures. Therefore, recent research focused on experimental approaches based on theoretical assumptions that could be derived from the process analysis presented earlier.

Geometry-invariant exposure strategies

For the realization of the previously discussed geometry-invariant temperature fields, two related approaches were followed. The first approach lies in the segmentation of larger cross-sections into smaller subsections, applying checkerboard-like exposure strategies, that could be transferred from metal processing [12]. It was found that thermal superposition effects could be affected by the temporal exposure sequence of respective checkerboard fields. Considering the inherent segmentation, high local temperature peaks are induced due to low scan vector lengths that promote unfavorable pore formation and shrinkage. Due to the temporal offset between the exposures of the chessboard fields, the mechanical connection between these fields is impaired. This can be initiated either by introducing an overlapping area or by exposing multiple contours. However, both approaches lead to the increase in the layer time and insufficient part quality [12].

On this account, a different approach for the adaptation of the scan vector layout was chosen for the generation of geometry-invariant temperature fields and part properties. Mathematically describable self-similar, space-filling curve patterns, such as the Peano curve [25], appear to be predestined for use as an exposure strategy in PBF (see Fig. 5.12). Through a variation of the scan vector orientation and low energy input, thermal superpositioning effects can be avoided, leading to homogeneous temperature distributions, a reduction of thermal shrinkage, and unified layer properties [20]. The use of the Peano curve as an exposure strategy at a material-specific energy level represents a solution for geometry and layer-invariant PBF.

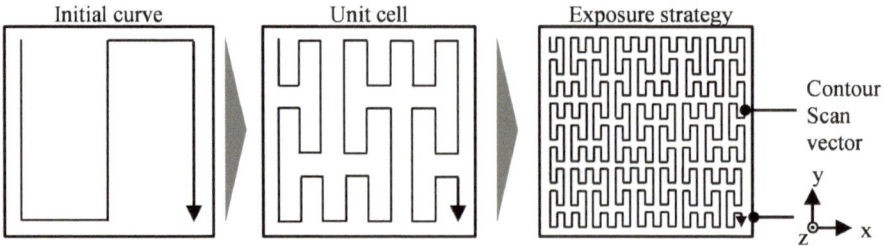

Fig. 5.12 Peano exposure strategy deducted from a self-similar mathematically describable Peano space-filling curve [11, 21]

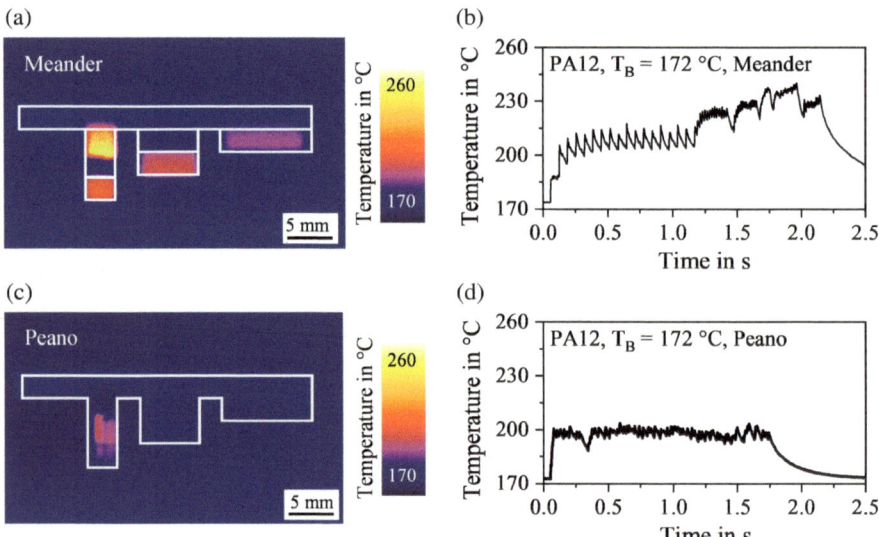

Fig. 5.13 **a** Thermograph and **b** temperature profile resulting from the meander exposure strategy and **c** thermograph and **d** temperature profile resulting from the Peano exposure strategy

In Fig. 5.13, thermographs of a geometrically complex structure with various scan vector lengths, using the standard meander exposure strategy and the Peano exposure strategy, are compared. Using the meander strategy, a geometry-dependent extent of thermal superposition is present, leading to geometry-dependent temperature levels and a temperature profile with high fluctuations. A considerably accelerated cooling of the exposed layers takes place using the Peano strategy, leading to a uniform temperature level that is independent of the macroscopic part geometry that is associated with the emergence of temperature fields merely on a mesoscale level. Furthermore, this strategy can be applied to influence material solidification to drastically reduce the build chamber temperature.

5.2.2.3   Consolidation

In PBF of polymers, the lowering of the build platform and the dosing of new powder for powder coating mainly characterize the consolidation step. In contrast to PBF of metals, material consolidation does not take place immediately after exposure but depicts a dependency on time, temperature, and the degree of crystallization. For PA12, isothermal crystallization is induced a few Kelvin below the build chamber temperature (see Chap. 4). Depending on the temperature level, crystallization-induced material solidification takes place within minutes to hours after exposure. Deriving an understanding of the flow, crystallization, and solidification processes is of great importance for PBF. These material transformations, which take place during the consolidation phase, can be used to develop new process strategies or for process monitoring under property prediction.

In the current research [31], a methodology based on optical deep learning was developed to predict material porosity based on the optical observation of the coalescence process. Based on the optical variability of emerging melt pools, the anisotropic radiance was shown to represent a significant indicator of the porosity of manufactured parts (see Fig. 5.14). Moreover, based on the wavelength-dependent Mie scattering occurring inside the melt pool, the coalescence of particles and the melting of spherulites is associated with a diminishing mesoscopic radiance of the exposed powder bed. By applying deep residual neural networks, a regression of the contactless-derived optical process properties and the corresponding part porosity can be obtained, allowing for the real-time prediction of morphological properties.

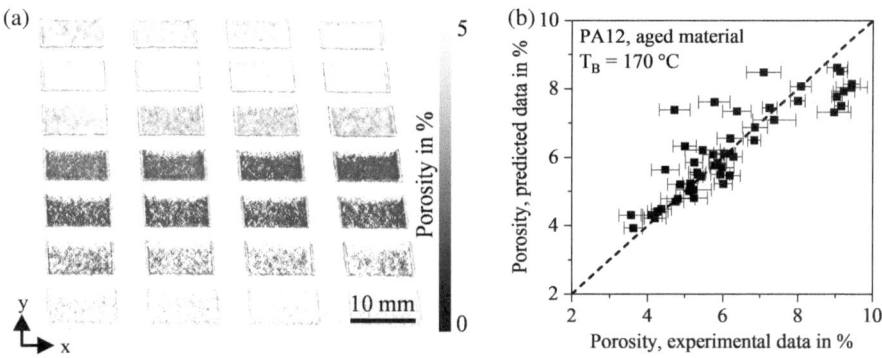

Fig. 5.14 a Spatial porosity distribution during the fabrication of one layer and **b** validation of the porosity using the implemented methodology

For the development of new process strategies, process monitoring and drawing the right conclusions are of great significance. Process-adapted material characterization allows for the prediction of material consolidation under processing conditions (section A3). For characterizing material consolidation, an experimental setup was developed that allows for the characterization of crystallization-induced material solidification within the PBF process chamber. The so-called stick-drop experiment facilitates the correlation of the material deformation to its consolidation state. In this setup, metal sticks with defined probe tips and defined weights fall from a defined height onto the molten material. It is observable that the resulting deformations are time-dependent and well-correlated to crystallization-induced consolidation as measured in process-adapted material characterization [9]. This finding is significant for the development of new processing strategies focusing on process acceleration and part property homogenization throughout the whole build chamber.

Accelerated consolidation through non-isothermal PBF

Based on the aforementioned use of the Peano exposure strategy for realizing a geometry-invariant exposure in the PBF process, a low-temperature process-variant was developed. Through quasi-simultaneous exposure—resembling the repeated exposure of distinct segments—it is possible to significantly reduce the build chamber temperature level without risking process abortion through crystallization-induced shrinkage. With the reduction in the build chamber temperature, single part production can be combined with the benefits of powder and beam-based additive manufacturing such as high dimensional accuracy or high mechanical properties. Furthermore, these exposure strategies allow for actively influencing the spherulite microstructure of the parts (see Fig. 5.15).

During quasi-isothermal PBF, the spherulite microstructure is a result of low cooling rates induced by the slow cooling process and isothermal crystallization at elevated temperature levels. Through the high cooling rates in non-isothermal PBF, it is, for the first time, possible to actively alter the part morphology with regard to the spherulite size and formed crystalline phases. The processing window (see Fig. 5.16a) is shifted to temperatures below the crystallization peak, rendering the processability of to-date critical materials possible for the PBF-LB/P process. Furthermore, materials that are difficult to process in isothermal PBF, such as glass fiber reinforced polyamide 6 (PA6-GF) or high-temperature materials, such as polyetheretherketone (PEEK), can be processed using standard PBF machinery.

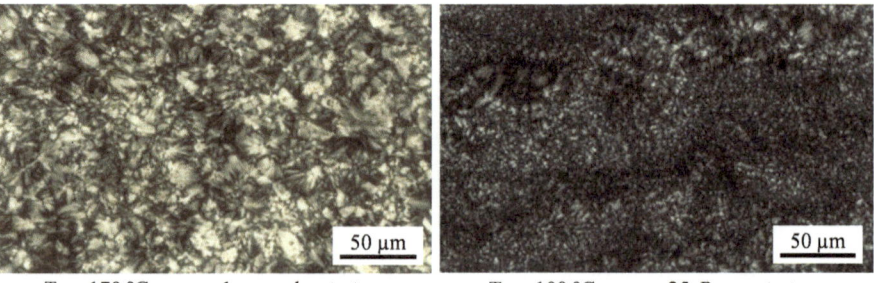

$T_B = 170\ °C,\ n_{Scan} = 1$, meander strategy $T_B = 100\ °C,\ n_{Scan} = 25$, Peano strategy

Fig. 5.15 Microscopic images of the typical part morphologies for the **a** standard PBF process and **b** non-isothermal PBF

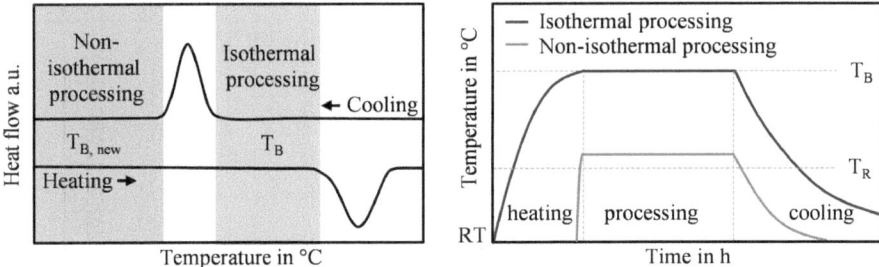

Fig. 5.16 a Schematic comparison of the process windows, **b** process times, and temperature profiles during PBF and non-isothermal PBF

Through the local and temporal discretization of exposure and rapid consolidation, shrinkage can be induced locally to an almost infinitesimally small scale. For PA12, the build chamber temperature could be reduced from 170 to 100 °C in the first step using multiple exposures [30]. Further material-specific process adaptations of the exposure strategy led to the reduction of the temperature build chamber temperature to room temperature level and the reduction of the processing time to a level comparable with quasi-isothermal processing [27]. Compared to the standard process, the preheating and cooling time, as well as the filling degree of the build chamber, no longer negatively influence the process times and part costs (see Fig. 5.16b), thereby allowing for the unimpeded recycling of unfused powder. Further advantages of non-isothermal PBF lie in the elimination of the thermal aging of the part cake during the build phase, allowing for material reuse without material degradation. Hence, an especially sustainable process variant was introduced. Furthermore, the process window is enhanced, rendering the fabrication of materials with previously improper structural or thermal properties possible. Transferability could, for example, be achieved for polypropylene (PP) [29] and for PP filled with polymers sensitive to heat degradation, such as agarose [28]. Both material systems could be processed to filigree specimens at room temperature instead of an elevated temperature of 135 °C.

5.2.3 Cooling Phase

In PBF of polymers, the cooling phase is decisive for the total process time and component availability. After the last layer has been produced during the build phase, the material cools down passively to a specified removal temperature, which is a few Kelvin above room temperature. Depending on the size of the build chamber and the volume of the parts produced, cooling times of several hours to days occur. This leads to high thermal exposure and position-dependent part properties, as they are dependent on the local temperature history. For this reason, adapting the cooling phase is a promising starting point for process optimization with respect to enhanced

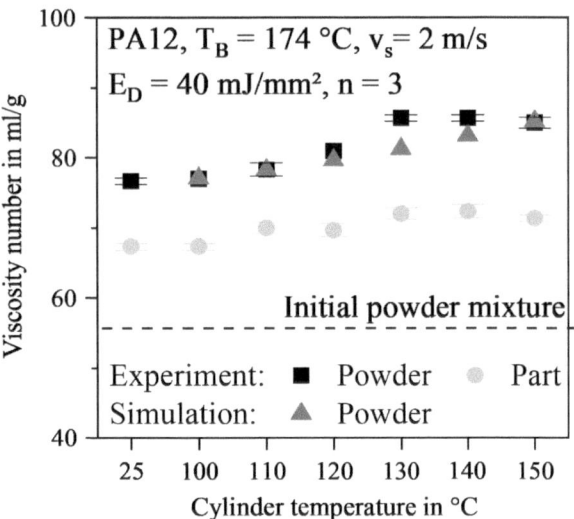

Fig. 5.17 Viscosity number of the powder bed and the part that is dependent on the cylinder temperature [17]

reproducibility and process time reductions. In [36], the cooling rates were enhanced by removing the parts from the powder bed during an early stage of the cooling phase, leading to slightly improved elongations at break but lower tensile strengths for higher cooling rates, in accordance with the mechanical properties of the non-isothermal processing. The elongation at break is one of the most sensitive part properties in PBF and therefore a suitable parameter for the evaluation of process optimization. The latter result is in good correlation with Gibson's findings [10], since the center of the build chamber, correlated with higher tensile strengths, is associated with lower cooling rates. In the current research [17], the focus is on the temperature control of the build cylinder temperatures during the processing phase to reduce the cooling time and impact on material aging. Using isothermal cylinder heaters, it is clearly visible that material aging can be reduced by the adaption of the cylinder temperature to a lower temperature level (see Fig. 5.17).

An increase in the viscosity number is an indication of thermally induced post-condensation. The results closely align with the model that was presented in [34] which allows for the prediction of the solution's viscosity of the part cake. In addition, a similar increase in the observed values derived for the viscosity number is observed in the generated part properties, indicating that material aging can be limited by the reduction of the cylinder temperature. The reduction of material aging is an important goal, as lower-aged material can be recycled more often and therefore drive the PBF technology further toward a resource-efficient manufacturing technology.

Apart from the impact on material aging, the influence of the cylinder temperature on part properties such as elongation at break is evident [17]. The optimum cylinder temperature depicts a dependency on the size and shape of the build cylinder (see Fig. 5.18a). The results show that the temperature optimum for the system used can be identified at 130 °C. At higher or lower temperature levels, a reduction of the

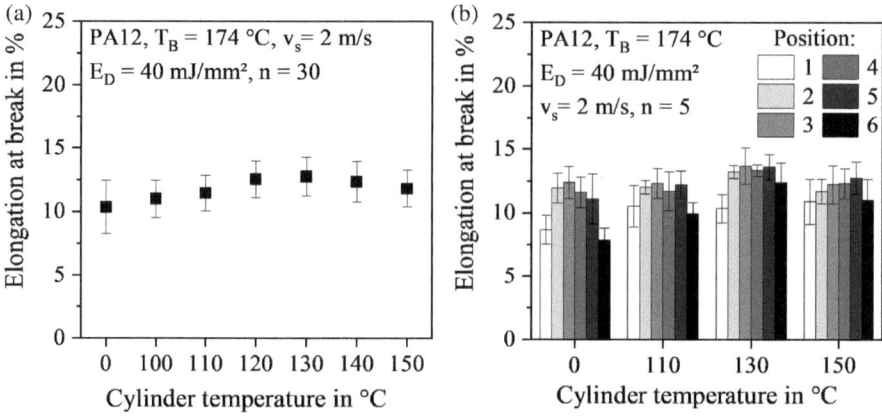

Fig. 5.18 **a** Influence of the cylinder temperature on the elongation at break and **b** the position-dependent elongation at break under variation of the cylinder temperature [17]

elongation at break is achieved. However, high standard deviations are present that can be correlated to the position of the part or rather the distance to the cylinder walls, indicating a correlation between the part position and the underlying reproducibility of the manufactured parts (see Fig. 5.18b). A spatial proximity to the cylinder wall is correlated with temporal and spatial inhomogeneities of the temperature fields that are linked to the formation of defects and reduced elongation at break. The closer the parts are to the cylinder walls, the higher the local cooling rates and therefore the lower the elongation at break.

However, at higher cylinder temperatures, the impact of local cooling rates on part properties can be reduced. These results directly lead to two different assumptions. First, the cylinder temperature has to be as high as possible to avoid local variations of the part properties within the build plane. Second, in order to achieve process acceleration through the adaptation of the cooling phase, a dynamic cooling system has to be developed. Therefore, in recent research [16], the impact of the geometrical and thermal properties of the build cylinder was analyzed. Through the implementation of layerwise build-up and the subsequent cooling through powder coating and layer-wise heating, simulations on the influence of different cylinder parameters were employed, allowing for studying the influence of boundary conditions such as the build cylinder dimensions on transient temperature distributions. In Fig. 5.19a, the temperature distribution is shown at the end of the process phase under variation of the build size between 50, 100 and 200 mm. The build cylinder temperature of 150 °C was reached within the deeper layers.

It is evident that the uniaxial size of the build cylinder is negatively correlated to the thermal homogeneity in a particular XY-plane, thereby further contributing to the accelerated cooling towards the cylinder temperature. Regarding the temperature development, the center of the build cylinder experiences a slowed cooling in

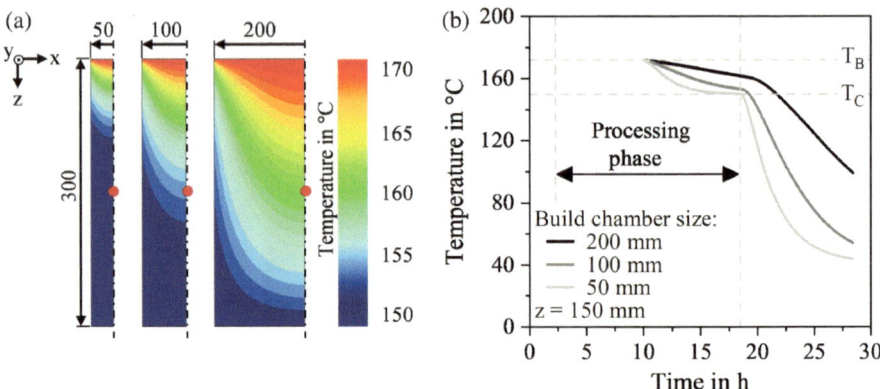

Fig. 5.19 **a** Temperature distribution within built chambers of variable build size at the beginning of the cooling phase and **b** temperature profile in the center of the build chambers, according to [16]

contrast to the cylinder walls. To avoid the distortion of the derived thermal characteristics, the center temperatures were evaluated in Fig. 5.19b. The results indicate that comparatively high cooling rates are correlated to small build cylinder dimensions, whereas slower cooling is induced through larger build cylinder dimensions. Due to those low cooling rates, large differences in the local thermal history occur, correlated with considerable alterations in the emerging mechanical part properties, as shown in Fig. 5.18.

Dynamic cooling strategies

In order to reduce the impact of the local temperature history on part properties, new temperature control concepts for the build cylinder were analyzed. For the homogenization of the temperature history, a concept for enabling dynamic temperature control was derived. Based on the ex situ findings obtained regarding material consolidation (see Chap. 4), high temperatures of the build cylinder close to the powder bed surface promote dense microstructures. After the material has started to solidify through isothermal crystallization, the reduction of the cylinder temperature is possible without risking crystallization-induced warpage.

In theory, tensions that are induced through crystallization can be compensated through high chain mobility at elevated temperature levels. Within deeper layers, higher cooling rates can be applied with regard to its feasibility in the field of heat transport processes. Due to the large volume of air between the unfused polymer particles, accessible cooling rates are inherently limited. The definition of a sectioned temperature control perpendicular to the build plane allows for identical thermal histories regardless of the part's position. In Fig. 5.20a, this new dynamic cooling strategy is compared to isothermal (constant) temperature control, representing the state of the art. It is evident that on the surface of the powder bed, minor thermal gradients can be observed, with the build chamber temperature and the applied heating

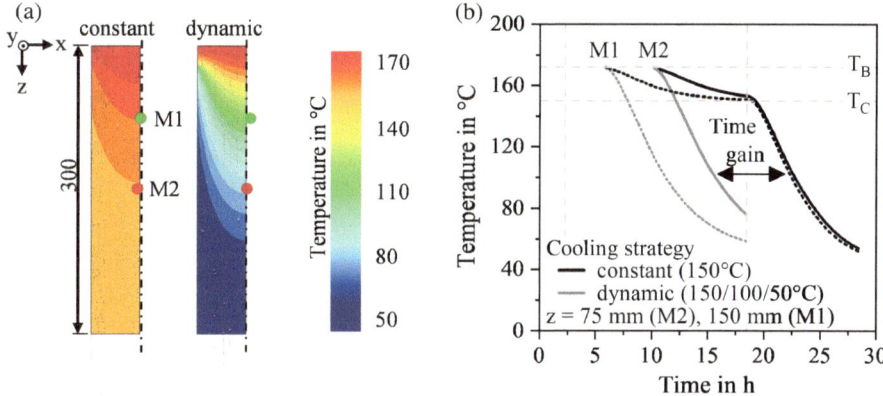

Fig. 5.20 **a** Temperature distribution within built chambers under variation of the cooling strategies at the beginning of the cooling phase and **b** temperature profiles in the center of the build chambers, according to [16]

system predominantly determining this thermal gradient. With regard to the temperature control of the build cylinder, low-temperature gradients in the build direction are prominent for the constant temperature control strategy. Using the dynamic cooling strategy, a large temperature gradient is visible at the end of the process, indicating completed crystallization and solidification of the majority of parts. Through the evaluation of two measuring positions (M1 and M2), it is possible to highlight the main differences between a constant and dynamic temperature control of the build cylinder (see Fig. 5.20). Using a constant temperature control, different initial cooling rates are present during the build phase, leading to the formation of locally varying component properties. After completion of the build phase and switching off all heating sources, maximum cooling rates, defined by heat transport within the part cake, are possible. However, a higher heat quantity is present which leads to longer cooling times.

In contrast to the state of the art, forced cooling mechanisms enabled by dynamic temperature control allow for almost identical temperature curves to form. This indicates that the temperature history can be unified in the height direction through the adaption of the cylinder system, significantly contributing to the homogenization of component properties in the height direction. These findings represent the foundation for a continuous PBF (endless process) with increased efficiency and reproducible properties. Furthermore, this new process strategy accelerates the part availability of individual and complex functional components and contributes to a lower aging impact on the powder to further increase the sustainability of the process.

5.3   Summary

In this section, the impact of thermal and temporal influences, respectively the temperature history of the PBF-LB/P process, and the resulting part properties, were displayed through in-depth process analysis. Based on the insights obtained, new processing strategies were derived that allow for enhanced process robustness, control over the resulting thermal history and part properties, and higher sustainability and productivity.

Precise information regarding the heating process and the powder coating process is beneficial since these process steps represent the relevant initial boundary conditions for the generation of parts with uniform properties and profiles at low process variations. However, adaptations of the exposure, consolidation, and cooling process are key factors as they directly affect component properties. A non-isothermal PBF process was introduced that allows for single part production, the processing of material systems that are currently unsuitable for PBF, and the targeted adaptation of morphological properties. For isothermal processes, the adaptation of the build cylinder through the integration of forced cooling is promising for the unification of component properties and for further enhancing the reproducibility and productivity in PBF-LB/P.

References

1. Bourell, D.L., Watt, T.J., Leigh, D.K., Fulcher, B.: Performance limitations in polymer laser sintering. In: Proceedings of the 8th International Conference on Laser Assisted Net Shape Engineering LANE 2014, vol. 56, pp. 147–156 (2014)
2. Drexler, M.: *Zum Laserstrahlschmelzen von Polyamid 12, Analyse zeitabhängiger Einflüsse in der Prozessführung.* Ph.D. thesis, Friedrich-Alexander-Universität Erlangen-Nürnberg (2016)
3. Drexler, M., Drummer, D., Wudy, K.: Effects of the powder bulk density on the porosity of laser molten thermoplastic parts. In: Proceedings of the 29th Annual Meeting PPS 29 (2013)
4. Drexler, M., Drummer, D., Wudy, K.: Einfluss des pulverauftragsprozesses auf den selektiven strahlschmelzprozess thermoplastischer kunststoffe. In: Proceedings of the 12th Rapid.Tech 2015 (2015)
5. Drexler, M., Greiner, S., Lexow, M., Lanzl, L., Wudy, K., Drummer, D.: Selective laser melting of polymers: influence of powder coating on mechanical part properties. J. Polym. Eng. **38**(7), 667–674 (2018)
6. Drexler, M., Kühnlein, F., Drummer, D.: Influence of powder coating on part density as a function of coating parameters. In: Proceedings of 4th International Conference on Additive Technologies—iCAT 2012 (2012)
7. Drexler, M., Lexow, M., Drummer, D.: Selective laser melting of polymer powder—part mechanics as function of exposure speed. In: *Proceedings of the 15th Nordic Laser Materials Processing Conference Nolamp* (2015)

8. Drummer, D., Drexler, M., Wudy, K.: Density of laser molten polymer parts as function of powder coating process during additive manufacturing. In: Proceedings of the 7th World Congress on Particle Technology (2014)
9. Drummer, D., Greiner, S., Zhao, M., Wudy, K.: A novel approach for understanding laser sintering of polymers. Add. Manuf. **27**, 379–388 (2019)
10. Gibson, I., Shi, D.: Material properties and fabrication parameters in selective laser sintering process. Rapid Prototyping J. **3**(4), 129–136 (1997)
11. Greiner, S.: *Bedeutung der geometrieabhängigen Belichtungstemperaturen für das Lasersintern von Kunststoffen*. Ph.D. thesis, Friedrich-Alexander-Universität Erlangen-Nürnberg, 2023
12. Greiner, S., Drummer, D.: Infrared monitoring of modified hatching strategies for laser sintering of polymers. In: Proceedings of the 11th CIRP Conference on Photonic Technologies (2020)
13. Greiner, S., Drummer, D.: Materialangepasste prozessstrategien für das lasersintern von kunststoffen. In: Proceedings of Technomer - Fachtagung über Verarbeitung und Anwendung von Polymeren (2021)
14. Greiner, S., Drummer, D.: Understanding aspect ratio effects in laser powder bed fusion of polyamide 12 by means of infrared thermal imaging. In: Proceedings of the 12th CIRP Conference on Photonic Technologies LANE 2022 (2022)
15. Greiner, S., Drummer, D.: Understanding geometry dependent temperature fields in laser powder bed fusion by means of infrared thermal imaging. In: Proceedings of the 8th International Conference on Additive Technologies—iCAT 2023 (2023)
16. Greiner, S., Jaksch, A., Cholewa, S., Drummer, D.: Development of material-adapted processing strategies for laser sintering of polyamide 12. Adv. Indus. Eng. Polym. Res. **4**(4), 251–263 (2021)
17. Greiner, S., Jaksch, A., Drummer, D.: Understanding cylinder temperature effects in laser beam melting of polymers. In: LMNE, Enhanced Material, Parts Optimization and Process Intensification. Springer, Berlin (2020)
18. Greiner, S., Lanzl, L., Wudy, K., Drexler, M., Drummer, D.: Untersuchungen zum potential der pulververdichtung durch den pulverauftrag beim selektiven laserstrahlschmelzen von kunststoffen. In: Proceedings of the 14th Rapid.Tech 2017 (2017)
19. Greiner, S., Lanzl, L., Zhao, M., Wudy, K., Drummer, D.: Influence of powder bed surface on part properties produced by selective laser beam melting of polymers. In: Proceedings of 7th International Conference on Additive Technologies (2018)
20. Greiner, S., Schlicht, S., Drummer, D.: Temperaturfeldhomogenisierung durch fraktale belichtungsstrategien im lasersintern von kunststoffen. In: Proceedings of the 17th Rapid.Tech 2021 (2021)
21. Greiner, S., Wudy, K., Wörz, A., Drummer, D.: Thermographic investigation of laser-induced temperature fields in selective laser beam melting of polymers. Opt. Laser Technol. **109**, 569–576 (2018)
22. Lexow, M., Drexler, M., Drummer, D.: Fundamental investigation of part properties at accelerated beam speeds in the selective laser sintering process. Rapid Prototyping J. **23**(6), 1099–1106 (2017)
23. Meyer, L., Wiedau, L., Wegner, A., Witt, G.: Innovative pulver-auftragsstrategien im lasersinter-prozess – einflussuntersuchung der packungsdichte in korrelation zur oberflächenrauheit. In: Proceedings of the 15th Rapid.Tech 2018 (2018)
24. Niino, T., Sato, K.: Effect of powder compaction in plastic laser sintering fabrication. In: Proceedings of the Solid Freeform Fabrication Symposium SFF (2009)
25. Peano, G.: Sur une courbe, qui remplit toute une aire plane. Mathematische Annalen **36**, 157–160 (1890)
26. Rietzel, D., Drummer, D., Kühnlein, F.: Investigation of the particular crystallization behaviour of semi-crystalline thermoplastic powders processed by selective laser sintering. In: 3rd International Conference on Additive Technologies-iCAT2010 (2010)
27. Schlicht, S., Cholewa, S., Drummer, D.: Process-structure-property interdependencies in non-isothermal powder bed fusion of polyamide 12. J. Manuf. Mater. Process. **7**(1), 33 (2023)

28. Schlicht, S., Drummer, D.: Accelerated non-isothermal powder bed fusion of polypropylene using superposed fractal exposure strategies. In: Proceedings of the 8th International Conference on Additive Technologies—iCAT 2023 (2023)
29. Schlicht, S., Drummer, D.: Thermal intra-layer interaction of discretized fractal exposure strategies in non-isothermal powder bed fusion of polypropylene. J. Manuf. Mater. Process. 7(2), 63 (2023)
30. Schlicht, S., Greiner, S., Drummer, D.: Low temperature powder bed fusion of polymers by means of fractal quasi-simultaneous exposure strategies. Polymers 14(7), 1428 (2022)
31. Schlicht, S., Jaksch, A., Drummer, D.: Inline quality control through optical deep learning-based porosity determination for powder bed fusion of polymers. Polymers 14(5), 885 (2022)
32. Schmachtenberg, E., Seul, T.: Model of isothermic laser-sintering. In: Proceedings of the 60th Annual Technical Conference of the Society of Plastic Engineers (ANTEC) (2002)
33. Soldner, D., Greiner, S., Burkhardt, C., Drummer, D., Steinmann, P., Mergheim, J.: Numerical and experimental investigation of the isothermal assumption in selective laser sintering of pa12. Add. Manuf. 37, 101676 (2021)
34. Wudy, K.: *Alterungsverhalten von Polyamid 12 beim selektiven Lasersintern*. Ph.D. thesis, Friedrich-Alexander-Universität Erlangen-Nürnberg (2017)
35. Wudy, K., Greiner, S., Zhao, M., Drummer, D.: Selective laser beam melting of polymers: In situ and offline measurements for process adapted thermal characterization. In: Proceedings of the 10th CIRP Conference on Photonic Technologies LANE 2018 (2018)
36. Zarringhalam, H.: *Investigation into crystallinity and degree of particle melt in selective laser sintering*. Ph.D. thesis, Loughborough University (2007)
37. Zhao, M., Wudy, K., Drummer, D.: Crystallization kinetics of polyamide 12 during selective laser sintering. Polymers 10(168) (2018)

Dietmar Drummer succeeded Prof. Ernst Schmachtenberg as Head of the Institute of Polymer Technology at Friedrich-Alexander-Universität Erlangen-Nürnberg on May 1, 2009. He was the spokesperson for the Collaborative Research Center 814 Additive Manufacturing, which focused on process understanding, development, and process-adapted material characterization. Furthermore, he heads the Polymer Group of Neue Materialien Fürth GmbH and 2 Keylabs at the Bavarian Polymer Institute.

Chapter 6
Three-Dimensional Multi-material Parts

Sebastian-Paul Kopp and Stephan Roth

6.1 Introduction

Generating multi-material polymer parts by means of laser-based additive manufacturing (PBF-LB/P) generally requires the realization of a defined, heterogeneous powder layer consisting of different materials. If the thermal processing properties of the polymers differ significantly from each other, the laser irradiation strategy may also need to be adapted [13, 15]. In this chapter, first the adapted laser irradiation strategy is described in detail, followed by an in-depth depiction of various strategies for realizing a defined, heterogeneous powder layer for multi-material PBF-LB/P.

6.2 Adapted Laser Irradiation Strategy for Generating Multi-material Parts by Means of Laser-Based Powder Bed Fusion

The realization of simultaneous, intensity-selective irradiation by means of infrared (IR) emitters and different laser beam sources is one of the key elements for successful multi-material PBF-LB/P. Beam-shaping elements offer a further degree of freedom to achieve the required three-dimensional temperature fields adapted to the melting and crystallization behavior of the different materials. For three-dimensional control of the temperature field within the powder bed, [12] showed that it is beneficial to select the second laser beam source in such a way that energy can penetrate deeper into the powder bed in comparison to CO_2 laser radiation. This was achieved by using a Thulium-laser emitting at a wavelength of 1.94 μm. In this way, the

S.-P. Kopp · S. Roth (✉)
Bayerisches Laserzentrum GmbH (blz), Konrad-Zuse-Straße 2-6, 91052 Erlangen, Germany
e-mail: s.roth@blz.org

© The Author(s) 2025
D. Drummer and M. Schmidt (eds.), *Progress in Powder Based Additive Manufacturing*, Springer Tracts in Additive Manufacturing,
https://doi.org/10.1007/978-3-031-78350-0_6

temperature in the build direction is selectively adjustable to prevent crystallization of the molten material during the process, which can lead to curling of the molten layer and consequently to failure of the build job.

6.2.1 Multi-material PBF-LB/P with Polymers Displaying Similar Thermal Property Profiles

To derive a general understanding of the material-dependent properties for combining polymer materials within one multi-material part, the temperature fields were analyzed during the process and compared with the part properties obtained, such as porosity and mechanical strength. In addition, material parameters relevant to the process and compatibility, such as the optical material properties or wetting behavior of the melts, were analyzed, leading to a comprehensive characterization methodology. This allows the selection of suitable materials for the realization of multi-material parts.

For this purpose, polyamide 12 (PA12, PA2200, EOS GmbH, Krailling, Germany), high-density polyethylene (HDPE, Coathylene NC 6454-F, Axalta Coating Systems GmbH, Pratteln, Switzerland), polypropylene (PP, Coathylene PD0580, Axalta Coating Systems GmbH, Pratteln, Switzerland), and a thermoplastic elastomer based on polyamide (TPE-A, LSS Laser-Sinter-Service GmbH, Holzwickede, Germany) were analyzed. These polymers were selected due to their wide range of properties with regard to the achievable mechanical strength, haptic properties, and chemical resistance. The starting materials were analyzed with regard to their mutual compatibility [11], flowability, and optical powder properties at room temperature [16]. Furthermore, these properties were characterized at temperatures up to the melting point [14] and modified using flow aids to achieve sufficient flowability for powder application by means of a doctor blade. Moreover, absorption-increasing particles were added in order to achieve high absorption for different laser wavelengths. A measurement setup was also realized and utilized to determine the thermal conductivities of the materials [16] that influence the temperature fields resulting from irradiation during the process. This setup allowed for generating multi-material parts consisting of two different materials with a centrally oriented boundary zone and a maximum size of 40 mm × 25 mm [15]. The simple component geometry was due to the powder application by means of a two-chamber doctor blade. However, it offered the possibility to analyze material and component properties, such as the wetting behavior of the melts and the achievable bond strength without influences arising from geometrically complex interfaces. .

In the first process step, two different powder materials were applied next to each other by means of a two-chamber doctor blade. Both materials were preheated with IR heaters to a temperature slightly below the melting temperature of the material with the lower melting temperature. The simultaneous preheating of the material with the higher melting temperature slightly below its respective melting temperature was realized by a CO_2-laser at a wavelength of $\lambda = 10.6$ μm. It irradiated the powder surface locally with homogeneous intensity distribution achieved by a diffractive optical element (DOE). This beam shaping method offers higher efficiency for the wavelength of a CO_2-laser compared to other approaches such as spatial light modulators (SLM). Due to the maximum laser power of 50 W of the CO_2-laser installed in the system, the size of the build area, and losses due to heat conduction, a maximum temperature increase of approx. 30 K compared to the preheating temperature provided by the IR emitters could be measured for the processed powder materials using a thermal camera. Since the DOE represents a geometrically rigid mask, only a constantly large area can be preheated by means of CO_2-laser radiation. In the final process step, a Thulium-laser ($\lambda = 1.94$ μm) was used to image a geometrically freely adjustable area in the powder bed by means of a mirror element and to melt both preheated materials simultaneously. The mirror element consists of a micromirror array in which each mirror can alternate between two mechanically defined tilt angles, with one tilt state deflecting the radiation incident on the mirror into the powder bed and the other tilt state deflecting the radiation into a beam trap. By varying the frequency of mirror tilting, different amounts of energy can be introduced into the material per unit time [12].This variable energy input enables flexibly adjustable temperature distributions in the powder bed through local variations of the time-averaged intensities to compensate for decreasing temperature profiles at material transitions or in the outer areas. For temperature control, a thermal camera was used to control the energy input provided by both the IR emitters and the CO_2 laser. In order to be able to determine the absolute temperatures of the powder bed surface required for temperature control, the material-dependent and temperature-dependent emission coefficients for different powder materials were first determined in the process-relevant temperature range [15].

The temperature fields were determined in the build plane by means of a thermographic camera and, perpendicular to the build plane, by means of thermocouples positioned in the powder bed. The measurement results showed that the spatial resolution of the achievable temperature field in the build plane mainly depended on the laser intensity and irradiation duration used for melting [15]. In case of long irradiation times of several seconds, the powder surrounding the irradiated powder surface also partially melted as a result of heat transport processes, which lead to deviations between the nominal geometry and the actual geometry of the component. In order to reduce the irradiation time, the maximum achievable intensity had to be increased. By using an additional DOE for homogenization of the Thulium-laser radiation instead of an optical fiber, the power losses could be reduced by 40%, so that a higher intensity could be achieved with the same achievable part size.

Utilizing the described experimental setup, multi-layer multi-material specimens were realized. The maximum achievable temperature increase of 30 K of the higher melting material by the CO_2 laser radiation compared to the preheating temperature provided by the IR emitters limited the material combinations to polymers showing similar thermal property profiles. As a further limitation, different compatibility principles of the materials, such as property compatibility and processing compatibility, were considered. Due to these incompatibilities, some of the materials used cannot be combined in other processes, such as multi-component injection molding, or only by using adhesion agents. Using the aforementioned multi-material setup for PBF-LB/P, however, composite parts consisting of previously incompatible materials were realized. Due to turbulences during powder application by means of the multi-chamber doctor blade, a transition zone was formed in which both powder materials were mixed [13]. This resulted in undercuts in the transition zone after melting and subsequent crystallization. Due to this form-fit joint, material composites could also be achieved with materials such as PA 12 and PP, which do not wet each other due to significantly different surface tensions of the melts. Hence, no diffusion takes place, which would lead to firmly bonded joints on molecular level. Overall, in addition to TPE-A/PE and PE/PP, PA 12 was also successfully processed together with PP [13]. For analyzing the mechanical component properties, tensile bars made of the individual materials and of the multi-material combinations were tested according to DIN EN ISO 527 [15]. In Fig. 6.1 the tensile strengths of the tensile test bars made of the individual materials are compared with the values determined for multi-material tensile bars. As can be expected from the fact that a form-fit joint is formed between areas of different materials, all multi-material parts fail at the transition zone between the two materials [15].

The processing compatibility with regard to the melting and crystallization temperatures of the materials, in combination with the temperature gradient between the different preheating temperatures at the interface, is decisive for the joint strength. If the crystallization temperature T_K of the material with the higher melting temperature is above the preheating temperature T_{VH} of the second material, as is the case with TPE-A ($T_{VH} = 124$ °C) in combination with PE ($T_K = 117$ °C), and in case of PP ($T_{VH} = 155$ °C) with PA 12 ($T_K = 150$ °C), a strongly connected interface is formed. Thus, the achievable tensile strengths are comparable to the tensile strength of the mechanically weaker single material. When combining PE with PP, the crystallization temperature of the material with the higher melting temperature is above the

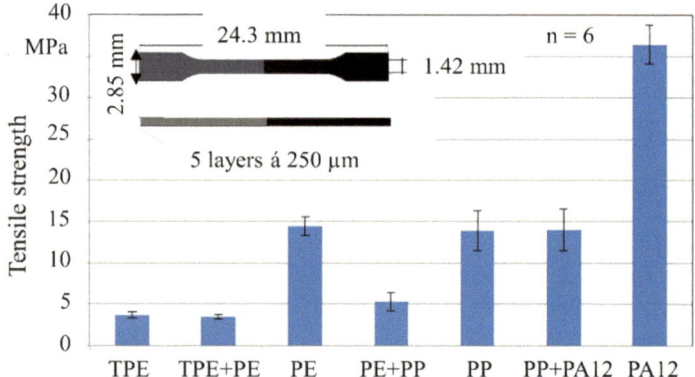

Fig. 6.1 Tensile strength of tensile test bars consisting of one material and multi-material test parts consisting of materials with similar thermal property profiles

preheating temperature of the second material. The temperature drop as a result of the heat flow in the interface zone of the two materials causes the higher melting material to crystallize at the interface. As a result of the crystallization shrinkage, the higher-melting material partially separates from the other material. As a result, defects are formed and the joint strength is reduced (cf. Fig. 6.1). The temperature gradient in the interface must be compensated for by higher energy input by means of a graded intensity profile in the transition region. As a result, crystallization and shrinkage can be controlled in time to prevent detachment of the crystallizing materials and increase the bond strength.

6.2.2   Simulation of Resulting Temperature Fields During Multi-material PBF-LB/P of Polymers Showing Similar Thermal Property Profiles

The objective of the finite element (FE) simulation model that was developed was to facilitate a numerical calculation of the spatial and temporal variations of the temperature fields in the powder bed. This helped to derive suitable process strategies in the context of the newly introduced simultaneous, intensity-selective irradiation strategy. Taking different material combinations into account, part densities are assigned to the materials according to the temperature history to obtain an enhanced understanding of the process, particularly of the consolidation behavior in the transition zones of multi-material distributions.

The simulation model included the subprocesses' energy inputs (large-area, simultaneous irradiation with laser energy input of discrete laser wavelength), dissipation of heat (heat conduction and convection), and melting (phase transition powder-melt). The FE model was created with the help of C++ and the FE library "Deal.ii" and extended by nonlinear material models, e.g., for the description of the temperature-dependent heat capacity with latent heat and heat conduction. Additionally, a suitable absorption model for the large-area simultaneous energy input by different radiation sources (e.g., CO_2- and Thulium-laser radiation) for multi-material layers was added. The FE model was used to calculate the resulting spatial and temporal behavior of the temperature field after simultaneous energy input into a powder bed as a function of different process parameters, such as preheating temperatures, intensities, or exposure times. The energy input was integrated into the simulation model as a volume source of the form $g(x, y, z) = (1 - R) - \beta - I(x, y) - exp[-\beta z]$. The attenuation of the irradiated intensity distribution $I(x, y)$ along the irradiation direction z was assumed according to the Lambert-Beer law. The required optical material parameters (reflectance R, absorption coefficient β) of different powder materials were determined from experimental measurements using an integrating sphere [16]. An important result of the simulation is that heat conduction and convection have a strong influence on the spatial and temporal temperature distribution and the developing gradients of the temperature field and thus also on the resulting layer geometry.

When irradiated with a defined intensity, a stationary state is reached after a certain time, in which the maximum temperature of the irradiated surface no longer changes. The duration of time and the resulting temperature for a given intensity depend on the size of the irradiation area [15]. This finding is of particular importance for irradiation surfaces of complex shape, where a corresponding temporal and spatial adjustment of the intensity becomes necessary to homogenize the temperature in all areas of the surface. Thus, the simulations show that for small irradiation surfaces, higher intensities are needed than for large irradiation surfaces to achieve a defined temperature inside the surfaces. In addition, heat conduction leads to an increase of the layer area and depth, resulting in "rounded" geometries, as well as inhomogeneous layer thicknesses (smaller thickness of the layers at the edges of the irradiation area).

The influence of heat conduction can be reduced with the aid of small temperature increases that are still sufficient for the melting process, and by short irradiation times. The simulations show that with irradiation times of less than 100 ms and correspondingly adjusted intensity, there is only a slight loss of heat from the irradiated surface due to thermal conduction. On the one hand, the temperatures reached are almost independent of the size of the irradiated area and, on the other hand, only minor heat conduction effects can be detected in the resulting layer geometry.

Furthermore, the simulation model was used to simulate the process of simultaneous laser beam melting for processing PA12 with PP (cf. Fig. 6.2) and compared with experimental measurements. The results show a qualitative agreement with respect to the resulting layer geometries and thicknesses, which are influenced by process parameter-dependent heat conduction effects, especially in the boundary zone of the two materials. Quantitative deviations of the simulated temperature distributions can be attributed to a systematic overestimation of the heat flow from the powder surface into the surrounding air. This is due to the fact that, within the framework of the simulation, the air temperature is set as a boundary condition equal to the preheating temperature, and an expected heating of the air directly above the powder bed is not represented by the simulation.

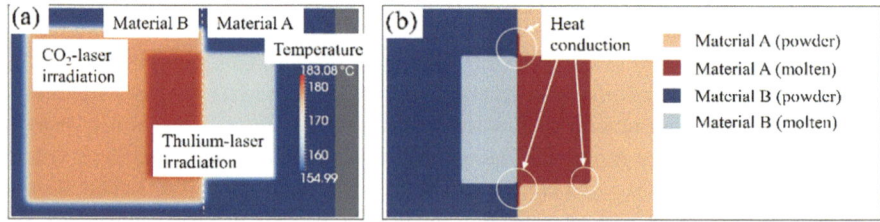

Fig. 6.2 **a** Simulated temperature field resulting from simultaneous laser irradiation (Material A: PP, Material B: PA12); **b** Phases of Materials A and B as a result of the temperature fields with heat conduction effects

6.2.3 Multi-material PBF-LB/P with Polymers Showing Significantly Different Thermal Property Profiles

In quasi-simultaneous preheating, the low-melting material is still preheated by means of IR radiant heaters, while the high-melting material is brought to the preheating temperature in a selective and quasi-simultaneous manner with the aid of a scanner-guided CO_2 laser. For this purpose, the DOE previously used for beam shaping of the CO_2-laser was replaced by a galvanometer scanner to set the complex temperature fields with high gradients required for multi-material processing, whereby the high deflection speeds of the scanner on the one hand, and the low thermal conductivity of the plastics on the other hand, are used for homogeneous preheating of the powder bed. It was shown that by implementing a scanner for irradiation by a CO_2-laser, a quasi-simultaneous energy input, depending on the illumination parameters (e.g., focus diameter, hatch distance, writing speed), and hence a homogeneous temperature field, could be realized [10].

When processing materials with significantly different thermal properties (e.g., PP/TPE-A), it was not possible to establish a bond between the materials when using a two-dimensional, homogeneous irradiation intensity (Thulium-laser) to melt the materials. Instead, curling of the melted PP region was observed near the TPE-A/PP boundary zone, which led to insufficient material connection in the boundary zone during the process. This is due to heat conduction losses arising from high temperature gradients in the boundary zone. Gap formation and curling of the PP region caused by premature crystallization can be prevented by locally adjusting the energy input in the boundary zone, as shown in Fig. 6.3 [10].

Here, the energy introduced by the Thulium-laser, which is necessary for the simultaneous melting of the materials, is increased by varying the tilting frequencies of the micromirrors used in a defined range on the side of the high-melting material in order to increase the temperature locally. For the material combination PP/TPE-A,

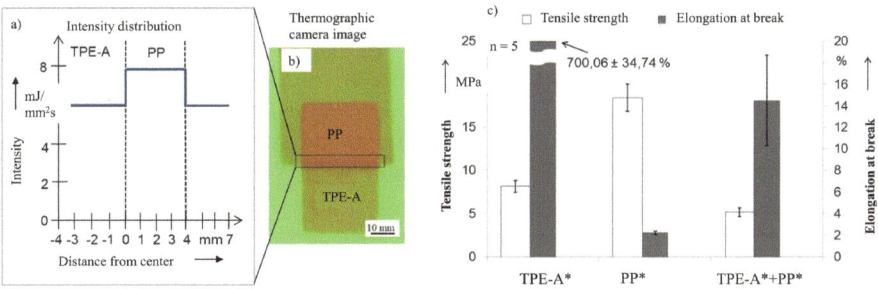

Fig. 6.3 **a** Intensity distribution of Thulium-laser irradiation; **b** Thermographic camera image (arbitrary units) during locally adapted energy input for processing PP and TPE-A; **c** Mechanical properties of the individual materials and the multi-material parts determined in tensile tests; all images are the property of the Bayerisches Laserzentrum GmbH and were first published in [10]

for example, a local increase in the intensity of the Thulium-laser by 2 mJ/mm^{2}s (base intensity: 6 mJ/mm^{2}s) in a range up to 4 mm relative to the boundary line was identified as suitable for generating a multi-material layer with an intact composite zone (cf. Fig. 6.3a). The bond strengths that were obtained were verified by tensile tests and the mechanical properties are shown in Fig. 6.3c. The sample designation (*) indicates the process- and material-dependent functionalization of the pure polymer powders with flow agents and absorption-enhancing carbon black in different amounts. The results indicate that the test parts are slightly weakened by the composite zone, as can be seen from the lower strength values of the multi-material parts compared to the characteristic values of the individual materials. The reason for this is a slightly increased porosity in the boundary zone.

6.2.4 Simulation of Resulting Temperature Fields During Multi-material PBF-LB/P of Polymers Showing Significantly Different Thermal Property Profiles

Here, the objective is to numerically calculate the spatial and temporal variations of the temperature fields in the powder bed in the context of (quasi-)simultaneous, intensity-selective irradiation. Based on this, suitable strategies for generating multi-material parts consisting of polymers with significantly different thermal property profiles were developed. The simulation model is based on a finite element model for the calculation of the temporal temperature distribution in the powder bed. It solves the nonlinear heat conduction equation numerically and uses temperature-dependent functions for heat capacity and heat conduction to account for the influence of the phase transition on the temperature history. By means of the simulation model, it was shown that the quasi-simultaneous irradiation (high-speed scanner-guided laser beam) of a $10 \times 10 \, cm^2$ area with a beam diameter of 2 mm can be very well approximated by simultaneous exposure from scan speeds of 10 m/s. This means that heat conduction can be neglected for 0.5 s for one exposure cycle (one complete scan of the layer). Thus, thermal conduction can be neglected for the duration of 0.5 s for one exposure cycle (single, complete scanning of the layer). By only considering simultaneous exposure in the modeling, a great deal of computational time can be saved, which facilitates simulating the process for manufacturing multilayered components in an acceptable amount of time in the first place. For example, a computation time of approx. 1.5 h is required for the process sequence of a layer with the size $30 \times 40 \, mm^2$. The simulation model was verified using experimental data in the processing of PA12 and PP by a simple test setup [27]. The process design and the resulting temperature profile in the generated part are shown in Fig. 6.4.

Good qualitative agreement was found between the experimental and calculated temperature data, with the results also allowing for conclusions to be drawn about the significance of various process and material parameters. For example, it can be seen

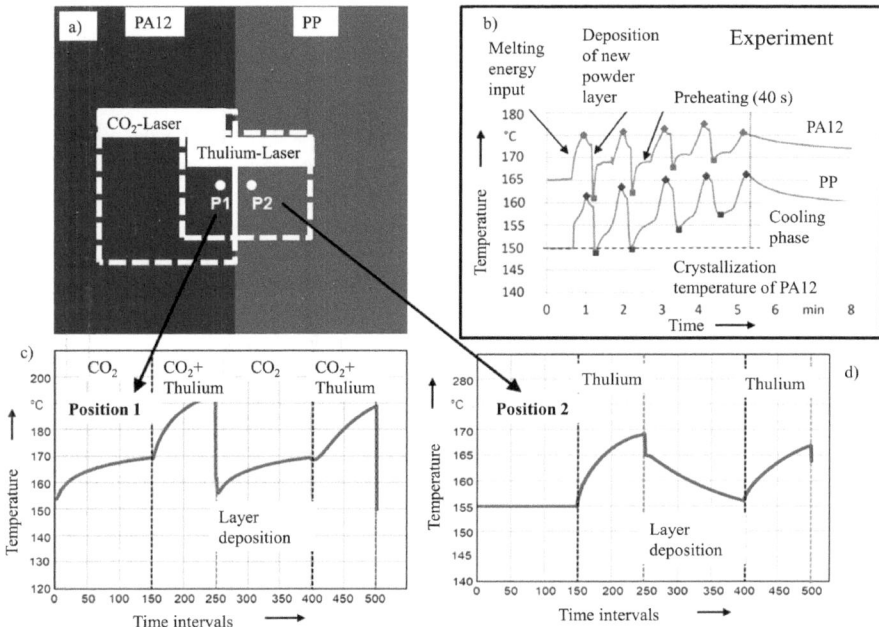

Fig. 6.4 Simulation setup with simple multi-material distribution (PA12/PP); **a** Irradiation areas and measuring points; **b** Experimental measurement of the temperatures in the PA12 and PP layers, respectively; **c** and **d** Calculated temperature curves at point 1 within PA12 and at point 2 within PP

that latent heat strongly influences the temperature history in the powder bed, which can be inferred from the different degrees of cooling of PA12 and PP after exposure to Thulium- and CO_2-lasers, respectively. PP has a significantly higher enthalpy of fusion (207 J/g) than PA12 (115 J/g), which results in a correspondingly high enthalpy of crystallization in the simulation model. A major goal was the derivation of strategies to compensate for heat transport losses. It was shown that, although their compensation by adjusting the irradiation intensity as a function of the irradiated area size is possible, the low structural resolution that occurs due to significant heat conduction at long exposure times cannot be prevented. The reason for this is that unwanted heat accumulation cannot be reversed, since no targeted heat dissipation by irradiation is possible. This ultimately leads to the unintentional coalescence of closely spaced exposed material zones into a common melt pool during the melting process.

6.3 Powder Application Strategies for Generating Multi-material Parts by PBF-LB/P

Generating multi-material polymer parts by means of PBF-LB/P generally requires the realization of a defined, heterogeneous powder layer consisting of different materials. A two-chamber doctor blade-based powder application device allows for applying two different powder materials simultaneously into the build chamber [13, 20]. However, although this enables fabrication of multi-material parts by means of PBF-LB/P, the complexity of these parts is strongly limited due to the simple and non-adaptable border zone between both materials. For this reason, a vibration nozzle-based powder application method was introduced in [26] as described in Sect. 6.3.1. A promising new approach for overcoming the limitations of conventional or nozzle-based powder application methods is the electrophotographic powder application known from two-dimensional printing of toner particles in the context of laser printers. This approach is described in Sect. 6.3.2, while the latest findings are presented in Sect. 6.4.

6.3.1 Vibration Nozzle System for Generating Multi-material Parts

In principle, nozzle-based powder application enables arbitrarily complex powder distributions with both lateral and vertical boundary zones. Based on the breaking of powder bridges inside the nozzle by a piezo actuator, the powder flow is stimulated for material deposition. Furthermore, a start-stop functionality is also provided. Figure 6.5a shows a schematic overview of the nozzle system for generating multi-material powder distributions within a PBF-LB/P process chamber.

Two additively manufactured metal nozzles with orifice diameters of 0.7 and 1.0 mm are attached to the standard coater of a modified PBF-LB/P system. Temperature control is achieved by means of water cooling or oil tempering, which are realized with the aid of fluid channels inside the nozzles. A function generator produces a sinusoidal output signal for the analog signal amplifier, which amplifies the signal by a factor of 100. The amplified sinusoidal signal induces vibrations in the piezo actuator, causing the powder bridges to break up, which allows for a flexible powder deposition.

A major challenge in the use of vibrating powder nozzles is the high build chamber temperature of up to 230 °C (when processing polybutylene terephthalate (PBT), for example), which affects the functionality of the nozzle-based powder application system. In particular, the application mechanism itself, i.e., the reversible formation of powder bridges, must be reproducibly ensured. Depending on the requirements of the powder material or the process, suitable temperatures can be set inside the nozzle. Figure 6.5b shows the dependence of the melt flow on the die temperature, using PA12 powder as an example.

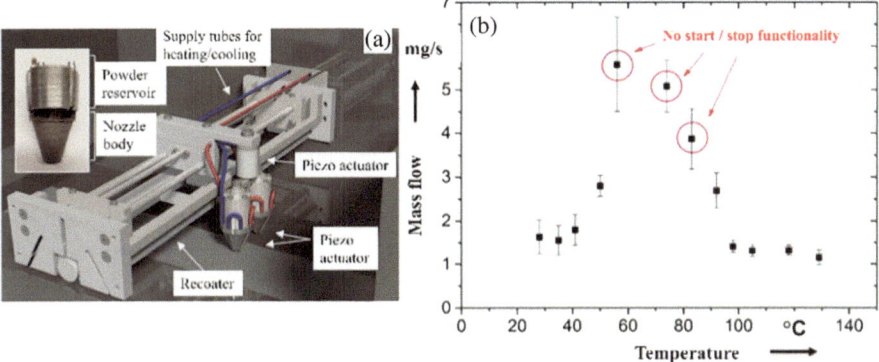

Fig. 6.5 a PBF-LB/P coater equipped with two vibration nozzles, a piezo actuator and supply for heating/cooling fluids; enlarged image: additively manufactured metal nozzle; **b** Average PA12 mass flow during vibration excitation (frequency = 300 Hz, voltage amplitude = 30 V) for a temperature range between 27 and 130 °C; all images are the property of the Bayerisches Laserzentrum GmbH and were first published in [10, 19]

It becomes apparent that both the voltage amplitude and the temperature have a great influence on the resulting mass flow. The temperature dependence of the mass flow (first strongly increasing, then strongly decreasing) is caused by two effects. Firstly, the initial increase of the mass flow is due to the increasing flowability due to the drying of the powder and thus the lower water adsorption leads to the reduced formation of liquid bridges between the powder particles. Due to the high flowability between 55 and 80 °C, no reversible powder bridge formation can take place, which means that the start-stop function of powder deposition does not occur. Furthermore, complete drying leads to improved particle-particle contact and an increase in Van der Waals forces, which is reflected in reduced mass flow to a constant level of approximately 1 mg/s between 90 and 130 °C. This shows that temperature control is essential for the function of a reproducible and uniform powder application. Since the translation speed is limited to 25 mm/s at a maximum possible mass flow of approx. 2 mg/s, the resulting layer preparation time for a line width of 1.2 mm, a layer thickness of 130 μm, and an area of 100 mm × 100 mm is over 5 min. For this reason, the efficient use of vibration nozzles for large-area powder layer preparation requires the additional use of other technologies, such as electrostatic powder transfer, which is used in electrophotography. Nevertheless, nozzle application can already be used efficiently for small, local material-side modifications (e.g., targeted filler addition for mechanical reinforcement in PBF-LB/P or local alloying in laser beam melting of metals (PBF-LB/M).

6.3.2 Electrophotographic Powder Application for Generating Multi-material Parts

Electrophotographic powder application (EPA) is based on electrostatic attractive and repulsive forces between powder particles and an entirely or selectively charged photoconductor. For this purpose, a first test setup was realized, whose essential element was an aluminum plate coated with a 100 μm thick, vapor-deposited layer of As_2Se_3 serving as photoconductive material [24, 25]. This plate was used to pick up powder from a powder bed (development step) for subsequent deposition (printing step) into a heatable chamber (cf. Fig. 6.6).

In general, the powder is developed outside the chamber and thus at comparatively low temperatures. A charge pattern can be generated on the photoconductor by means of a charging unit (Corona or Scotron) and subsequent exposure by a mirror element with a radiation source emitting at a wavelength of 632 nm. The powder in the powder bed needs to be electrostatically charged beforehand by a further charging unit. The

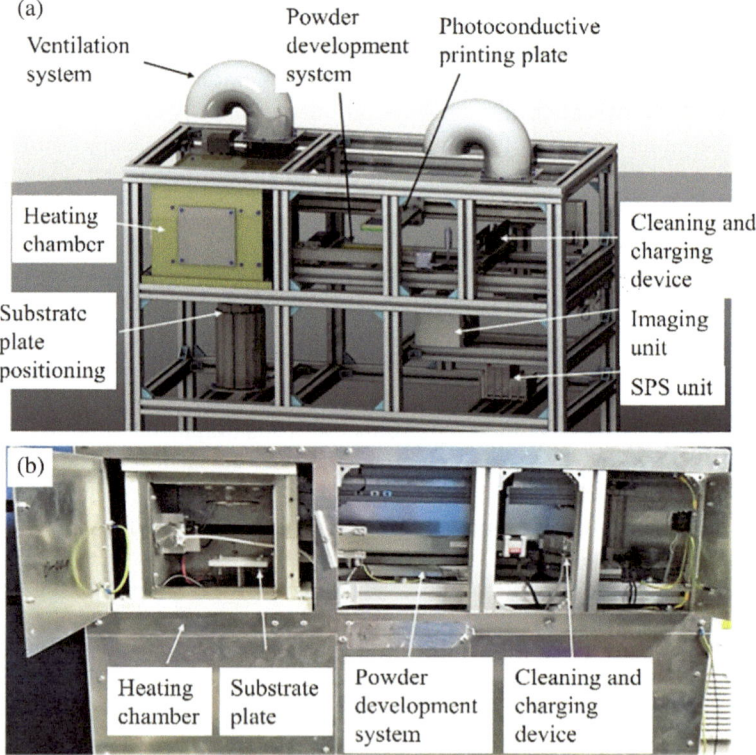

Fig. 6.6 a Rendering image; **b** Realized first electrophotographic powder application setup; all images are the property of the Bayerisches Laserzentrum GmbH and were first published in [25]

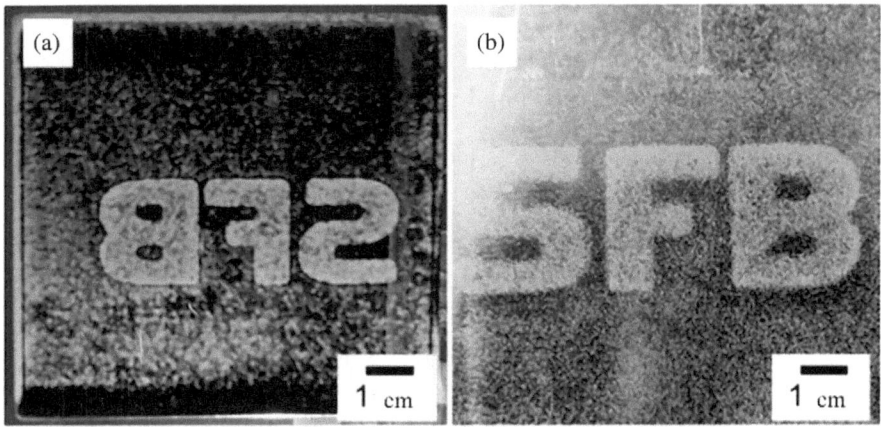

Fig. 6.7 **a** Polypropylene (PP) powder pattern developed on a photoconductive plate (PCP); **b** PP deposited onto a metal substrate plate; all images are the property of the Bayerisches Laserzentrum GmbH and were first published in [25]

powder deposition on a height-adjustable platform in the chamber was realized by a transfer grid in the first test setup (cf. Fig. 6.6).

However, as shown in Fig. 6.7b, the deposited powder layer (letters "SFB") only shows poor coverage in the range of 70% and, additionally, a strong false print (powder particles deposited into unintended areas) of approximately 50% surrounding the entire powder pattern is visible.

Furthermore, the subsequent electrophotographic deposition and laser-based fusion of several powder layers to generate three-dimensional (3D) parts has not been conducted to date.

6.4 Current Findings from the Research

Electrophotographic powder application is a promising approach for overcoming the limitations of conventional or vibration nozzle-based powder application methods. EPA for laser-based powder bed fusion of polymers (EPA-PBF-LB/P) consists of six main process steps, based on [7, 9, 23]. A schematic overview of the process steps is shown in Fig. 6.10a. First, the powder and a photoconductive plate (PCP, aluminum plate coated with photoconductive material) are homogeneously charged [17, 18, 22]. In EPA, powder particles can be charged via gas discharge (Corona or Scorotron charging) or triboelectrically [2, 3]. While gas discharge-based powder charging does not require any further functionalization of the particles in terms of their charging behavior, for triboelectric charging the particles need to be functionalized with charge control agents (CCAs) in order to tune their charging tendency. As shown in Fig. 6.8a, polymer particles such as PP typically display a

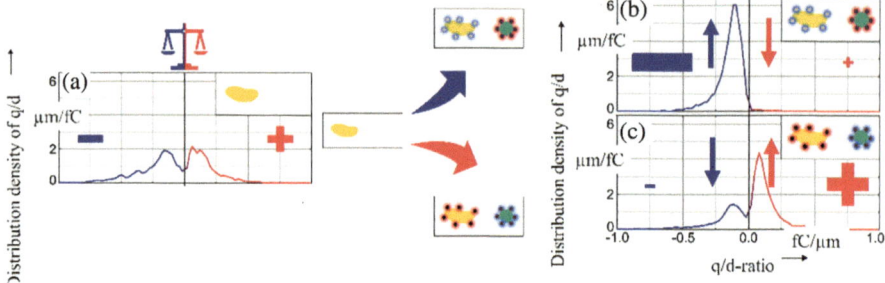

Fig. 6.8 Functionalization of polymer particles (yellow potato-shaped particles) with charge control agents (CCAs, blue and red circles surrounding the polymer particles) and use of suitable carrier particles (cyan circular particles with their respective surface functionalization) for tuning the triboelectric charging behavior of polymer particles from bimodal (**a**) to monomodal negative (**b**) or positive (**c**); all images are the property of the Bayerisches Laserzentrum GmbH and were first published in [3]

bimodal triboelectric charge distribution, which is unsuitable for EPA. Only the synergy of both functionalizing the particle surface with CCAs and choosing suitable carrier particles known from two-component toners allows for controlling crucial parameters of triboelectric charging, namely the polarity, amount of charge, and resulting surface potential. This enables the triboelectric charging mechanism to be used for electrophotographic powder application in PBF-LB/P [3].

As described in [2, 3], triboelectric charging offers considerable advantages compared to gas discharge-based (Corona or Scorotron) powder charging. In particular, in the case of triboelectric powder charging, the powder flowability, which in many cases renders the use of certain powders for PBF-LB/P impossible, does not play any role in the powder application. This allows for pharmaceutical polymers or drugs, which often show poor flowability, to be processed in EPA-PBF-LB/P [6].

Another remarkable advantage of triboelectric charging instead of gas discharge-based powder charging is the fact that the layer thickness can be adjusted by changing the development field strength (cf. Fig. 6.9). This is due to the fact that not only the uppermost powder layer, which in the case of gas discharge-based powder charging is charged by incorporated ions, is charged via triboelectric charging. Instead, the entire, several hundreds of micrometers thick powder layer is homogeneously charged [3]. Hence, a thicker powder layer can be developed and deposited, which offers the possibility of significantly increasing the productivity of EPA-PBF-LB/P. In the case of conventional doctor blade- or roller-based powder application methods, the layer thickness is typically in the range of 100–150 μm and strongly depends on the particle sizes of the powder used [1, 2, 4, 8]. Considering the different possible layer thickness values shown in Fig. 6.9, it can be concluded that utilizing triboelectric charging in combination with EPA enhances the flexibility of tailoring the layer thickness of powder layers deposited onto the build platform for PBF-LB/P [3].

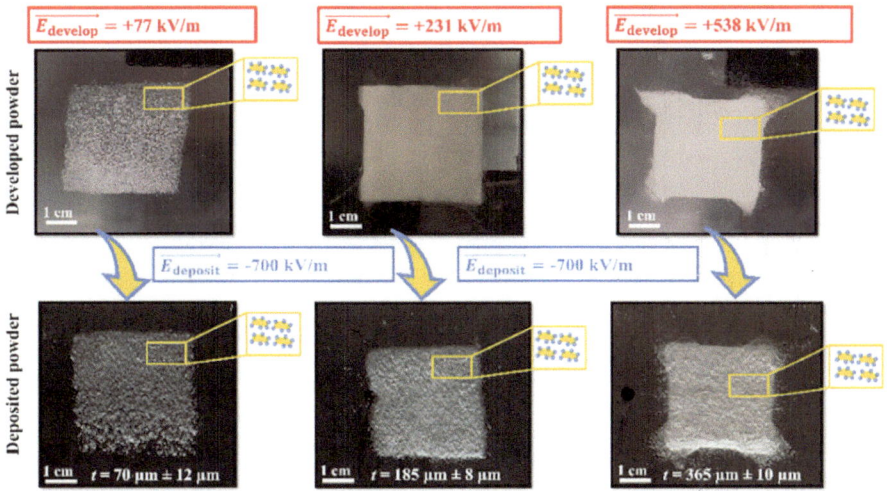

Fig. 6.9 PP functionalized with 0.5 wt% silica(−) as CCA and mixed with 40 wt% positively charging carrier (MF83-100) for triboelectric charging developed to PCP using different electric development field strengths and deposited onto the build platform; number of measurements n = 20 for layer thickness; all images are the property of the Bayerisches Laserzentrum GmbH and were first published in [3]

In the second step of EPA-PBF-LB/P, the PCP is selectively discharged, thereby creating a latent charge pattern by selective illumination using a DLP at a wavelength of 632 nm. In step three, the so-called development step, charged powder particles are attracted towards the PCP due to electrostatic forces. Depending on the polarity of charging the powder particles in step two, charged particles can either be developed into charged (charged area development, CAD) or discharged (discharged area development, DAD) regions of the latent charge pattern of the PCP. In step four, the printing step, the PCP is moved laterally to the build chamber and the powder pattern is deposited onto the build platform located inside the build chamber. In contrast to the heating chamber of the first electrophotography setup shown in Fig. 6.6, the build chamber of the EPA-PBF-LB/P setup used in [3–5] offers the possibility of conventional CO_2-laser based powder fusion. Thus, in step five, irradiation by a laser beam takes place to fuse the deposited powder particles. Finally, in step six, the PCP is cleaned.

The general demonstration of the working principle of EPA in the context of PBF-LB/P on a single-layer basis was shown in [21, 23, 25].

However, printing more than one layer by EPA-PBF-LB/P introduces some challenges, which are mostly related to the electrostatic phenomena taking place during EPA. As demonstrated in [7], controlling the electric field applied in terms of field strength and especially shape is crucial for powder deposition to achieve high dimensional accuracy and coverage of the deposited powder pattern. For tailoring the electric deposition field, a transfer frame was developed and tested in [7] and replaced the

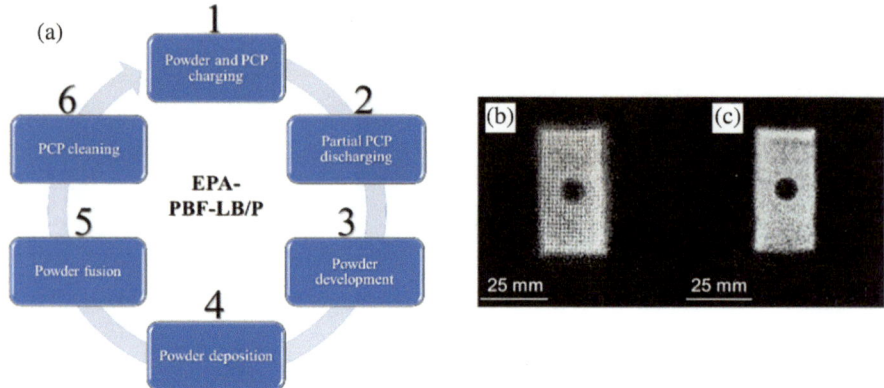

Fig. 6.10 a Schematic overview of the main process steps of EPA-PBF-LB/P; comparison of electrophotographic powder deposition of PP using a transfer grid (**b**) and a transfer frame (**c**); all images are the property of the Bayerisches Laserzentrum GmbH and were first published in [3, 5]

transfer grid used so far (Sect. 6.3.2). In Fig. 6.10b, c, the power deposition results of both transfer structures are compared.

In addition, for successful multi-layer EPA-PBF-LB/P, the accumulation of charges accompanied by a progressive decrease of the electric field strength of the powder deposition field needs to be compensated [4]. This necessitates a strategy for compensating for the charge accumulation independently of the already generated part thickness. In case of constant environmental conditions, this should lead to a net charge of zero by neutralization of contrarily charged ions. In fact, after depositing a powder layer with a certain surface potential, a powder layer with contrary charging but the same magnitude of surface potential should be deposited on top of the previous layer. This means that the operating modes of EPA-PBF-LB/P need to be regularly changed from CAD to DAD. Although changing the polarity after each powder deposition would be even more efficient for compensating charge accumulation, a certain duration is needed for the reversion of polarity by the high-voltage power supply. In [4], changing the operation mode from CAD to DAD after every second powder deposition was demonstrated to be a good trade-off between the duration for the reversion of the polarity and maintaining significant part height growth. In Fig. 6.11a left, already for the fifth PP layer deposition, an increase of the magnitude of the surface potential to 1100 V accompanied by a part height growth of 0 μm can be observed. Applying the charge compensation strategy developed in [4] allows for maintaining the surface potential values (cf. Fig. 6.11b) at which significant part height growth is achieved.

A further aspect of increasing the powder deposition efficiency is the development and implementation of a piezoelectric excitation device in [5]. By reducing the van der Waals interaction forces due to agitation of the powder particles by means of a piezoelectric excitation, their flow behavior and, thus, the powder deposition efficiency was significantly enhanced.

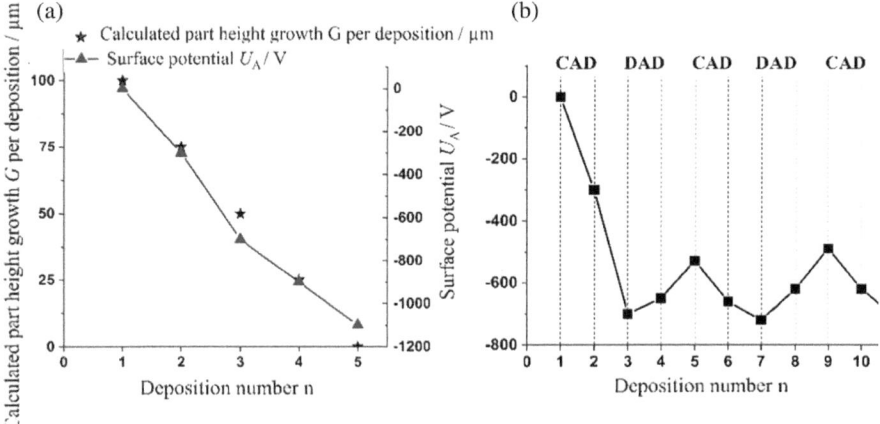

Fig. 6.11 **a** Surface potential of a PP powder layer deposited into the build chamber without a charge compensation strategy; **b** Surface potential with an applied charge compensation strategy; all images are the property of the Bayerisches Laserzentrum GmbH and were first published in [4]

It is important to note that in the case of conventional powder application methods, only a small portion of the powder in the build chamber is used to generate the actual part [4, 20] and the remaining powder must be recycled at great expense. Since EPA offers the possibility of selectively applying only the powder volume necessary for generating the desired part, EPA-PBF-LB/P can significantly reduce the environmental impact of PBF-LB/P. Beyond this, the number of powder materials applicable for PBF-LB/P can be increased due to the independence of the powder flowability—especially in the case of triboelectric charging. In [4], the parts generated by EPA-PBF-LB/P could be validated to show at least the same or better mechanical properties compared to parts generated by conventional PBF-LB/P (cf. Fig. 6.12a, b). The selective and precise powder deposition enabled by EPA furthermore allows for the generation of multi-material parts with graded transition zones between the single materials (cf. Fig. 6.12c).

6.5 Conclusion

Generating multi-material polymer parts by means of laser-based additive manufacturing (PBF-LB/P) requires an adaptable energy input and the realization of powder layers consisting of different materials. Particularly in the case of multi-material parts made from polymers with significantly different thermal property profiles, the use of a second laser source allows the individual thermal requirements of the materials to be met. In this context, a Thulium-laser emitting at a wavelength of 1.94 μm is especially beneficial due to the deeper penetration into the powder bed.

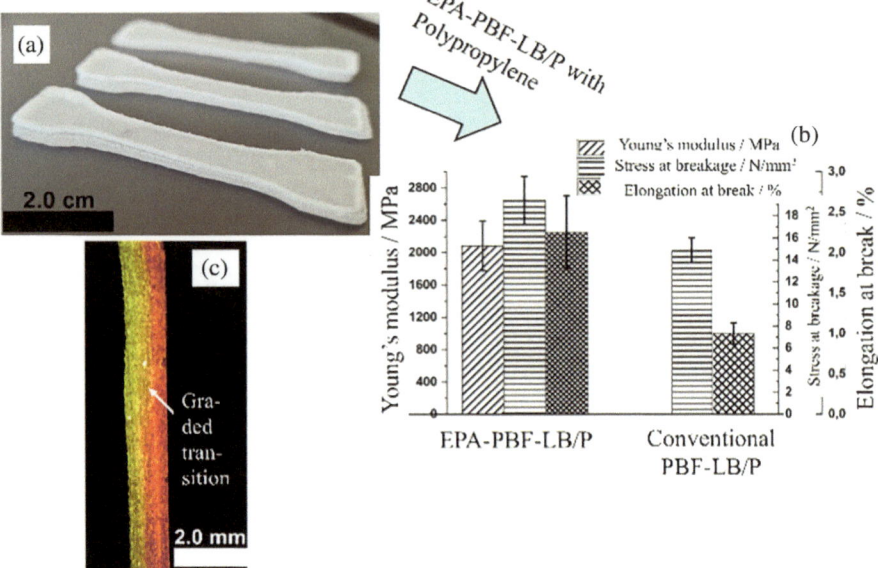

Fig. 6.12 **a** Polypropylene (PP) tensile test specimen generated by EPA-PBF-LB/P; **b** Mechanical properties of the tensile test specimen shown in (**a**) (error bars indicate a standard deviation of 3 measurements; values for conventional PBF-LB/P from [28]; **c** Multi-material part generated by EPA-PBF-LB/P with a graded transition between two different PA12 blends; all images are the property of the Bayerisches Laserzentrum GmbH and were first published in [4]

For generating tailored powder layers consisting of more than one material, both the nozzle-based and the electrophotographic approach are suitable powder application methods for PBF-LB/P. However, while a nozzle-based powder application inevitably results in a comparatively high surface roughness of the powder layer and is relatively slow, electrophotography can overcome the shortcomings of conventional powder application methods for PBF-LB/P. It enables the deposition of powder layers consisting of various materials with a high degree of geometrical flexibility and accuracy. Due to the selective powder deposition, only the amount of powder required to generate the actual parts is applied, which significantly increases the powder's efficiency. In addition, the dependence of the powder deposition on flowability is greatly reduced, which facilitates the use of new powder materials in PBF-LB/P. By combining the advantages of an adapted laser energy input in combination with selective electrophotographic powder deposition, highly challenging powders in terms of thermal degradation and flowability were successfully manufactured [6].

References

1. Dechet, M.A., Gómez Bonilla, J.S., Grünewald, M., Popp, K., Rudloff, J., Lang, M., Schmidt, J.: A novel, precipitated polybutylene terephthalate feedstock material for powder bed fusion of polymers (pbf): Material development and initial pbf processability. Mater Des. **197**, 109265 (2021)
2. Düsenberg, B., Kopp, S.-P., Tischer, F., Schrüfer, S., Roth, S., Schmidt, J., Schmidt, M., Schubert, D.W., Peukert, W., Bück, A.: Enhancing photoelectric powder deposition of polymers by charge control substances. Polymers **14**(7), 1332 (2022)
3. Kopp, S.-P., Düsenberg, B., Eshun, P.M., Schmidt, J., Bück, A., Roth, S., Schmidt, M.: Enabling triboelectric charging as a powder charging method for electrophotographic powder application in laser-based powder bed fusion of polymers by triboelectric charge control. Add. Manuf. **68**, 103531 (2023)
4. Kopp, S.-P., Medvedev, V., Frick, T., Roth, S.: Expanding the capabilities of laser-based powder bed fusion of polymers through the use of electrophotographic powder application. J. Laser Appl. **34**(4), 042032 (2022)
5. Kopp, S.-P., Medvedev, V., Roth, S.: Targeted vibration excitation for increasing the powder deposition efficiency in electrophotographic powder application for laser-based powder bed fusion of polymers. Procedia CIRP **111**, 55–60 (2022)
6. Kopp, S.-P., Medvedev, V., Tangermann-Gerk, K., Wöltinger, N., Rothfelder, R., Graßl, F., Heinrich, M.R., Januskaite, P., Goyanes, A., Basit, A.W., Roth, S., Schmidt, M.: Electrophotographic 3d printing of pharmaceutical films. Add. Manuf. 103707 (2023)
7. Kopp, S.P., Stichel, T., Roth, S., Schmidt, M.: Investigation of the electrophotographic powder deposition through a transfer grid for efficient additive manufacturing. Procedia CIRP **94**, 122–127 (2020)
8. Kruth, J.P., Wang, X., Laoui, T., Froyen, L.: Lasers and materials in selective laser sintering. Assembly Autom. **23**(4), 357–371 (2003)
9. Kumar, A.V., Dutta, A., Fay, J.E.: Electrophotographic printing of part and binder powders. Rapid Prototype J. **10**(1), 7–13 (2004)
10. Laumer, T., KG, M.: Erzeugung von Thermoplastischen Werkstoffverbunden Mittels Simultanem, Intensitätsselektivem Laserstrahlschmelzen. Ph.D. thesis, Friedrich-Alexander-Universität Erlangen-Nürnberg (FAU) (2017)
11. Laumer, T., Koopmann, J., Stichel, T., Amend, P.: Generation of multi-material parts with alternating material layers by simultaneous laser beam melting of polymers (2014)
12. Laumer, T., Stichel, T., Amend, P., Roth, S., Schmidt, M.: Analysis of temperature gradients during simultaneous laser beam melting of polymers. PhPro **56**(C), 167–175 (2014)
13. Laumer, T., Stichel, T., Amend, P., Schmidt, M.: Simultaneous laser beam melting of multimaterial polymer parts. J. Laser Appl. **27**(S2), S29204 (2015)
14. Laumer, T., Stichel, T., Bock, T., Amend, P., Schmidt, M.: Characterization of temperature-dependent optical material properties of polymer powders. AIP Conf. Proc. **1664**, 160009 (2015)
15. Laumer, T., Stichel, T., Riedlbauer, D., Amend, P., Mergheim, J., Schmidt, M.: Realization of multi-material polymer parts by simultaneous laser beam melting. JLMN-J. Laser Micro/Nanoengineering **10**(2) (2015)
16. Laumer, T., Wudy, K., Drexler, M., Amend, P., Roth, S., Drummer, D., Schmidt, M.: Fundamental investigation of laser beam melting of polymers for additive manufacture. J. Laser Appl. **26**(4), 042003 (2014)
17. Matsusaka, S., Masuda, H.: Electrostatics of particles. Adv. Powder Technol. **14**(2), 143–166 (2003)
18. Pai, D.M., Springett, B.E.: Physics of electrophotography. Rev. Mod. Phys. **65**(1), 163 (1993)
19. Schuffenhauer, T., Stichel, T., Kopp, S.-P., Roth, S., Schmidt, M.: Process route adaption to generate multi-layered compounds using vibration-controlled powder nozzles in selective laser melting of polymers (2019)

20. Setter, R., Hafenecker, J., Rothfelder, R., Kopp, S.-P., Roth, S., Schmidt, M., Merklein, M., Wudy, K.: Innovative process strategies in powder-based multi-material additive manufacturing. J. Manuf. Mater. Process. **7**(4), 133 (2023)
21. Setter, R., Stichel, T., Schuffenhauer, T., Kopp, S.P., Roth, S., Wudy, K.: Additive manufacturing of multi-material polymer parts within the collaborative research center 814. In: Lecture Notes in Mechanical Engineering, pp. 142–15 (2021)
22. Shahin, M.M.: Mass—spectrometric studies of corona discharges in air at atmospheric pressures. J. Chem. Phys. **45**(7), 2600 (2004)
23. Stichel, T., Brachmann, C., Raths, M., Dechet, M.A., Schmidt, J., Peukert, W., Frick, T., Roth, S.: Electrophotographic multilayer powder pattern deposition for additive manufacturing. Jom **72**(3), 1366–1375 (2020)
24. Stichel, T., Brandl, T., Hauser, T., Geißler, B., Roth, S.: Electrophotographic multi-material powder deposition for additive manufacturing. Procedia CIRP **74**, 249–253 (2018)
25. Stichel, T., Geißler, B., Jander, J., Laumer, T., Frick, T., Roth, S.: Electrophotographic multi-material powder deposition for additive manufacturing. J. Laser Appl. **30**(3), 032306 (2018)
26. Stichel, T., Laumer, T., Baumüller, T., Amend, P., Roth, S.: Powder layer preparation using vibration-controlled capillary steel nozzles for additive manufacturing. Phys. Procedia **56**, 157–166 (2014)
27. Stichel, T., Laumer, T., Schmidt, M.: Simulation des (quasi-)simultanen laserstrahlschmelzens zur herstellung von multi-material-bauteilen aus polymeren. In: Rapid.Tech + FabCon 3.D— International Trade Show + Conference for Additive Manufacturing, pp. 312–329 (2018)
28. Tan, L.J., Zhu, W., Sagar, K., Zhou, K.: Comparative study on the selective laser sintering of polypropylene homopolymer and copolymer: Processability, crystallization kinetics, crystal phases and mechanical properties. Add. Manuf. **37** (2021)

Stephan Roth has been serving as the Managing Director of the Bavarian Laser Center located in Erlangen since 2010. He obtained his doctoral degree in Mechanical Engineering in 2008. In Additive Manufacturing, the research facility deals not only with process technology but also with the optical properties of powder materials and new powder application systems.

Chapter 7
Processing Strategies for Electron Beam Based Powder Bed Fusion

Christoph Breuning, Jakob Renner, Matthias Markl, and Carolin Körner

7.1 Introduction

Electron beam powder bed fusion (PBF-EB) is an additive manufacturing process that uses a high-energy electron beam for the local consolidation of metal powder. The layer-wise nature of the process enables the tool-free fabrication of complex geometries with few geometrical restrictions. Moreover, the combination of high vacuum conditions with high process temperatures and the ability to control the spatio-temporal energy input not only enables the processing of a large variety of metal alloys but also the local tailoring of microstructure and properties [1, 2]. However, the reliable, defect-free fabrication of complex geometries with predefined, homogeneous properties remains a major challenge in PBF-EB. Local defect formation and material properties are determined by the local thermal conditions that are governed by the cumulative heating effect and are based on the combination of the part geometry, process parameters, processing conditions, and scanning strategy. Variation of the cross-section and return time within each layer, as well as variations of the geometry in the building direction, change the local thermal conditions and can lead to local defect formation if processing boundaries are violated [3]. Suitable processing strategies that control the process parameters and scanning strategy have to be developed to tailor the local thermal conditions based on the features of the underlying geometry, prevent defect formation, and tailor the local properties. Since the local thermal conditions are not directly experimentally accessible, thermal simulations are utilized to investigate the spatio-temporal melt pool evolution and identify suitable processing strategies. Previous research showed the ability of simplified thermal models—that neglect fluid convection, latent heat release, radiation, and

C. Breuning · J. Renner · M. Markl · C. Körner (✉)
Friedrich-Alexander-Universität Erlangen-Nürnberg, Chair of Materials Science and Engineering for Metals, Martensstraße 5, 91058 Erlangen, Germany
e-mail: carolin.koerner@fau.de

© The Author(s) 2025   127
D. Drummer and M. Schmidt (eds.), *Progress in Powder Based Additive Manufacturing*, Springer Tracts in Additive Manufacturing,
https://doi.org/10.1007/978-3-031-78350-0_7

vaporization and assume constant material properties—to predict the spatio-temporal melt pool evolution with sufficient accuracy for predictive simulation [4–6]. These numerical approaches, however, simplify the underlying physics of the process and are not able to precisely describe the stochastic processes during powder application and in the active melt pool [7, 8]. Therefore, reliable layer-wise process monitoring is required for quality assurance and to guarantee the defect-free fabrication of complex geometries. In addition to the aforementioned stochastic processes, uncertainties in the estimation of local processing conditions during processing, such as local preheating temperature or beam characteristics, prevent the precise description and prediction of the process when solely using numerical simulations. In order to reduce simulation uncertainties and adapt developed processing strategies to the actual processing conditions of the current process, additional process information is required. In-situ process monitoring provides the necessary insight into the process and information for either the adaption of the process through control or to establish the relationships between local thermal conditions and final properties during the process. The harsh process environment at high temperatures in vacuum conditions prevents the use of common optical process monitoring solutions due to metallization. Electron optical (ELO) process monitoring in PBF-EB can overcome these limitations, and enables the in situ detection of surface defects and melt contours [9–12]. This chapter introduces a framework combining predictive simulation and in-situ process monitoring using ELO imaging that enables the reliable, defect-free fabrication of complex geometries in PBF-EB with predefined, homogeneous properties using line-based hatching strategies.

7.2   Framework

The proposed framework extends the PBF-EB process and integrates both feed-forward and feedback control strategies based on a shared process database, as detailed in Fig. 7.1. The process database contains information about the relationships that enable the initial process parameter selection and process control. The first relationship comprises the interactions between the process parameters (power P, velocity v), processing conditions (beam diameter σ, preheating temperature T_p), and the emerging melt pool geometry and thermal conditions. The second relationship connects the melt pool geometry and thermal conditions with the emerging structure and properties. The emerging structure and properties can be further subdivided into the primary objective of a defect- and surface-structure and the secondary objectives of microstructure and mechanical properties. Both relationships are further detailed in Sect. 7.5.

The goal of the feed-forward compensation prior to fabrication is to provide an optimal processing strategy under defined processing conditions to achieve specific, homogeneous thermal conditions according to predefined objectives in an arbitrary geometry. For this purpose, the specific thermal conditions that correspond to the

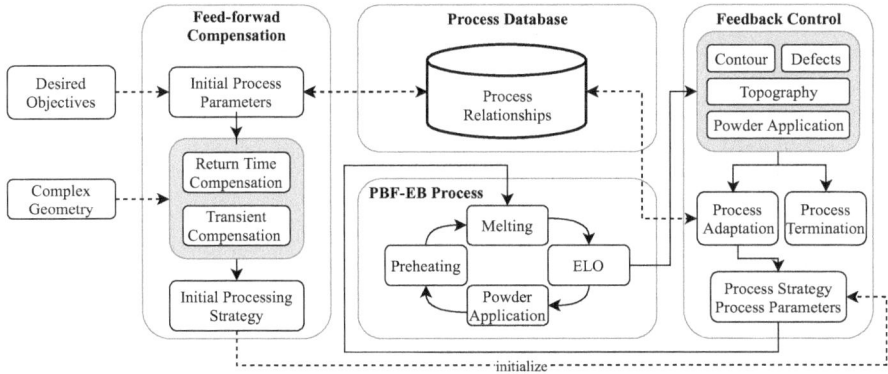

Fig. 7.1 Schematic of the proposed framework with three main components: process database that contains all process relationships, feed-forward control that provides the initial process parameters and processing strategy to reach the desired objectives for complex geometries, feedback control during processing based on the layer-wise ELO information which adapts the process strategy and process parameters when deviations from the optimal process are detected

defined objective have to be identified from the process database. Additionally, suitable processing strategies have to be developed and applied that enable the fabrication of complex geometries with constant thermal conditions, as outlined in Sect. 7.6. During the process, in-situ process monitoring provides information about the melt surface contour and topography, as well as the powder application in between layers. From this layer-wise information, deviations of the process from the ideal process and potential defect initiation sites can be identified. Based on these deviations, the feedback control system is able to update the processing strategy during the process to prevent defect formation and to guarantee the desired homogeneous thermal conditions, which is detailed in Sect. 7.7. Suitable strategies for process adaptations are based on the relationships between the processing and thermal conditions that are contained in the process database. If a defect is nevertheless detected, its influence on the desired objectives and the overall process stability has to be assessed. For instance, if the process stability is at risk and the termination criteria based on surface topography and powder application are exceeded, the process can be aborted safely. In this chapter, the functionality and application of this framework is illustrated using Ti–6Al–4V as a model material.

7.3   Electron-Optical Process Monitoring

ELO process monitoring provides the necessary information to monitor and control the process and accelerate the establishment of the process database. To obtain information about the current build surface, the electron beam scans the build surface and generates an ELO image, similar to a raster-electron microscope. The backscattered electron (BSE) signal at each position of the raster scan is based on the local BSE behavior, which is governed by the underlying material and the local surface topography. Differences in the BSE signal are utilized to distinguish the melt surface and powder surface, identify surface defects, determine powder applications, and quantify the melt surface topography [9–11].

Figure 7.2a shows an exemplary ELO image from the cross-section of a caliper brake. The difference in the BSE signal from the molten area with high signal and the unconsolidated powder bed with low signal is based on the different backscatter behavior, as indicated in Fig. 7.2b. The large difference leads to a high image contrast between the powder bed and melt surfaces, which enables the clear identification of melt contours during the process. Precise information about the melt contours are crucial to identify deviations from the target contour and the dimensional accuracy of the final geometry, which is detailed further in Sect. 7.7.

The characteristic geometric features of surface pores also lead to a distinct change of the BSE signal, as shown in Fig. 7.2 (right), and result in a reduction of the BSE signal at the pore location. This characteristic signal reduction within molten surfaces enables the clear identification of pores [9]. In order to automatically identify defect and pore formation for quality assurance and process control, a neural-network-based segmentation approach is used to reliably identify pores based on their characteristic signal change. This approach is based on the *U-Net* architecture [13], which takes the original ELO image as input and outputs segmentation maps according to predefined

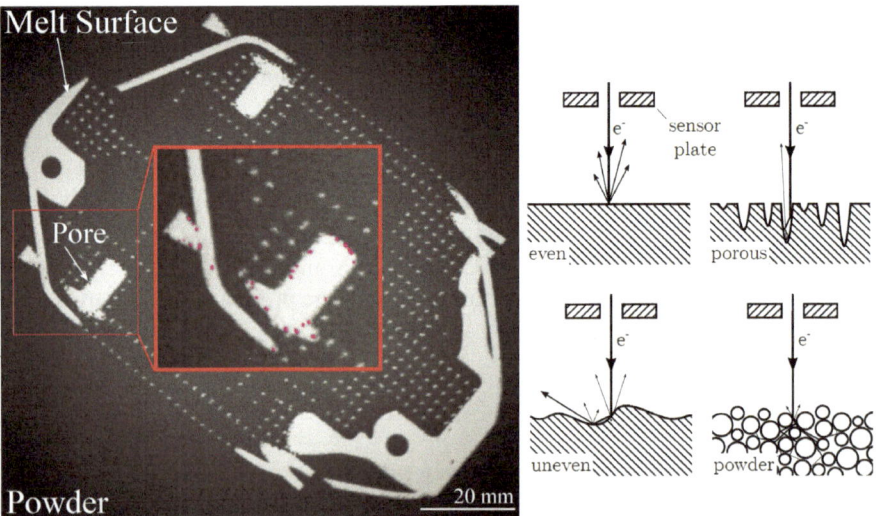

Fig. 7.2 Exemplary ELO image (**a**) of a caliper brake with powder, melt surfaces, and pores. The inset plot shows the indicated section of the ELO image, with identified pores. Schematic backscatter behavior and BSE signal at the sensor plate for different surface topographies common in PBF-EB (**b**) according to [10]

categories. The inset in Fig. 7.2a highlights the identified pores in a section of the original ELO image. ELO process monitoring setups with a single BSE detector are limited to a qualitative description of the surface topography. However, for a comprehensive description of the process state and the subsequent process adaptation, quantitative information about the surface topography is required. A multi-detector setup enables the determination of the local surface orientation based on the BSE signal of four opposite detectors and opens up the possibility to reconstruct the melt surface [11]. Based on these height maps, quantitative measurements, and a comprehensive description of the melt surfaces is possible. For this purpose, a multi-detector setup based on four opposite detectors was designed considering the total BSE signal yield and the signal difference of opposite detectors to extract sufficient gradient information. Figure 7.3a shows a top view of the final multi-detector design and two exemplary ELO images obtained for a complex geometry from two opposite detectors. The shading effect based on the local surface normal is clearly visible. Based on this information, the local gradient in one dimension can be determined, as shown in Fig. 7.3 (right).

In addition to the detector design, the complete computation chain that generates surface information from ELO images, including the surface tilt and solid angle contrast correction, was developed and implemented. Further information about the detector setup and computation chain can be found elsewhere [11]. Ultimately, the final melt surface can be obtained by integrating the normal vector fields, which are defined by the gradients shown in Fig. 7.3c, over the build surface. Figure 7.4a shows an exemplary surface reconstruction of a complete build surface and the comparison of the surface topography (b) obtained with the surface height map determined using a laser scanning microscope.

The process cycle can be further modified by inserting an additional ELO imaging step after the powder application step, which enables the identification of local variations in the powder application. Figure 7.5 shows a series of ELO images obtained

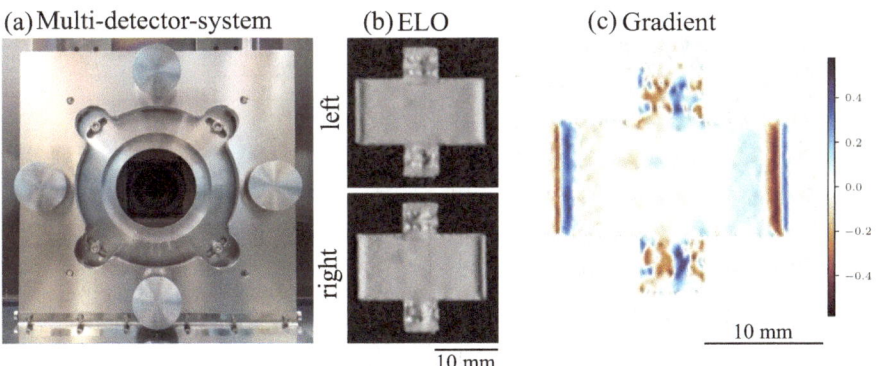

Fig. 7.3 Top view of the multi-detector system (**a**) with four opposite detectors, exemplary ELO images of the left and right detector (**b**), and surface gradient derived in the x direction (**c**)

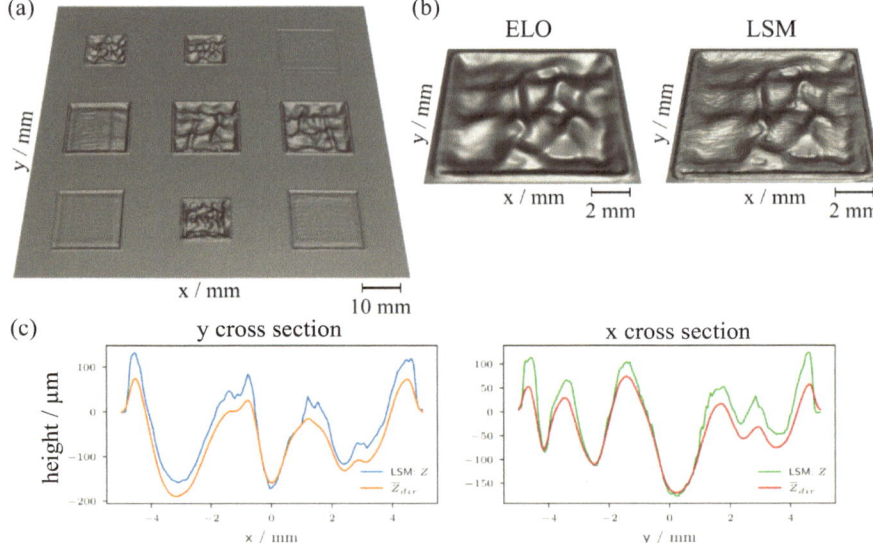

Fig. 7.4 Surface reconstruction of the whole build surface (**a**); comparison between the surface reconstruction obtained from ELO imaging and the surface obtained from a laser scanning microscope (**b**) with two cross-sections in the x and y directions (**c**). Reproduced from [11]

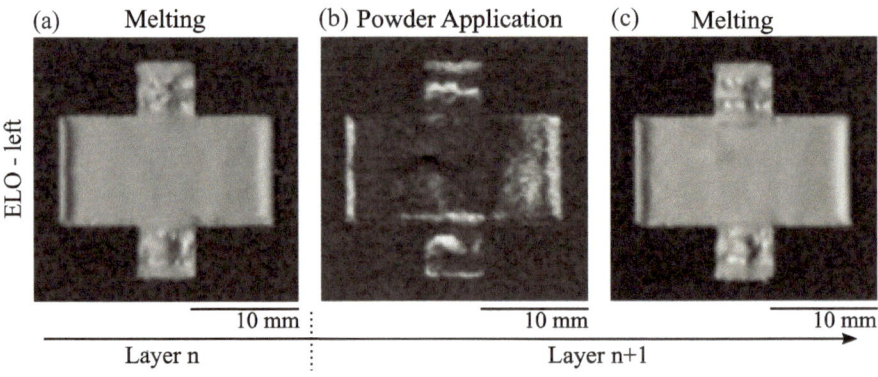

Fig. 7.5 ELO images obtained by the left detector in the modified process cycle. ELO image after the melt step of layer n (**a**), ELO image after the powder application of layer $n + 1$ (**b**), and ELO image after the melt step of layer $n + 1$ (**c**)

from the left ELO detector starting with the melting step of layer n and the subsequent powder application step from layer $n + 1$. This approach facilitates the detection of the uneven powder application and the formation of an uneven powder layer due to the uneven surface morphology of the previous melt surface which can lead to the propagation of defects to subsequent layers.

The information about the melt contours, defect and pore formation, and local surface topography can be combined not only for a quality assurance of the build, but also to determine the relationships underlying the process database and the control of the process, which is further detailed in Sect. 7.7.

7.4 Process Model

A simplified thermal model is used to obtain the spatio-temporal melt pool evolution in complex geometry to develop new processing strategies and determine the relationships governing the process. Two different solution approaches to solve the heat equation in three dimensions with a Gaussian surface heat source are utilized. The first approach solves the heat equation using an explicit finite difference solution on the graphics processing unit (GPU) and enables the efficient calculation of the detailed thermal evolution of complex geometries. Since the relationships between process parameters and thermal conditions are established in the quasi-stationary state of the hatch, where the cumulative heating effect is completely developed, the detailed thermal evolution is not necessary. Therefore, a semi-analytical solution of the heat equation is implemented, which enables the computation of the temperature at any given point in space and time without the need for the computation of the whole thermal history at each point of the simulation domain. This solution enables the efficient computation of a large number of process parameter combinations that are required to establish the process parameter database. Both models neglect fluid convection, latent heat release, radiation, and vaporization, and further details on the implementation are provided in the literature [3, 4, 14].

7.5 Process Database

The key component of the framework is the process database, which contains the relationships necessary to initialize and control the process according to predefined objectives. In order to enable the selection of the initial process parameters under specific processing conditions and the control of the process, the process database establishes the connection between process parameters (p, v), thermal conditions (σ, T_p), and final properties. The properties of interest include the defects and surface structure as the primary objective, and the microstructure and mechanical properties as the secondary objective. These relationships are established based on the quasi-stationary thermal conditions of primitive, cuboid geometries and can be transferred to complex geometries using customized processing strategies, which are further detailed in Sect. 7.6. The establishment of these relationships using numerical simulations and experimental approaches is detailed in the following subsections.

Processing Parameter—Thermal Conditions

The first relationship relates to the processing parameters with the emerging quasi-stationary melt pool geometry and thermal conditions, as shown in Fig. 7.6. Since the melt pool geometry and thermal conditions are not directly experimentally accessible, this relationship is determined using the process model described in Sect. 7.4. To better relate machine-specific processing parameters (p, v) to physical properties, the melt pool geometry and thermal conditions are calculated as a function of the area energy $E_a = P/v \cdot l_o$ and the lateral velocity $v_{lat} = v \cdot l_o/l_m$, which govern the energy input of the current hatch line and the magnitude of the cumulative heating effect [15]. Both parameters, namely the line offset l_o and line length l_m, contain information about the scanning strategy and the geometry. The emerging melt pool geometry can be characterized by different quantities that govern the defect formation, including the melt pool depth, lateral melt pool extension, and melt pool lifetime. For each process parameter combination of area energy and lateral velocity, each quantity describing the melt pool geometry is calculated and can be visualized in the process parameter space. As a result, each combination of area energy and lateral velocity yields a unique combination of energy input and cumulative heating effect. Therefore, a unique melt pool geometry and unique thermal conditions arise for each parameter combination. The melt pool size and melt pool lifetime increase with a higher total energy input at a constant lateral velocity, and with larger lateral velocity at a constant energy input.

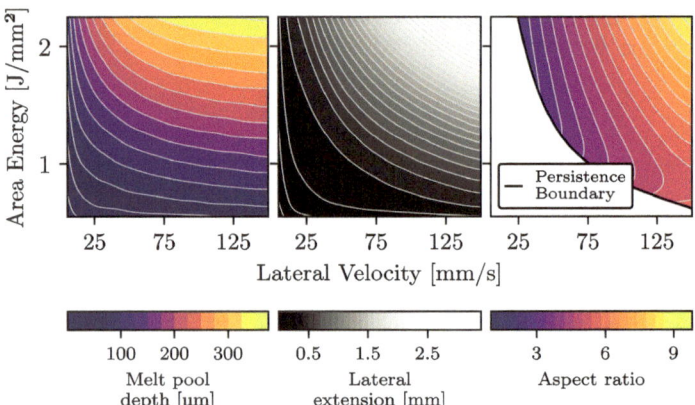

Fig. 7.6 Relationships between the processing parameters and thermal conditions for 1200 process parameter combinations calculated for area energies of 0.6–2.2 J/mm² and lateral velocities of 2.5–150 mm/s with a line offset of 100 μm, line length of 15 mm, preheating temperature of 1023 K, and a beam diameter of 200 μm

Thermal Conditions—Defect and Surface Structure

The second relationship relates specific thermal conditions to their corresponding emerging defect and surface structure. Through a combination of the calculated melt pool geometries and thermal conditions with experimentally observed defect formation and surface evolution, both after and during the process using process monitoring, it is possible to identify the thermal conditions and melt pool geometries underlying defect formation. Based on these relationships, processing boundaries can be established, and thermal conditions can be identified that satisfy the primary objective and enable defect-free fabrication. Two main defect types with different defect characteristics are common during PBF-EB processing and are related to their underlying thermal conditions and melt pool geometries in the following [16, 17].

The first defect type is characterized by the emergence of connection faults and pore formation, and results from insufficient consolidation of the powder bed [1, 18]. To avoid the formation of pores and consolidation defects, sufficient connection between the subsequent layers has to be established. The quantity governing the powder consolidation is the penetration depth of the melt pool into the powder bed. When the melt pool is not deep enough, no sufficient connection between subsequent layers can be established and connection faults occur and pores form. The necessary melt pool depth for sufficient connection is determined by the actual local layer thickness, which is highly dependent on the previous melt surface, but can be determined in approximation from the effect layer thickness of the process. For a sufficient connection between two subsequent layers, the melt pool depth has to exceed the effective layer thickness [18]. To consider local fluctuations of the powder bed and variations of the initial surface topography, an additional safety factor is included. Based on the combination of the established relationships between the processing parameters and melt pool geometry and the geometric criterion for sufficient consolidation, the consolidation boundary can be established, which follows the melt pool depth curve at the necessary depth and as shown in Fig. 7.7.

The second defect type is characterized by the emergence of an uneven surface topography at higher energy inputs and lateral velocities. There are, however, two different types of surface unevenness, namely the formation of distinct surface bulges and the formation of a labyrinth-like surface structure, as already showcased in Fig. 7.4. Based on experimental results, the surfaces of cuboid samples showed distinct surface bulges for specific thermal conditions, while other combinations did not show any bulge formation. The formation of surface bulges was related to a specific melt pool feature, i.e., the melt pool regime, that locally tracks if the melt pool is still liquid when the beam returns [5, 6]. This property can be calculated for each position in a hatch contour. In the case of the underlying cross-snake strategy, the melt pool can be characterized into three distinct melt pool regimes. The first melt pool regime is characterized by a typical teardrop-shaped, trailing melt pool that follows the beam movement. The remaining regimes are different forms of a persistent melt pool, where the melt pool is still liquid when the beam returns on the adjacent hatch line. This is the case when the melt pool lifetime exceeds the local return time of the beam. As a consequence of the cross-snake strategy, the persistence condition is fulfilled at the turnings points of the hatch, and a temporary persistent melt pool emerges. With increasing melt pool lifetime at higher energy inputs and decreasing return times, the temporary persistent regions from each turning point interleave at the center of the hatch and a permanent persistent melt pool emerges, where the melt pool is liquid when the beam returns from either side of the hatch. Any hatch surface can be characterized by a combination of these melt pool shapes and each of these shapes exhibits a different material transport characteristic [6]. Based on the different material transport characteristics, the onset of a permanent persistent melt pool leads to heterogeneous material transport at different locations of the hatch and ultimately results in the formation of surface bulges over many layers. Therefore, the necessary requirement for the formation of an even surface is homogeneous material transport conditions over the course of the hatch. Based on this requirement, the persistence boundary can be established as shown in Fig. 7.7, where a partial persistent melt pool forms and different material transport regimes coexist. Below the persistence boundary, the combination of trailing and temporary persistent melt pools enable the fabrication with homogeneous material transport conditions. Above the persistence boundary, a line-like persistent melt pool that covers the whole line length also ensures homogeneous material transport conditions.

The second characteristic, i.e., uneven surface morphology that displays a labyrinth-like surface structure, was identified for process parameter combinations in the permanent persistent melt pool region with high lateral velocities, as shown in Fig. 7.4. At constant energy input, a higher lateral velocity results in the elongation of the permanent persistent melt pool in the lateral direction. This results in the formation of a shallow melt pool over a large area, which is similar to a thin melt film [4]. Under the action of the surface tension driving to reduce the surface-to-volume ratio,

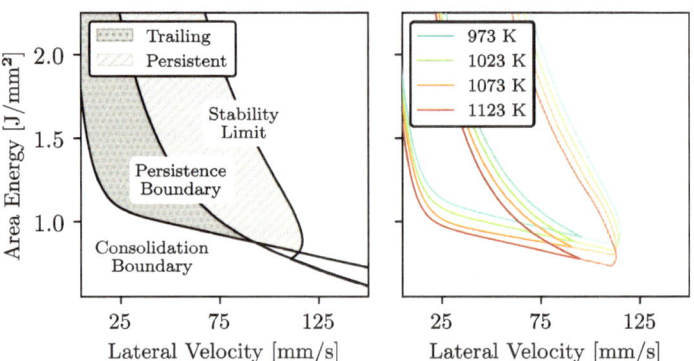

Fig. 7.7 Process space based on Fig. 7.6 with a line offset of 100 μm, line length of 15 mm, preheating temperature of 1023 K, and beam diameter of 200 μm together with the identified processing boundaries (left) based on [3]: consolidation boundary, persistence boundary, and melt pool stability limit. The influence of the preheating temperature on the location of the processing boundaries is also shown (right)

the liquid melt pool starts to aggregate into ridges over the course of the solidification process and results in the characteristic labyrinth-like surface structure. The threshold for the onset of the liquid melt aggregation was determined based on the combination of experimental data during the process and numerical calculations at an aspect ratio of 4.7 ± 0.3 between the lateral extension and the melt pool depth [4]. The aggregation of liquid melt not only requires a specific driving force, but also the necessary time for the process to take place. Therefore, this criterion only applies to the permanent persistent melt pool regime with a sufficient melt pool lifetime, and can be combined into the melt pool stability limit in Fig. 7.7.

Based on the established relationships between the melt pool geometry and the emerging defect structures, processing boundaries can be established that limit the thermal conditions under which defect-free fabrication is possible.

Thermal Conditions—Microstructure and Mechanical Properties

As described previously, each point in the process parameter space is defined by a unique melt pool geometry and thermal conditions. Since there are different locations in the process parameter space that comply with all processing boundaries, the fabrication of dense samples with even surfaces is possible with different melt pool geometries and thermal conditions. Different thermal conditions, melt pool geometries, and solidification conditions affect the development of microstructures, alloy compositions, and ultimately, mechanical properties and enable the deliberate selection of specific thermal conditions matching the desired microstructure and mechanical properties of interest. To determine the specific thermal conditions necessary to reach specific properties, the underlying relationships have to be established. Relationships relating thermal conditions to microstructure and mechanical properties are material-specific and have to be experimentally determined. In the following, three exemplary relationships are described for two different materials, Ti–6Al–4V and IN718, covering the alloy composition, microstructure evolution, and tailoring of mechanical properties.

The first example focuses on the influence of the thermal conditions on the final alloy composition, whereby different energy inputs and melt pool lifetimes affect the evaporation of volatile elements [2, 16, 19]. In the case of the model material Ti–6Al–4V, the aluminum content changes significantly as a function of energy input and lateral velocity. Figure 7.8a shows the aluminum loss in wt.% for different process parameter combinations within the processing boundaries. Higher volumetric energy densities lead to higher peak temperatures and a higher melt pool lifetime, resulting in a significantly higher aluminum loss [2, 19].

Figure 7.8b shows the corresponding aluminum element mappings for constant energy input and lateral velocity with various line offsets, which reveal a distinct difference. Lower line offsets with lower peak temperatures lead to a lower aluminum

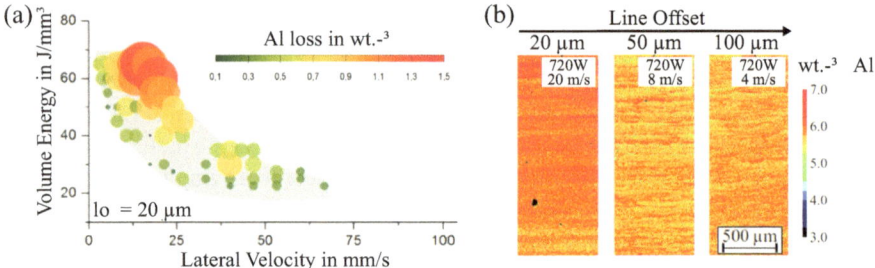

Fig. 7.8 Aluminum loss of dense, even samples fabricated within the processing boundaries (**a**) based on [2], and aluminum element mapping of Ti–6Al–4V (**b**) for different process parameter combinations with the same energy input but different line offset

evaporation. These variations in composition as a function of the underlying thermal conditions can be leveraged to tailor the properties and achieve a desired secondary objective, while always facilitating the fabrication of a dense, even sample. The thermal conditions also directly influence the solidification conditions and the solidification behavior, enabling microstructure control and, consequently, the control of the mechanical properties. Figure 7.9a illustrates the relationship between thermal conditions and the emerging α-lamella thickness in Ti–6Al–4V. With increasing energy input, a constant lateral velocity, the α-lamella thickness, increases as detailed in Fig. 7.9b.

The relationship between the emerging α-lamella thickness and the mechanical properties (yield strength and the ultimate tensile stress) of the final material are shown in Fig. 7.9c [17]. Based on these relationships, specific thermal conditions can be selected according to the desired mechanical properties. Materials without

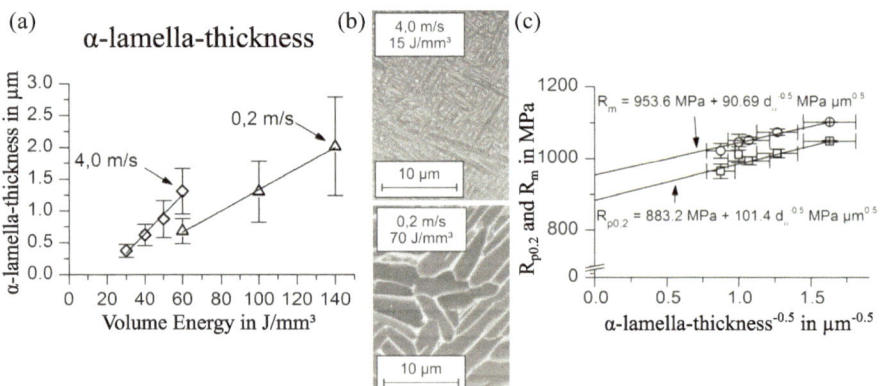

Fig. 7.9 Relationship between thermal conditions on the α-lamella thickness in Ti–6Al–4V (**a**), exemplary microstructure of two selected thermal conditions (**b**), and the relationship between α-lamella thickness and mechanical properties (**c**). Reproduced from [17]

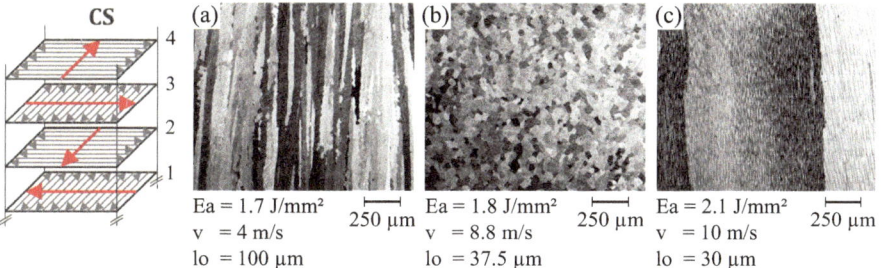

Fig. 7.10 Microstructure variations in IN718 using the cross-snake hatch strategy under different thermal conditions based on [20]: columnar (**a**), equiaxial (**b**), and coarse columnar microstructure (**c**) can be achieved through adaptation of the thermal conditions

solid state transformations allow for further microstructure control and modifications, since the solidified microstructure is preserved during cooling. Figure 7.10 shows the relationship between different thermal conditions and the emerging microstructure of cuboid samples fabricated using IN718 with the standard cross-snake strategy with a 90° rotation after each successful layer. Variation of the thermal conditions within the processing boundaries, through changes in the area energy, velocity, and line offset, enable the fabrication of sample with a columnar (a), equiaxial (b), or coarse columnar microstructure (c). Each individual microstructure variation leads to differences in the mechanical properties.

Tailoring of the microstructure is not only limited to the change of processing parameters in each layer, as the hatch sequence over multiple layers can also be used to influence the final microstructure. Deviating from the 90° rotation after a successful layer and only changing the hatching direction every 10 layers by 90° opens the possibility for a tilt of the columnar grains in the hatching direction. This can be especially interesting for aligning mechanical properties along loading directions.

Relationships correlating the thermal conditions within the processing boundaries enable the tailoring of mechanical and functional properties without defect formation. Based on exhaustive relationships, the inverse problem can be solved and thermal conditions can be identified that correspond to a specific, desired material property.

Processing Conditions—Thermal Conditions

In order to transfer the initially selected processing parameters to the actual processing conditions and to provide the database for process adaptation, the relationships between processing conditions and the emerging thermal conditions have to be established. The two main sources of uncertainty are the local preheating temperature and the exact beam diameter or the energy distribution in the beam. Therefore, the relationship between the thermal conditions and these processing conditions is determined numerically. With higher preheating temperature the energy necessary to reach the melt pool geometry decreases, which also results in a shift of the processing boundaries towards lower energy inputs and smaller lateral velocities, as shown in Fig. 7.7 (right). Changes in the beam diameter also significantly influence the thermal conditions [21]. The feedback control system utilizes these relationships to determine the right process parameter adaptations to achieve the same melt pool geometry and thermal conditions under changing processing conditions, which will be further detailed in Sect. 7.7.

7.6 Feed-Forward Control

The objective of the feed-forward control is to provide an optimal initial processing strategy for a geometry to reach defined homogeneous thermal conditions corresponding to a predefined objective which are based on the relationships established in Sect. 7.5. In order to achieve this goal, suitable processing strategies have to be developed that can provide homogeneous thermal conditions in complex geometries. To determine geometry-independent compensation strategies that enable the transfer of predefined thermal conditions, the influence of geometric features on the local evolution of thermal conditions and the melt pool geometry has to be identified. The spatio-temporal melt pool evolution and local thermal conditions are based on the energy input of the current hatch line and the remaining energy from previous hatch lines [3, 16]. With constant processing parameters, the energy input from the current hatch line remains constant over complex geometries. The energy remaining from previous hatch lines is determined by their individual energy input and their respective thermal loss until the current hatch line, which is governed by the local return time of the beam. Based on this dynamic, different cases can be identified in which complex geometries lead to local variations of the thermal conditions and melt pool geometry. With constant processing parameters, a variation in the scan length leads to the local change in the return time and influences the local contribution of energy from previous hatch lines. This results in a larger melt pool for shorter scan lengths and can even lead to violations of processing boundaries, as shown in Fig. 7.11.

In addition, turning point effects that result from the cross-snake hatch strategy result in local return time variations in each hatch line. Both effects are especially pronounced for large scan lengths and large changes in scan length with constant processing parameters and are addressed by the return time compensation strategy in Sect. 7.6.1. Since the thermal conditions are governed by the energy input of the current line and the energy remaining from previous hatch lines, a special case emerges for locations in complex geometries without prior hatch lines. At these positions, there is no contribution of previous hatch lines to the thermal conditions, and therefore, different thermal conditions emerge and a smaller melt pool develops. In order to address these special cases, a strategy to compensate for transient effects is detailed in Sect. 7.6.2.

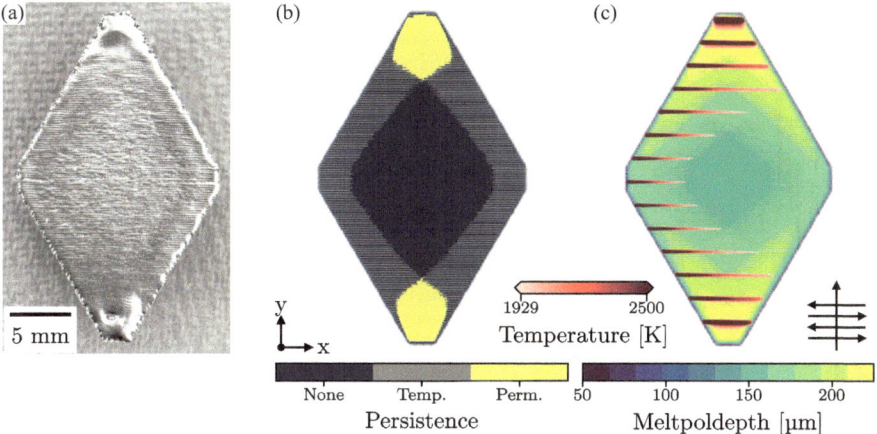

Fig. 7.11 Final melt surface of a model geometry fabricated using the cross-snake strategy (**a**) with P = 530 W, v = 3.3 m/s, lo = 100 μm, and Tp = 1023 K based on [3]. Calculated persistent regions (**b**) and local melt pool depth, (**c**) (Thermal Conditions—Microstructure and Mechanical Properties) according to the temperature fields. Representative melt pool geometries facing the negative x direction are overlaid when the beam reaches the end of one hatch line. Hatching direction along x-axis, traverse direction along the positive y-axis

7.6.1 Return Time Compensation

The goal of the return time compensation is to create defined thermal conditions in complex geometries independent of the line length by tailoring the spatio-temporal energy input [3]. With constant processing parameters, the energy input is constant for each position in a complex geometry. However, the line length directly influences the local return time and, consequently, lower line lengths lead to a higher cumulative heating effect and larger melt pools as shown in Fig. 7.11c. To decouple the local return times from the line length and create a geometry-independent scanning strategy, idle segments are introduced into the hatching sequence. Idle segments do not deposit any additional energy but enable the tailoring of the local return time through adjustment of their idle time t_{idle}. To create the same thermal conditions with constant energy input and homogeneous cumulative heating contribution, idle segments have to be positioned in between melting segments and the respective return time has to be adjusted to match the return time of a reference geometry. Figure 7.12 showcases the effect of the return time compensation with the minimum bounding rectangle as reference geometry.

The constant return time independent of the local line length leads to the same thermal conditions for melt pool formation and a homogeneous melt pool geometry and persistence regime over the course of the hatch. This prevents violations of processing boundaries and enables the fabrication of a homogeneous sample without surface defects (Fig. 7.12a) and with constant melt pool geometries (Fig. 7.12c). The return time compensation can be further extended to consider turning point effects.

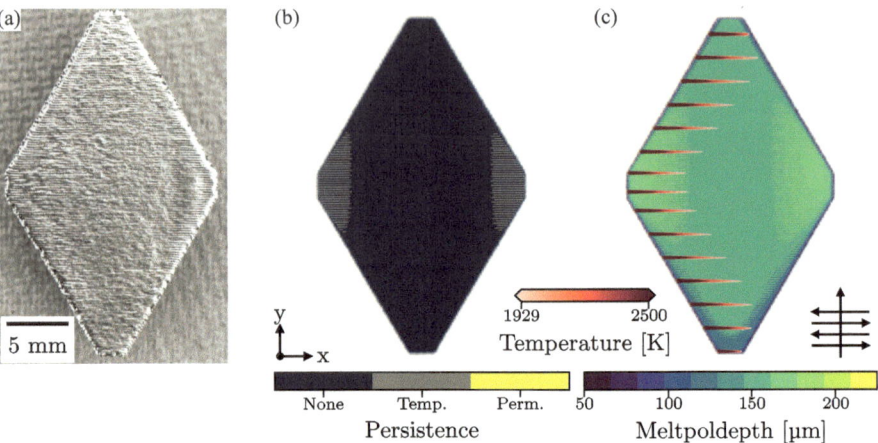

Fig. 7.12 Final melt surface of a model geometry fabricated using the return time compensation (**a**) with P = 530 W, v = 3.3 m/s, lo = 100 μm, and Tp = 1023 K based on [3]. Calculated persistent regions (**b**) and local melt pool depth (**c**) according to the temperature fields. Representative melt pool geometries facing the negative x direction are overlaid when the beam reaches the end of one hatch line. Hatching direction along x-axis, traverse direction along the positive y-axis

Since turning point effects occur based on the low return times at the turning points of the cross-snake strategy, a combination of reference geometry and processing parameters can be selected for fabrication, where the turning point effects are located in the idle segments of the hatch and not in the melting segments. This enables the fabrication of complex geometries with a homogeneous melt pool geometry and homogeneous thermal conditions. Since the return time compensation can emulate any thermal condition by the selection of a proper reference geometry and processing parameters, it is possible to achieve the same thermal conditions in any complex geometry corresponding to the relationships defined in the process database.

7.6.2 Compensation of Transient Effects

To compensate for the transient effects that arise at positions in complex geometries without prior hatch lines and ensure homogeneous thermal conditions, additional energy has to be deposited to compensate for the lack of contribution of prior hatch lines. For this purpose, a suitable universal processing strategy has to be identified that enables the deposition of additional energy at the right locations without disrupting the thermal conditions of the remaining hatch. While there are different possibilities to deposit additional energy at these locations, e.g., increasing the beam power or decreasing the beam velocity, they have to match the underlying thermal conditions in the quasi-stationary state which emerge based on a step-wise energy input in specific intervals with thermal loss in between. In contrast to the adaptation of the processing

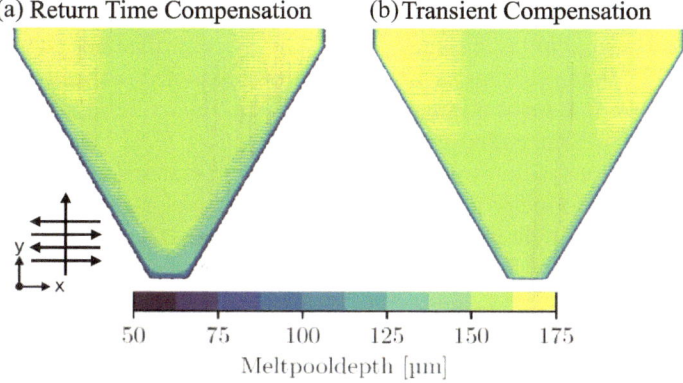

Fig. 7.13 Comparison of the melt pool depth distribution of the return time compensation strategy (**a**) with the transient compensation strategy (**b**) with P = 530 W, v = 3.3 m/s, lo = 100 μm, and Tp = 1023 K

parameters of the first hatch line—which do not conserve the thermal conditions—the adaptation of the line offset of the first hatch lines is a valid option to deposit additional energy at the start of the hatch that preserves the energy input intervals. Since the thermal conditions in the quasi-stationary state and the additional energy required to reach the thermal conditions are different for each process parameter combination, the optimal distribution of line offsets is determined using an optimization approach. The objective of the optimization approach is the minimization of the difference between the quasi-stationary melt pool depth distribution and the melt pool depth distribution at the start of the hatch as a function of the line offset of the first hatch lines. Based on the results of the optimization, the optimal solution emerges for line offset combinations where the first hatch line is repeated for n times before the hatch continues with the original line offset. To transfer this basic strategy to complex geometries, we have to adapt the hatching sequence. Since each line segment without prior hatch lines is preceded by idle segments, to achieve constant return times, the idle time corresponding to the segment can be used for the required additional energy input at the corresponding location. This enables the required energy input without disrupting the return times of future hatch lines and results in a reduction of transient effects and the formation of a homogeneous melt pool and thermal conditions, as detailed in Fig. 7.13.

7.7   Feedback Control

The goal of the feedback control system is to monitor the process state based on the layer-wise ELO data obtained and control the process by updating the processing strategy when necessary. This chapter focuses on the defect-free fabrication and control of dimensional accuracy. Since the local preheating temperature at the start and during the build job is currently neither closely controlled nor precisely known, the initial processing strategy is selected under an assumed preheating temperature that can also be measured at the bottom of the start plate. In addition to the initial uncertainty of the preheating temperature assumption, local heat accumulation in hot spots, or regions with insufficient preheating temperatures, can occur based on a combination of the selected preheating theme, local geometric features, and the melting sequence of multiple geometries within the process. Depending on the initially selected thermal conditions and processing parameters, a variation of the preheating temperature shifts the emerging melt pool geometry and the processing boundaries according to the relationships determined in Fig. 7.7 (right) and can lead to the violation of processing boundaries and defect formation for previously stable processing parameters. Since the shift of the preheating temperature in either direction is typically a continuous systematic process, slight changes in the surface topography can be identified during the process and appropriate countermeasures can be taken. Based on the unambiguous relationship between each processing boundary and its characteristic defect features, the underlying defect origin can be clearly identified and modified process parameters can be chosen to achieve the desired thermal conditions under updated processing conditions according to the relationships determined in Sect. 7.5. An alternative approach is the local adaptation of the preheating theme to control the local preheating temperature, removing the need for process parameter adaptation. However, further research is necessary to develop suitable adaptive preheating strategies.

The comparison of the melt surfaces and their extracted melt contours with the desired model contours enables the control of the dimensional accuracy. Figure 7.14 shows the comparison of melt surfaces with the original desired CAD-Contour for four consecutive layers, whereby the areas of excess and missing material are indicated. Based on the hatching direction and the location of the transient compensation, the melt surface shows characteristic excess material. Moreover, based on the magnitude of the mismatch between the melt contour and original contour, the offset of the scan strategy in each direction can be adapted during the process. In this case, the top row shows the initial processing strategy, where excess material can be identified at each start of the hatch due to the compensation of the transient phase, as well as an overall increase of the melt area due to the beam characteristic. The bottom row in Fig. 7.14 shows the melt contours of the adapted scanning strategy, where the extent of the hatch area is decreased in each direction, leading to a significantly higher dimensional accuracy. This information concerning the dimensional accuracy for a specific thermal condition and its respective adaptation to match the desired contour is further fed back into the process database for a better initialization of subsequent build jobs with the same thermal conditions. Furthermore, it is not only possible to feedback information about the dimensional accuracy at every layer during the process, but also after the process based on a 3D reconstruction of the final part based on the melt contours of each layer.

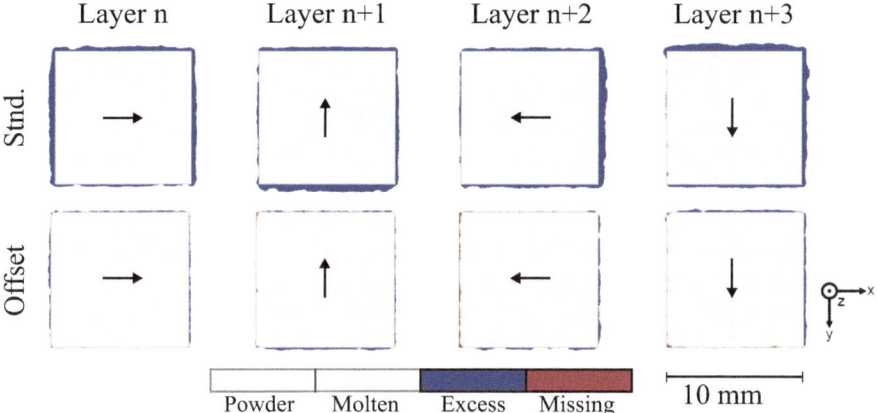

Fig. 7.14 Adaptation of the contour offset for a specific process parameter based on ELO images of four consecutive layers fabricated with Ea = 2.25 J/mm^2, vlat = 27 mm/s, lo = 100 μm, and Tp = 1023 K. ELO images showing excess material (top row) for four consecutive layers of cuboids with 10 mm line length using the compensation of the transient effects. ELO images of the adapted process strategy (bottom) leading to a significant reduction of the over-melting and a higher dimensional accuracy

Termination Conditions

If it is not possible to adapt the process to prevent defect formation or a sudden change in processing conditions occurs and the adaptation of the process is no longer possible, suitable termination criteria have to be implemented to partially or completely abort the process and prevent machine damage. For this purpose, the maximum amplitude in the surface topography from the powder bed to the highest surface elevation is monitored. When a maximum threshold is reached and previous process adaptations have not been successful, this geometry is prone to further defect formation and is aborted. Another important aspect is the powder application, as without sufficient powder application the build cannot progress. Especially for long build processes, powder application over long time spans can lead to heterogeneous powder applications depending on the powder level in the powder hoppers. Therefore, the powder application of the build platform is monitored. If the powder application is insufficient for multiple subsequent layers, melting of the same melt surface without additional power can lead to excessive energy input and surface swelling. Therefore, the build is terminated when the powder application is not sufficient for a number of subsequent layers.

7.8 Summary and Conclusion

This chapter introduces a framework for the fabrication of complex geometries with homogeneous predefined properties in PBF-EB based on a combination of feed-forward and feedback control. The framework operates based on a shared process database, that contains all the necessary information to initialize and control the process. The database relates the process parameters and processing conditions to emerging thermal conditions, and thermal conditions to the respective properties. Based on the process database and the processing strategies that were developed to enable the transfer of arbitrary thermal conditions to complex geometries, optimal initial process parameters can be identified. During the process, ELO process monitoring with a multi-detector system provides the necessary comprehensive information about the melt contours, surface topography, and powder application for quality assurance and to control the process. Within the scope of this chapter, the process model underlying the process database and the processing strategies to transfer thermal conditions to complex geometries is limited to the melting step of the process with constant preheating temperature and one material. This approach is, however, universally applicable independent of the underlying material and PBF-EB machine and can be further extended to include a comprehensive process model considering heat accumulation in each geometry over multiple layers.

References

1. Körner, C.: Additive manufacturing of metallic components by selective electron beam melting-a review. Int. Mater. Rev. **61**(5), 361–377 (2016)
2. Scharowsky, T., Bauereiß, A., Körner, C.: Influence of the hatching strategy on consolidation during selective electron beam melting of Ti–6Al–4V. Int. J. Adv. Manuf. Technol. **92**, 2809–2818 (2017)
3. Breuning, C., et al.: A return time compensation scheme for complex geometries in electron beam powder bed fusion. Add. Manuf. 103767 (2023)
4. Breuning, C., et al.: A multivariate meltpool stability criterion for fabrication of complex geometries in electron beam powder bed fusion. Add. Manuf. **45**, 102051 (2021)
5. Pistor, J., Breuning, C., Körner, C.: A single crystal process window for electron beam powder bed fusion additive manufacturing of a CMSX-4 type Ni-based superalloy. Materials **14**(14), 3785 (2021)
6. Breuning, C., et al.: Basic mechanism of surface topography evolution in electron beam based additive manufacturing. Materials **15**(14), 4754 (2022)
7. Markl, M., Körner, C.: Multiscale modeling of powder bed-based additive manufacturing. Ann. Rev. Mater. Res. **46**, 93–123 (2016)
8. Scharowsky, T., et al.: Melt pool dynamics during selective electron beam melting. Appl. Phys. A **114**, 1303–1307 (2014)
9. Arnold, C., et al.: Layerwise monitoring of electron beam melting via backscatter electron detection. Rapid Prototyping J. **24**(8), 1401–1406 (2018)
10. Arnold, C.: Fundamental Investigation of Electron-Optical Process Monitoring in Electron Beam Powder Bed Fusion. Friedrich-Alexander-Universität Erlangen-Nürnberg (FAU), Diss (2023)

11. Renner, J., et al.: Surface topographies from electron optical images in electron beam powder bed fusion for process monitoring and control. Add. Manuf. **60**, 103172 (2022)
12. Pobel, C.R., et al.: Immediate development of processing windows for selective electron beam melting using layerwise monitoring via backscattered electron detection. Mater. Lett. **249**, 70–72 (2019)
13. Ronneberger, O., et al.: U-net: Convolutional networks for biomedical image segmentation. In: Medical Image Computing and Computer-Assisted Intervention–MICCAI 2015: 18th International Conference, Munich, Germany, Proceedings, Part III 18. Springer International Publishing (2015)
14. Arnold, C., Breuning, C., Körner, C.: Electron-optical in situ imaging for the assessment of accuracy in electron beam powder bed fusion. Materials **14**(23), 7240 (2021)
15. Pobel, C.R., et al.: Processing windows for Ti-6Al-4V fabricated by selective electron beam melting with improved beam focus and different scan line spacings. Rapid Prototyping J. **25**(4), 665–671 (2019)
16. Juechter, V., et al.: Processing window and evaporation phenomena for Ti-6Al-4V produced by selective electron beam melting. Acta Materialia **76**, 252–258 (2014)
17. Scharowsky, T.: Grundlagenuntersuchungen zum selektiven Elektronenstrahlschmelzen von TiAl6V4. FAU University Press (2017)
18. Rausch, A.M., et al.: Predictive simulation of process windows for powder bed fusion additive manufacturing: influence of the powder bulk density. Materials **10**(10), 1117 (2017)
19. Klassen, A., Scharowsky, T., Körner, C.: Evaporation model for beam based additive manufacturing using free surface lattice Boltzmann methods. J. Phys. D: Appl. Phys. **47**(27), 275303 (2014)
20. Pobel, C.R., et al.: Innovative processing strategies for selective electron beam melting - Influence of scan line spacings on composition of Ti–6Al–4V and microstructure of IN718. In: Proceedings of the 6th International Conference on Additive Technologies iCAT 2016 (2016)
21. Reith, M., et al.: Impact of the power-dependent beam diameter during electron beam additive manufacturing: a case study with γ-TiAl. Appl. Sci. **12**(21), 11300 (2022)

Matthias Markl completed his doctorate in 2015 in the field of simulation of electron beam powder bed fusion under the supervision of Prof. Dr.-Ing. habil. Carolin Körner. Matthias Markl then took over as head of the Numerical Simulation working group, which focuses on the simulation of metal additive manufacturing processes.

Prof. Carolin Körner has headed the Chair of Materials Science and Technology of Metals at Friedrich-Alexander-Universität Erlangen-Nürnberg since 2011. She is primarily active in additive manufacturing (electron beam metals), casting technology (investment casting, die casting), process and microstructure simulation, and alloy development. She also heads a working group at the Central Institute for New Materials and Process Technology ZMP and Neue Materialien Fürth GmbH in Fürth.

Chapter 8
Laser Beam Melting of Metals

Florian Nahr and Michael Schmidt

8.1 Introduction

In the powder bed fusion of metals applying a laser beam (PBF-LB/M), components are built layer-wise by melting a thin bed of powder using a laser beam as an energy source. The build chamber is filled with shielding gas using purified Argon or Nitrogen to prevent oxidation and allow for efficient heat conduction and convective cooling of the build surface. Although laser beam melting operates at ambient temperatures, most of the PBF-LB/M machines are capable of substrate preheating. The powder particles absorb the photons in the first microseconds after the impact of the laser beam and generate a melt pool. The photons are subsequently further absorbed by the melt pool as the latter moves ahead, melting the powder layer to form the desired geometry [13]. It supports a wide variety of metals, ranging from steels over titanium alloys to nickel-based super alloys, just to name a few. Since PBF-LB/M shows great potential for functionally optimized and lightweight designs, the processing of aluminum alloys is of particular interest [9].

8.2 Processing of Aluminum Alloys in PBF-LB/M

At first, only casting alloys of the aluminum-silicon system close to the eutectic composition with excellent weldability were successfully used for PBF-LB/M. Processing aluminum alloys with a laser beam is generally more difficult due to their high reflectivity and high thermal conductivity. Furthermore, high strength aluminum

F. Nahr · M. Schmidt (✉)
Friedrich-Alexander-Universität Erlangen-Nürnberg, Institute of Photonic Technologies,
Konrad-Zuse-Straße 3/5, 91052 Erlangen, Germany
e-mail: michael.schmidt@fau.de

© The Author(s) 2025
D. Drummer and M. Schmidt (eds.), *Progress in Powder Based Additive Manufacturing*, Springer Tracts in Additive Manufacturing,
https://doi.org/10.1007/978-3-031-78350-0_8

149

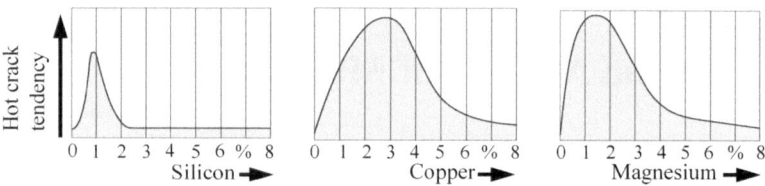

Fig. 8.1 Hot crack tendency of aluminum alloys [10]

Table 8.1 Chemical composition of aluminum alloys according to DIN EN-573-3 [2]

Material	Si	Fe	Cu	Mn	Mg
EN AW-2024	0.50	0.50	3.80–4.90	0.30–0.90	1.20–1.80
EN AW-2219	0.20	0.30	5.80–6.80	0.20–0.40	0.02
EN AW-2618	0.15–0.25	0.90–1.40	1.80–2.70	0.25	1.20–1.80
EN AW-4046	9.00–11.00	0.50	0.03	0.40	0.20–0.50
EN AW-5083	0.40	0.40	0.10	0.40–1.00	4.00–4.90
EN AW-6082	0.70–1.30	0.50	0.10	0.40–1.0	0.60–1.20

alloys exhibit hot cracking for laser processes, which are formed when columnar grains grow during solidification. Shrinkage opens spaces between the solidifying melt with insufficient interdendritic liquid at the interface to close them. The hot cracking sensitivity depends on the alloy's content as depicted in Fig. 8.1 [17].

Based on the alloying components from the different aluminum alloys (Table 8.1), it is evident that even among the wrought alloys some are more difficult to produce in PBF than others [10]. For complete melting of the powders, the radiation is focused by single-mode fiber lasers in continuous wave operation with rotationally symmetrical Gaussian distributions. It has been demonstrated that high-strength wrought alloys from the aluminum copper (Al-Cu) system, namely EN-AW-2024 [3], 2219 [4], and 2618 [1] can basically be processed in PBF-LB/M with relative densities above 99.5%. The correct adjustment of processing parameters such as laser power, scan velocity, hatch distance, or layer thickness has a significant impact and Fig. 8.2 shows the effect of different hatch distances (a–c) and the use of support structures (d) on the part quality. Supports lead to a smaller cross-section area compared to parts melted directly on the substrate, while reduced heat transfer through the supports could lead to a slower consolidation which allows for better filling of the interdendritic spaces with the liquid melt. Furthermore, heat accumulation through the supports might improve the complete melting of powder at locations where aluminum oxide from air/powder interactions might have piled, leading to processing defects [1].

However, there are many additional factors that contribute to a successful fabrication of aluminum alloys in PBF-LB/M. As shown in the following sections, powder properties such as the particle size, shape, and distribution, as well as the adhesion of nanoparticles, can have a significant impact on the process. Furthermore, the challenges of in situ alloying of Al-Cu alloys are discussed. Apart from the powder

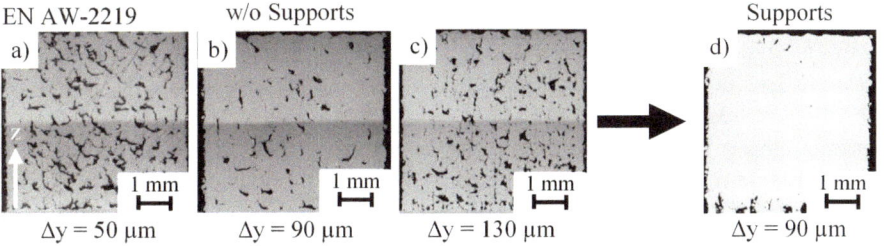

EN AW-2219   w/o Supports   Supports

$\Delta y = 50\ \mu m$ $\Delta y = 90\ \mu m$ $\Delta y = 130\ \mu m$ $\Delta y = 90\ \mu m$

Fig. 8.2 EN AW-2219 parts without support structures and different hatch distances (**a–c**), and with support structures (**d**) [1]

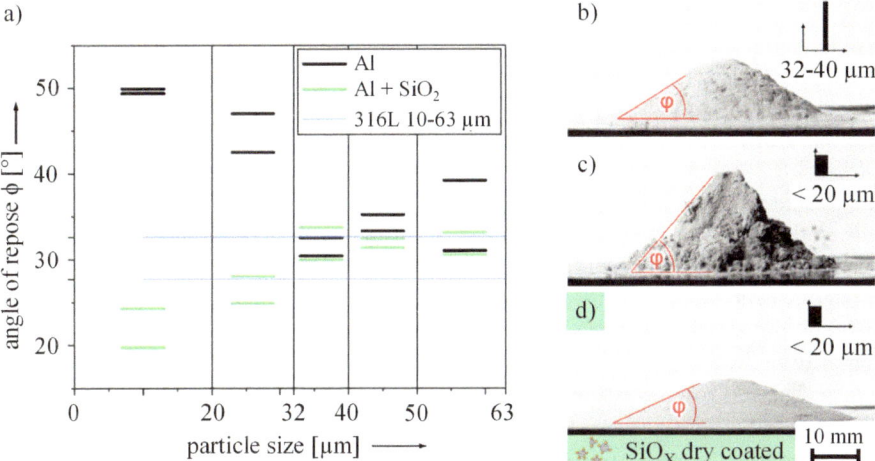

Fig. 8.3 Reducing the angle of repose ϕ by dry coating with SiO$_x$ **a** ϕ of the five sieved particle size ranges with and without SiO$_x$ compared to stainless steel 316L powder with a size distribution between 10 and 63 μm as a baseline. **b** $\phi = 32°$ particles 32–40 μm Al-Si, **c** $\phi = 50°$ particles <20 μm Al-Si, and **d** $\phi = 24°$ particles <20 μm with dry coated SiO$_x$ mixed for 1 [6]

quality, the laser beam profile is a key factor for the consistent fabrication of dense parts and beam shaping can improve the produceability of wrought aluminum alloys.

8.2.1   Functionalization of Aluminum Powder for PBF-LB/M

The successful production of parts from aluminum alloys in PBF-LB/M starts with the powder properties, since they affect the flowability and thus the layer deposition. Particles atomized with Argon (Ar) are spherical, but have the most satellite particles, which are smaller particles attached to larger ones. Particles atomized with Nitrogen (N) have ovoidal shapes more closely resembling potatoes than spheres. Particles atomized with N and air have spattered shapes that are mostly non-spherical [6]. Particle size distributions with powder fine proportions <20 μm achieve higher relative densities for standard parameters. Additionally, the process window extends. Identical trends are found for nitrogen atomized powder. The use of finer particles is thus more robust against varying process parameters in PBF-LB/M. However, although fine powders with particle sizes <20 μm are usually classified as unusable for the PBF-LB/M process because of their low flowability, mixing the powder with flow aids proves to be a viable option as shown in Fig. 8.3. For this purpose, the powder and the flow aid are shaken under an Ar atmosphere with a tumbling unit for a defined set amount of time [6].

As a reference, spherical stainless steel 316L is compared to a particle size distribution (PSD) of 10–63 μm. Aluminum particles <20 μm have by far the largest angle of repose ϕ and aluminum particles ranging from 32 to 40 μm comprise the only powder fraction without SiO_x that reaches a similar ϕ to 316L with both measured angles. An indicator of the effect of dry coating SiO_x is offered by the strong decrease in ϕ of Al particles <20 μm from 50° in (c) to 24° in (d). While (c) shows steep cliffy shapes and flanks that are so jagged that ϕ cannot be measured unambiguously, the pile with SiO_x in (d) is wider, flatter, and smoother. Like the flowability, the angle of repose does not directly reflect the behavior during powder coating in the PBF-LB/M process, but it indicates processability. Therefore, Fig. 8.4 shows optical images of the layer homogeneity of powders inside the PBF system with different particle morphologies, PSD densities, and SiO_x addition. The images are acquired via a digital camera on a tripod monitoring the powder bed through a protective glass. The top row shows layers created from sieved powders with narrow PSD of 32–40 μm without SiO_x nanoparticles. The smoothest layers were created from the most spherical, Ar atomized particles. The ovoidal, N atomized particles result in few grooves while N/air has the most inhomogeneous surface. The middle row (d–f) includes fine particles <20 μm but no SiO_x, which results in a more irregular surface. The bottom row shows layers including 15–17 wt% fine particles <20 μm dry coated with SiO_x nanoparticles. With Ar and N atomized powders in (g) and (h), the results are very similar to those without fine particles and without nanoparticles in (a) and (b) [6]. It can be summarized that dry coating SiO_x improves the PBF-LB/M of fine aluminum powders, whereby argon atomization results in more homogeneous layers than nitrogen. The nitrogen/air atomized powder with fine particles <20 μm and SiO_x results in smoother layers than the nitrogen/air 32–40 μm, indicating that dry coating SiO_x also improves the PBF-LB/M of non-spherical particles. Since the powder layers with SiO_x nanoparticles exhibit a better surface quality, parts with higher density can consequently be produced. Furthermore, the addition of SiO_x nanoparticles has no effect on the microstructure and the mechanical properties. The use of the fine powder fraction increases the material efficiency because such particles cannot be avoided in powder atomization and are usually discarded for PBF-LB/M [6]. Comparing the adhesion of air-atomized powder with and without adhered nanoparticles shows a reduction of tensile strength by an order of magnitude with the addition of nanoparticles, which correlates with a greatly improved powder deposition in PBF. This behavior can be observed for various metallic nanoparti-

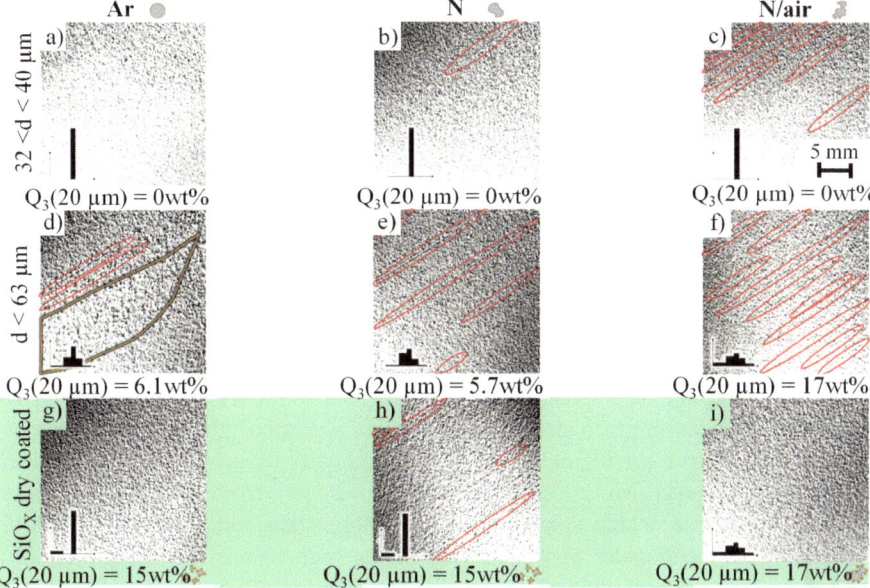

Fig. 8.4 Optical images depicting the layer homogeneity of powders with different atomization gas and particle morphology (columns) and particle size distribution densities (rows). Inhomogeneities are marked as red areas [2]

cles, as well as for fumed silica and carbon black, with the latter two being much less expensive and thus more attractive for later industrial use [5]. The tendency to agglomeration decreases with increasing particle diameter, since the weight force of the particles exceeds the interparticle forces. Therefore, dry coating is particularly effective when the powder tends to form agglomerates due to interparticle forces. If fine powder is mixed with SiO_x, the nanoparticles act as spacers and simultaneously reduce the interparticle forces [7].

8.2.2 In-Situ Alloying

For the investigation of aluminum alloys in PBF, powders from prealloyed feedstocks are typically used. However, another approach is the mixing of powders with different chemical compositions. The components for mixtures, elementally pure powders, and master alloys are atomized in large scales. They are available in a more reliable quality in terms of particle shape and impurity content than custom atomized small batches of prealloyed powders. In the following, three approaches to produce powder mixtures suitable for PBF-LB/M of aluminum alloys are presented [8].

8.2.2.1 Mixture Based on Microscale Powder Components

There are two fundamental challenges for a successful in situ alloy formation based on microscale powder particles. Firstly, homogeneous mixing of the different alloying constituents must take place in the melt pool. On the other hand, the segregation of the heterogeneous powder mixtures during powder feeding must be avoided for particles with large density differences. The production of powder mixtures prior to PBF-LB/M based on aluminum (Al) and copper (Cu) powders in a Turbula shaker leads to a strong segregation of the Al-Cu powder system. This is due to the significant density differences and the low interparticle adhesion forces, which are mandatory for the application of a homogeneous powder layer on the previously melted and already re-solidified layer. This leads to an inhomogeneous powder application and hence inhomogeneous powder layer. For a reduction of the segregation in the powder, a special application mechanism can be applied that mixes the starting powders by means of an active stirring element until immediately before the actual application of the powder layer. Moreover, fine particles <20 μm from the lighter aluminum powder can counteract the sinking of the larger copper particles by gravity. In addition to mixing in the powder booth, mixing in the melt also offers, in principle, the potential for homogenizing the material composition. Depending on the process parameters such as laser power or scan speed, this results in a melt pool width of between 100 μm and 250 μm with a weld penetration depth of approx. 50–100 μm. Taking into account the particle sizes used for copper of between 20 and 60 μm and the low volume fraction of approx. 2% for technically interesting wrought alloys, homogeneous, controllable mixing of the melt remains a challenge [8]. The influence of the particle size distribution on the segregation behavior can be examined by using different powder mixtures. Elemental copper and aluminum powders are mixed for 1 h in a ratio of 94.7 wt.-% Al, 5 wt.-% Cu, each with different powder size fractions. Furthermore, 0.3 wt.-% SiO_x is added, due to its positive influence described in Sect. 8.2.1. The alloying element copper is, on the one hand, the main alloying element of the EN AW-2xxx alloys (see Table 8.1) and on the other hand, the element with the highest segregation tendency due to its density. Figure 8.5 shows the de-mixing behavior for binary Al-Cu powder mixtures, whereby samples with no visible de-mixing are categorized as mixed and represented in green squares [8].

Samples with highly visible de-mixing of reddish Cu powder on the test tube bottom, top, or both, are categorized as de-mixed. De-mixed samples are represented by red squares. Samples that reveal local accumulations of Cu particles throughout the height of the test tube are categorized as intermediate, and represented in beige squares. All mixtures with particle sizes Al > Cu de-mix and only samples including Al particles <20 μm remain mixed. These are the mixtures with both Al and Cu particles <20 μm and those with Cu particles >40 μm. The remaining samples fall into the intermediate category. Thus, it is shown that the production of a stable heterogeneous powder mixture is challenging yet possible [8]. A transferability of the powder mixture to the powder layer application within a PBF-LB/M system is given. Furthermore, dense specimens can be produced for the varying powder fractions as shown in Fig. 8.6 [8].

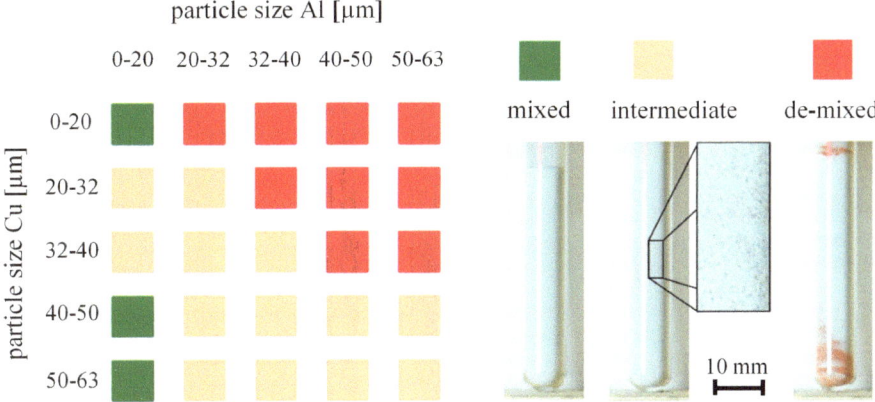

Fig. 8.5 De-mixing of Al-Cu powder mixtures under variation of particle size fractions from test tube experiments, 94.7 wt-% Al, 5 wt-% Cu, and 0.3 wt-% SiO_x [8]

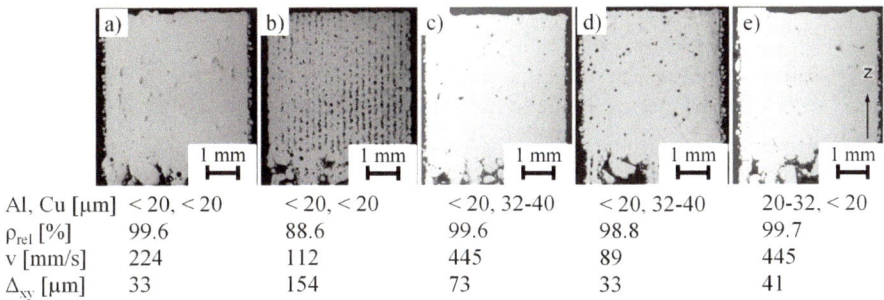

Al, Cu [μm]	< 20, < 20	< 20, < 20	< 20, 32-40	< 20, 32-40	20-32, < 20
ρ_{rel} [%]	99.6	88.6	99.6	98.8	99.7
v [mm/s]	224	112	445	89	445
Δ_{xy} [μm]	33	154	73	33	41

Fig. 8.6 Exemplary micrographs of Al-Cu specimens; **a** few defects; **b** ΔTXY too large; **c** few defects; **d** many spherical gas pores because of too high energy input; **e** few defects [8]

Figure 8.7 shows backscattered electron (BSE) images of the manufactured parts compared to those of a conventional extrusion profile made of EN-AW 2024. Cu, in particular, is in focus due to it yielding the highest segregation tendency. While slight accumulations of Cu (white arrows) and spherical pores (yellow arrows) are still visible in the backscattered electron (BSE) contrast image of the PBF-LB/M samples in the as-built condition (a–b), the corresponding BSE image of the T4 heat-treated sample appears completely homogeneous with only a few pores left (c). In contrast, the reference sample shows clear Cu accumulations in the BSE images (d). The BSE analyses prove that samples with a homogeneous alloy composition can be produced by means of in situ alloy formation in PBF-LB/M [8].

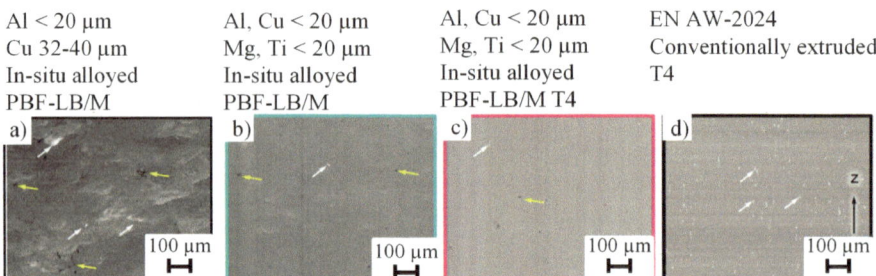

Al < 20 μm | Al, Cu < 20 μm | Al, Cu < 20 μm | EN AW-2024
Cu 32-40 μm | Mg, Ti < 20 μm | Mg, Ti < 20 μm | Conventionally extruded
In-situ alloyed | In-situ alloyed | In-situ alloyed | T4
PBF-LB/M | PBF-LB/M | PBF-LB/M T4

Fig. 8.7 BSE images of samples; **a** LBM of binary Al-Cu, Al particles <20 μm and Cu 32–40 μm as built; **b** quaternary as built; **c** quaternary T4; **d** EN AW-2024 T4 conventionally manufactured [8]

8.2.2.2 Generation of Cu-Coated Al Particles by Vapor Deposition

The second possible approach for in-situ alloying consists in coating Argon-dusted Al particles (20 μm < Ø particle < 60 μm) with copper nanoparticles by vapor deposition. An increase in reflectance on thin films of these modified powder systems with increasing copper content can be achieved. However, the copper content is still far below that of technical wrought alloys.

8.2.2.3 Addition of the Alloying Elements as Metallic Nanoparticles

In the third approach, the microscale Al host particles are mixed with nanoscale alloying powders resulting in different Al-Cu powder systems with different mixing ratios. To prevent oxidation of the nanoparticles, the modified powder systems have to be glued in a glovebox under an inert gas atmosphere with only a few ppm of reactive oxygen. Depending on the powder system investigated, structures with a relative density of more than 98.30% can be generated that are largely without defects. However, metallic nanoparticles exceed the cost of inert gas atomized microparticles by several orders of magnitude.

8.2.3 Influence of PBF Systems on Processing

As mentioned in Sect. 8.2, wrought Al-Cu alloys, which are regarded as difficult to weld and susceptible to hot cracking, can be successfully produced with relative densities above 99.5% However, the influences and restrictions of the PBF systems used play an important role in this context. For example, no defect-free horizontal cylindrical specimens can be produced if the installation space is too small, which is due to locally variable thermal boundary conditions caused by an inhomogeneous protective gas flow. In addition, an effective removal of spatter by the gas flow is not possible at all installation positions. However, transferring established processing parameters from one system to another comes with its own difficulties. The process gas flow, available laser power, and especially the minimum beam respectively the maximum achievable intensity differ significantly between the various systems that are currently available. If the laser power is changed, the exposure speed must also be adjusted for a comparable line energy. However, this change is also accompanied by a change in the fluid dynamics and the solidification conditions. Accordingly, a process parameter adjustment is always necessary [4].

8.2.4   *Influence of the Intensity Profiles on the Process Dynamics*

So far, it was shown that modifications of the powder properties and processing parameters lead to the fabrication of wrought aluminum alloys in PBF. As already indicated in Sect. 8.2.3, the process dynamics in all laser-based melting processes are significantly influenced by the beam profile and the associated intensity distribution. According to the state-of-the art PBF-LB/M, Gaussian or top-hat beam profiles are used. However, a large variety of beam profiles can be generated using diffractive optical elements. Figure 8.8 shows three modified beam profiles. The first profile, RING, corresponds to a ring with an outer diameter of 900 μm and a width of 110 μm. The integrated intensity is higher at the sides than in the center and is thus intended to counteract the heat flow from the melt pool and to ensure a more uniform temperature distribution. The objective of the second beam profile, LINE, is to keep both the intensity and the integrated intensity (line power) constant and orthogonal to the traverse movement in the processing point and thus to make the energy input more uniform compared to DEF. This is contrasted by the third profile, POINTS, which has approximately twice the peak intensity as LINE but leaves areas free between the individual beam points. These areas are not exposed and are therefore hardly affected by evaporation, allowing an unhindered melt flow in these areas. The measured intensity distribution and the section through the processing spot are shown in Fig. 8.8 [16].

Analysis of the process dynamics by means of high-speed camera recordings provides information about the process mode as a function of the selected process parameters. The classification into heat conduction welding (HCW, stable melt pool with no keyhole or turbulences), transition zone (TZW, increasing turbulences keyhole starts to form), as well as deep penetration welding (DPW, stable keyhole) can be made according to Fig. 8.9. Even at high powers in the processing spot, the weld pool is consistently calm when the RING profile is used (a). With the other beam profiles considered, as well as with DEF (b and c), the first turbulences in the melt pool already occur at significantly lower laser powers or higher exposure speeds [16]. LINE shows the lowest threshold at the transition between HCW and

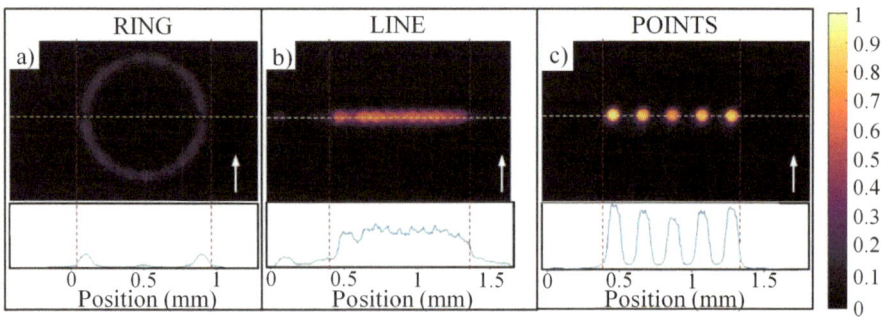

Fig. 8.8 Different beam profiles RING (**a**), LINE (**b**), and POINTS (**c**), as well as their respective intensity distribution [16]

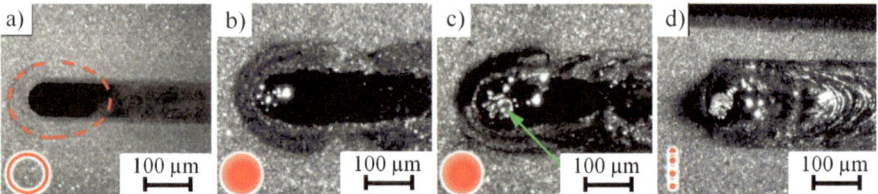

Fig. 8.9 **a** HCW. Highly stable melt pool with RING profile: P = 2.7 kW, v = 25 mm/s. The dashed ellipse indicates the size of the melt pool which is partly hidden by an oxide layer. **b** Early TZW. The first turbulences in the melt pool occur. DEF with P = 2.6 kW, v = 400 mm/s. **c** Late TZW. The keyhole is about to form but collapses right afterward. DEF with P = 2.6 kW, v = 200 mm/s. **d** A stable keyhole is formed. LINE with P = 2.0 kW, v = 100 mm/s [16]

TZW. As already suspected, this effect is due to the impeded melt flow caused by the evaporation-induced melt pool deformation and the associated higher temperature of the melt pool. Simulations support this conjecture. DEF has a peak intensity that is 40% higher than POINTS, although the low-deformation region of the melt pool is larger. These opposing trends account for the comparable boundary between HCW and TZW [16]. The weld cross-section and, in particular, the weld base shape change at different travel speeds. While the cross-section has the shape of a lens at slow speeds regardless of the beam profile used, the accumulated intensities are reflected at higher speeds. Thus, at high speeds, melt pool cross sections can be generated that resemble the shape of a rectangle. In PBF-LB/M, the necessary overlap can be reduced by selecting the appropriate beam profile. As a result, the efficiency can be increased and the total energy input can be reduced [16].

8.3   Current Findings from the Research

The previous section demonstrates the possibility for the basic processing of various wrought aluminum alloys with commercial PBF-LB/M equipment under certain conditions. By modification of the powder with various additives, a homogeneous powder layer can be fabricated, thereby increasing the produceability. Furthermore, going away from a Gaussian intensity distribution to a ring profile using beam shaping proves to be a valid method for a more stable melt pool. However, even established materials for laser beam melting such as the casting alloys AlSi10Mg still offer many interesting possibilities for investigation and especially the development of tailored microstructures in PBF-LB/M is an important research topic as shown in the following. Furthermore, functional strategies for the hot-crack-free processing of wrought aluminum alloys with an exceptionally high hot cracking sensitivity are presented.

8.3.1   AlSi10Mg in PBF-LB/M—An Old and Boring Material?

AlSi10Mg is one of the most investigated aluminum alloys in PBF-LB/M because of its wide process window. The composition close to the eutectic allows for a wide variety of different processing parameters and a microstructure with excellent mechanical properties. However, there are areas that still need a deeper understanding and one of these is the correlation between geometry, microstructure, and mechanical properties for adding tailored properties to a part. A possibility for achieving this is by influencing the solidification, which then directly affects the local part properties. Therefore, different part structures are presented. The first structure is a large cube that consists of various smaller cubes with three parameter sets allowing different solidification conditions during PBF-LB/M. Through these adjusted parameter sets, the goal is to obtain a specimen with locally defined mechanical properties. Secondly, a complex structure consisting of narrow cylinders between broader conic shapes is presented. Alternating between the different geometries affects the heat flux in the part and thus influences the microstructure and mechanical properties.

8.3.1.1   Large Cube

The cube consists of $6 \times 6 \times 6$ ($X \times Y \times Z$) small cubes with an edge length of 7 mm and a 2 mm contour surrounding the arrangement of small cubes (total size 46 mm $\times$ 46 mm $\times$ 42 mm). Three different types of parameter sets (PS) are used as depicted in Table 8.2. PS 1-3 are alternating in the X and Y directions, as well as in the build direction Z, as shown in Fig. 8.10. Furthermore, the layer thickness of 30 μm and the scan vector rotation of 37° are kept constant for all three PS. For varying the beam

Table 8.2 Cubic parameters [14]

Setting	Power P_L in W	Scan speed v_s in mm/s	Hatch in μm	Beam profile	Beam diameter (1/e2) dL in μm
PS 1	300	1650	130	Gaussian	78
PS 2	360	3250	20	Gaussian	78
PS 3	800	400	350	Top-Hat	650

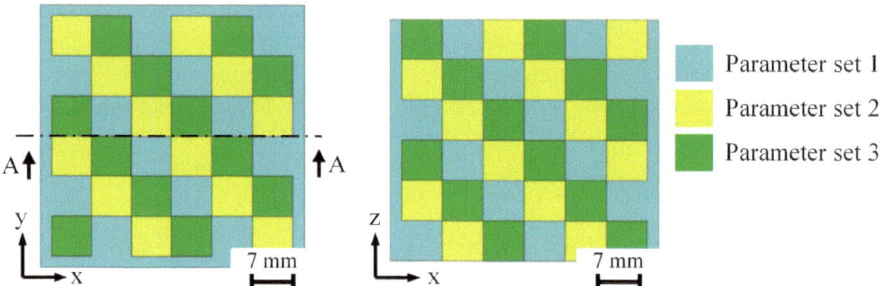

Fig. 8.10 Schematic representation of the cubes and the varied parameter sets. The build direction is in Z [14]

profile between cubes, a single-mode laser with a Gaussian intensity distribution and a multi-mode laser with Top-Hat profile was used [14].

During the build job, it takes almost 5 h until a steady temperature state of approximately 145–155 °C is reached, while after the end (17.8 h) of the build job, the temperature drops quite rapidly due to the cooling effects caused by the gas stream and heat conduction.

Figure 8.11 shows the grain structure of the different parameter sets. The biggest grain and sub-grain size are obtained by the use of PS 3, whereby the application of the single mode laser leads to a significantly smaller structure. Additionally, the grain size can be varied by 25% between the two parameter sets with Gaussian beam profiles. The parameter sets also have an impact on the grain orientation. For PS 1, the grains point perpendicularly to the weld seam, whereas PS 2 has a mixture of short columnar grains and some smaller epitaxial grains at the melt pool border. Finally, the grains of PS 3 are strictly oriented in the build direction. The orientations can be attributed to the higher (PS 2 and 3) or lower melt pool overlap (PS 1). Furthermore, the distribution of the alloy elements of silicon and magnesium depend on the parameter set after an in-situ aging of approximately 9 h at 150 °C. For all three parameter sets, the silicon is concentrated at the grain boundaries, and magnesium is spread equally over the analyzed cross-section. This indicates, on the one hand, that by in-situ aging no significant amount of microscale Mg_2Si secondary

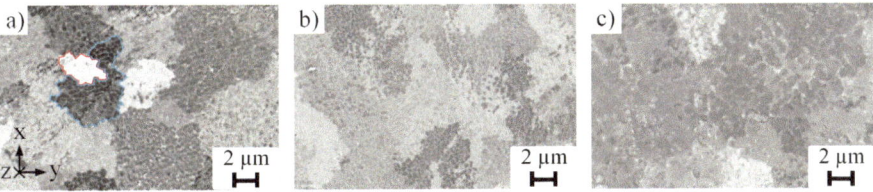

Fig. 8.11 SEM images of PS 1 (**a**), PS 2 (**b**), and PS 3 (**c**) taken at the xy-plane [14]

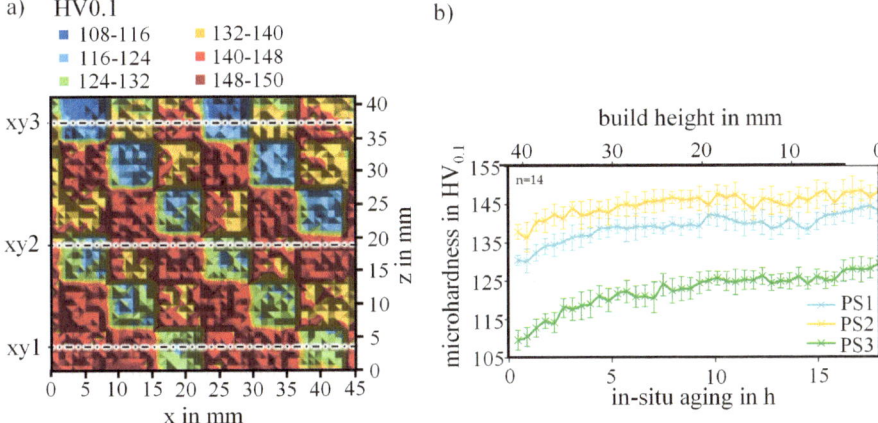

Fig. 8.12 Microhardness HV0.1 in the xz-cross-section (**a**) and the corresponding effect on the hardness along the build height (**b**) [14]

precipitations is formed. On the other hand, the impact time of PS 3 is still too short to significantly degenerate the silicon network, despite the huge laser spot and slow scanning speed [14].

The different parameter sets show strong variations in their respective micro-hardness. The use of PS 2 leads to the highest microhardness values, whereas PS 3 exhibits the lowest microhardness and PS 1 is settled in-between but closer to PS 2. Comparing the microhardness between xy-cross-sections and the xz-cross-section, anisotropic mechanical properties become visible due to the columnar grain growth of AlSi10Mg during PBF-LB/M. For the microhardness in the xz-direction, the effect of the temperature history over the build height becomes evident, leading to an in-situ aging. The microhardness increases for all parameter sets used, which is most probably caused by the formation of secondary Mg_2Si nano-precipitation. The in-situ aging effect is further shown by the evolution of the mean microhardness per parameter set along the build direction [14] (Fig. 8.12).

8.3.1.2   Influence of Complex Geometries

The cube already alluded to the significance of the part structure on the temperature history by exhibiting an in-situ aging effect leading to different microstructures and hardness values throughout the part. This effect can further be shown by varying the part geometry for constant parameters. For this purpose, a complex structure consisting of three conic shapes with narrow cylinders in between hindering heat flux is built. The cross-section of the part is shown in Fig. 8.13 in addition to detailed images of the porosity and the energy-dispersive X-ray spectroscopy (EDX) images of the microstructure. The relative density decreases along the build height for both cylindric and conic shaped areas. Furthermore, the narrow cylindric areas exhibit less porosity compared to the conic sections.

The microstructure also heavily depends on the geometry. Similar to the cubic structure, the narrow cylindric sections exhibit a silicon network with magnesium equally spread over the analyzed area. With increasing cross-section diameter, the silicon network decomposes and the diffusion-controlled formation of Mg2Si precipitates takes place. These effects correlate directly with the surface temperature of the part during the build job. The temperature history is shown in Fig. 8.14 and matches the findings from previous investigations conducted with Ti6Al4V [12]. As the exposed cross section is enhanced, the temperature increases, reaching a first peak at the largest diameter and before decreasing again as the cross-section diminishes again. At the bottom of the second narrow cylinder, the inter-layer temperature almost reaches the initial value. The trend repeats itself for the next two sections while the peak temperature in both sections rises slightly. A larger solid cross-section accelerates the heat conduction in the negative building direction compared to powder material. The heat accumulates before the transition to the cylindric section, thereby leading to in-situ heat treatment affecting the microstructure. The effects of the temperature on the microstructure directly correlate with the hardness along the build

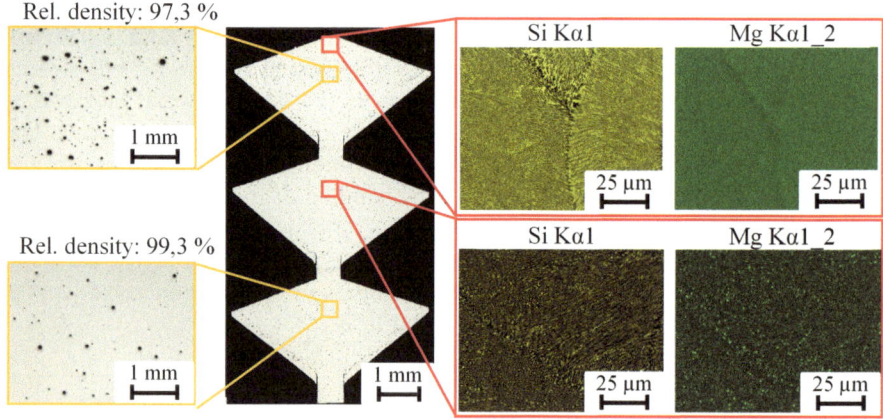

Fig. 8.13 Influence of the geometry on the relative density and microstructure

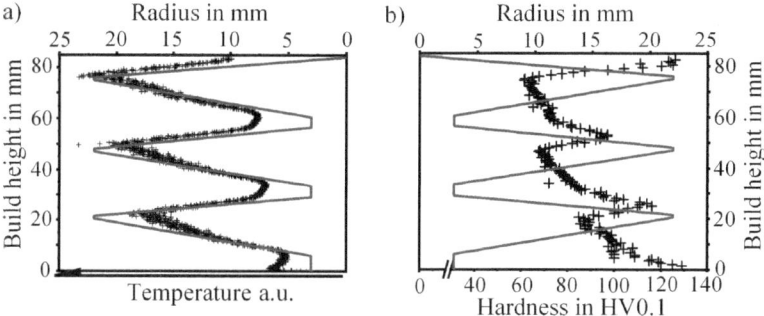

Fig. 8.14 Temperature history (**a**) and hardness (**b**) along the build height for the complex shaped part

height. The strengthening silicon network in the radially decreasing part of the conic shape leads to the highest hardness, whereas the hardness decreases in the conic sections as the network decomposes.

8.3.2 Processing Methods for High-Strength Aluminum Alloys

8.3.2.1 Grain Structure Evolution of EN AW-2024

To minimize hot cracking, different combinations of temperature gradients and solidification rates can be used to achieve specific solidification conditions in order to influence the resulting density and microstructure, as well as internal stresses. Figure 8.15 shows a comparison of two specimens, with one exhibiting hot cracks while the other one is sound. Furthermore, the resulting processing window is depicted. A scanning speed of 160 mm/s seems to be a hard border for the appearance of hot cracks. At a hatch distance of 60 μm, gas porosity is clearly visible, further decreasing for hatch distances of up to 80 μm, and for even higher distances, hot cracks start to form. It appears that the local temperature gradient is lowered by narrow hatching, whereby cracks can be avoided [15].

Depending on the processing parameters, different microstructures evolve as shown in Fig. 8.16. Figure 8.16a and d are characterized by extremely elongated grains, whereby the increased width of the grains in Figure (d) compared to Figure (a) can be attributed to lower cooling rates during processing. In both cases, grain lengths exceeding the size of the image are visible. Figure (b) shows a grain width in the range between Figures (a) and (d). However, the average grain length is significantly shorter in contrast to the other samples mentioned and a reasonable explanation for this effect might be the re-melting of lower layers during processing. This leads to the assumption that the grain orientation is not only influenced by the thermal

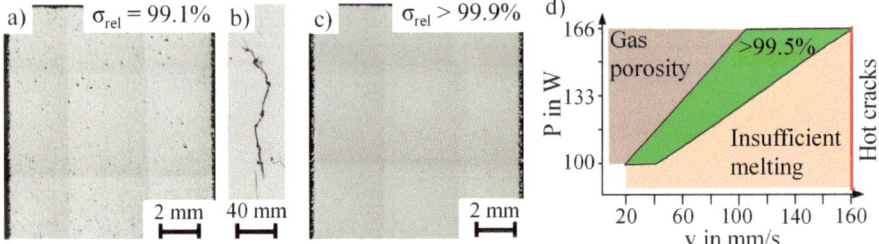

Fig. 8.15 Cross-section of different specimens with the highest relative density PL = 400 W, vs = 6500 mm/s, ΔTy = 20 µm, Tpre = 200 °C. **b** Magnification of a hot crack. **c** PL = 133 W, vs = 80 mm/s, ΔTy = 60 µm, Tpre = 20 °C. **d** Process window [15]

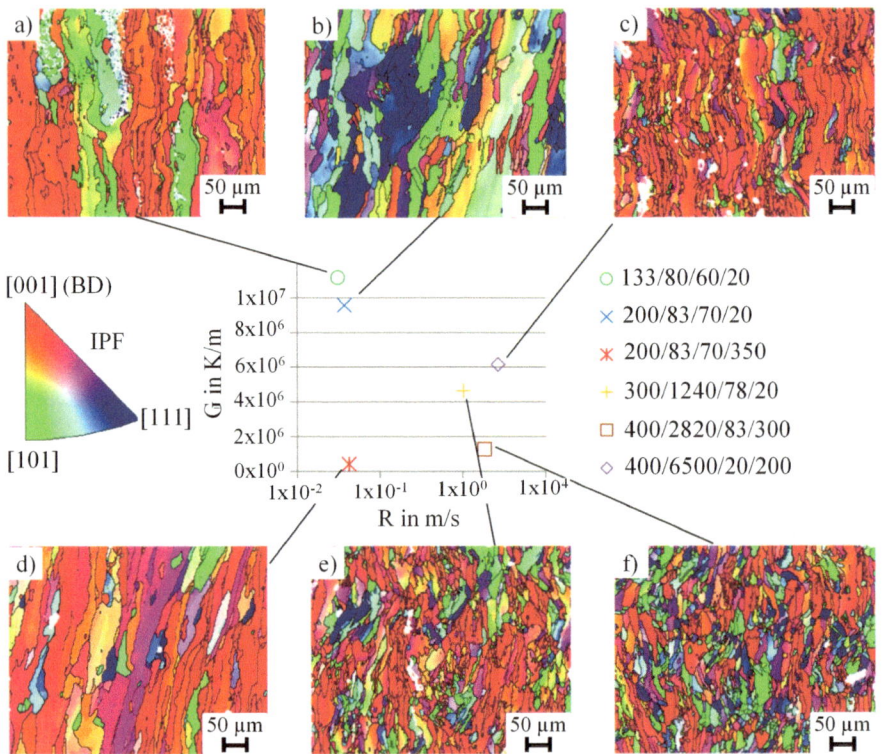

Fig. 8.16 Characteristic grain structure dependent on the thermal gradient (G) and the travel speed of the liquid-solid interface (R). The microstructures are compared for varying G at low R (**a**, **b**, and **d**) and high R (**c**, **e**, and **f**). Nomenclature of the legend: $(P_L/v_S)/(\Delta y/T_{pre})$ [15]

gradient in the bottom part but also at the outer boundaries of the molten pool. Large deviations between the orientation of the thermal gradient at the solidification front and the alignment of the grains lead to an interruption of the epitaxial grain growth in the early stages during solidification. The results of the parameter sets depicted in Figures (c), (e), and (f) show a decreased grain size due to higher cooling rates. For Figures (e) and (f), equiaxed grains can be found to a lesser extent, whereby the detectable amount is larger for Figure (f). Both parameter sets reached the columnar to equiaxed transition even though this line is not crossed completely [15]. Figure (c) shows the formation of very thin and elongated grains aligned in a zig-zag orientation proving a continuous change in temperature gradients. This might be due to the high scanning speeds causing a thermal gradient in hatch orientation perpendicular to the scanning vectors and their direction. In the present case, scanning vectors are rotated by 37° after each layer, while always starting at the point that is the furthest on the left of the sample. Correspondingly, the rotation of this layer is interrupted every third or fourth layer, thus being retrieved in the distance between two peaks in the zig-zag pattern. However, even though the ratio of the temperature gradient and the solidification rate G/R of Figure (c) are below half of Figure (e)'s, almost no globular grains can be found. This might be explained by the formation and shape of the molten pool, as high scanning speeds are applied during processing while the laser power is kept at a relatively low value, leading to a low line energy density. As a result, an elongated molten pool that is particularly shallow at the same time is formed, thereby supporting the epitaxial growth parallel to the building direction [15]. Stress-strain curves are shown in Fig. 8.17 for as-built samples (a) and T4 heat-treated samples (b). The T4 treated PBF-LB/M samples in both build directions exceed the values of the as-built samples. Taking a closer look at the stress-strain curves, it is evident that samples that are built vertically have a reduced elongation at break and also a lower yield strength and tensile strength. This is surprising, as these samples are characterized by highly elongated grains in the build direction, indicating fewer weak spots at the grain boundaries. In the as-built condition, the vertical samples display an even lower elongation compared to the heat-treated ones. The reason for this can be due to an ultra-thin layer that is not always completely remolten by subsequent exposure and might grow on top of the weld seams during PBF-LB/M. However, subsequent heat treatment appears to heal these defective layers, resulting in a completely ductile fracture for both vertically and horizontally built samples [15].

8.3.2.2   Addition of Nanoparticles in EN AW-6082

The acquired processing window shown in Fig. 8.15 for EN AW-2024 can successfully be transferred to EN AW-6082. However, it is still restricted by the scan speed. One possibility to further reduce hot cracking lies in the additives of the powder with TiB_2 nanoparticles as shown in Fig. 8.18. This effect can be traced back to several factors positively influencing the suppression of cracks. First, it appears that the nanoparticles can increase the absorption, which reduces the cold cracking tendency. Secondly, TiB_2 acts as a heterogeneous nucleation agent, thus enabling faster

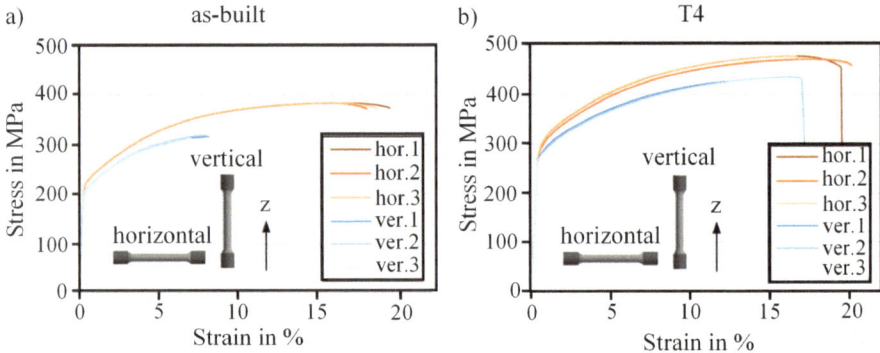

Fig. 8.17 Stress-strain curves for EN AW-2024 in the as-built condition (**a**) and heat treated condition (**b**) [15]

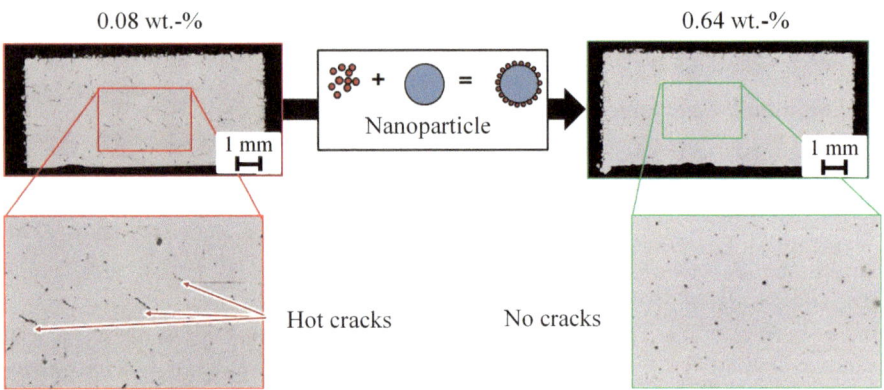

Fig. 8.18 Comparison of the hot cracking for 0.08 wt.-% and 0.64 wt.-% TiB$_2$

nucleation. This effect is shown in Fig. 8.19 by comparing the grain morphology with and without the addition of nanoparticles. A higher content of nanoparticles leads to an increased number of fine grains, which additionally share the same orientation. The smaller grain size delays the onset of strength development in the semi-solid. Furthermore, the higher nucleation rate in the melt pool affects the growth direction of the dendrites, which thus do not only grow in front of the solidification front. The last effect could be a positive influence on the thermal history leading to more evenly rising temperature throughout the layers and therefore a smaller thermal gradient.

8.3.2.3 Qualification of In-House Beam Shaping for PBF-LB/M

There are different approaches to change the intensity profile of the laser. The beam profiles in Fig. 8.8 were realized by diffractive optical elements. However, this approach has two drawbacks. Firstly, it comes with the necessity to redesign

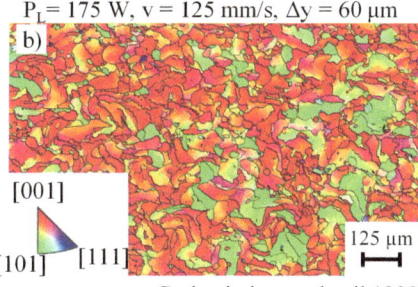

EN AW-6082 without TiB₂   **EN AW-6082 with 0,64 wt.-% TiB₂**

P_L = 175 W, v = 125 mm/s, Δy = 60 μm P_L = 175 W, v = 125 mm/s, Δy = 60 μm

IPF Coloring ∥ z Grains in image detail 508   Grains in image detail 1239

Fig. 8.19 EBSD images for EN AW-6082 without addition of TiB₂ nanoparticles (**a**) and with 0.64 wt.-% TiB₂ (**b**)

Table 8.3 Power ratio and spot diameter measurements

Index	Power ratio core/ring	Spot diameter (1/e2) in μm
0	100/0	113
1	70/30	163
2	60/40	227
3	50/50	267
4	40/60	292
5	20/80	326
6	10/90	334

the optical path for implementation of DOEs and secondly, DOEs can be damaged by too high intensities. Recent developments in the industry led to the introduction of multiple core fiber lasers, which offer the capability for in-source dynamic beam shaping. The AFX-1000 laser from nLight enables switching from a Gaussian intensity profile to different point/ring shaped beam profiles. This is realized by shifting the intensity distribution between an inner and outer core, leading from a full Gauss (Index 0) to a ring profile (Index 6) as shown in Table 8.3.

The melt pool quality and surface roughness acquired with the point/ring profiles show the qualification of the AFX-1000 laser for further use in the production of wrought aluminum alloys in PBF-LB/M. The point/ring shaped profiles exhibit fewer cracks at higher weld speeds when compared to the Gaussian beam profile as shown in Fig. 8.20. This could be traced back to the larger spot diameter and possible slight preheating and postheating effects during melting. All point/ring profiles exhibit wider melt pools compared to the purely Gaussian beam profile. Furthermore, the width of the Gaussian beam profile is far more dependent on the beam power. The point/ring profiles have surprisingly higher penetration depths than the Gaussian beam for low beam powers although the purely Gaussian beam overcomes this effect at higher beam powers. Index 5 exhibits still higher depths than indices 3 and 6.

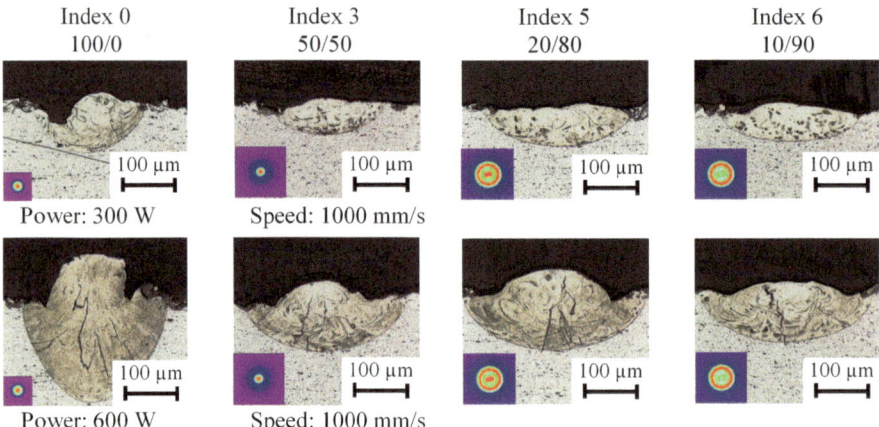

Fig. 8.20 Comparison of the melt pools for the different indices at 1000 mm/s and 300 W respectively 600 W beam power [11]

Furthermore, the melt pool of the pure Gaussian beam converges more sharply while the ring profiles create a more oval melt pool with a higher aspect ratio. In combination with the general larger melt pool width, the hatch distance can be varied in a broader area resulting in two positive effects for further processing in PBF. First of all, a sufficient melt pool overlap is provided, even at high hatch distances, leading to an increased build rate. For smaller hatch distances on the other hand, the adjacent track is more evenly remelted, which could result in less pore and crack formation. Going from Index 0 to higher indices leads to further improvements, as shown in the processing windows for the different profiles in Fig. 8.21. The point/ring profiles exhibit no keyhole formation at higher laser powers. Penetrations depths, which reach too deep into the part can lead to extensive pore formation. Another advantage of shifting the intensity distribution from core to ring is the generation of more stable weld tracks over the scan length. The formation of humps is significantly reduced at high weld speeds and beam powers. Based on these results, the assumption could be made that ring-based beam shapes result in a reduction of the balling effect in PBF-LB/M. Furthermore, the lack of unwanted elevations in the generated layer leads to a more uniform powder layer and reduced damage on the recoater. Both effects result in less flaw formation. These results give a strong indication towards an improved processing of aluminum alloys like EN AW-5083, EN AW-6082, or EN AW-7075 [11].

8.3.2.4 Two-Color Thermography

A quantitative evaluation of process control, beam shaping, or nanoparticle additives with respect to their hot cracking tendency requires the use of a spatially and temporally high-resolution temperature determination method to acquire information (e.g.,

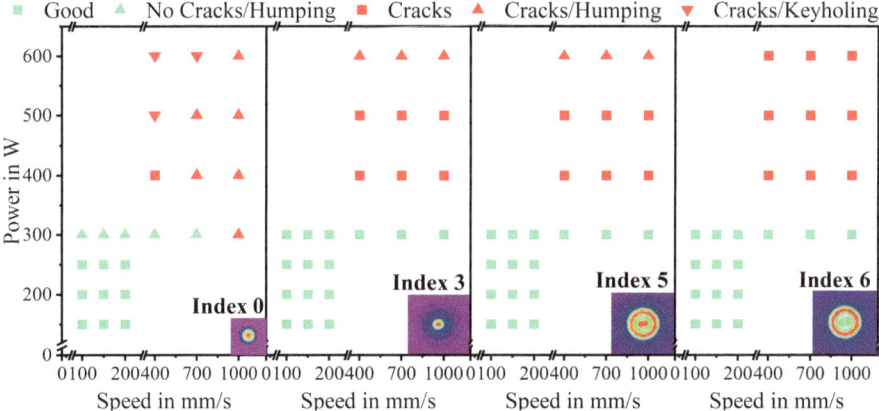

Fig. 8.21 Processing window for the single weld tracks according to the melt pool cross-sections [11]

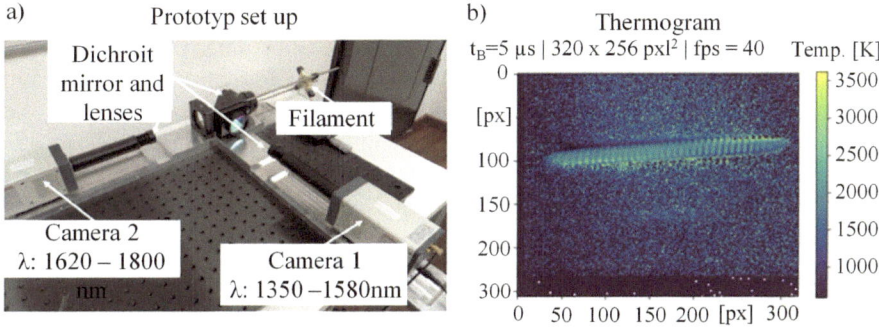

Fig. 8.22 Setup of the two-color thermography and an exemplary thermogram of a filament [18]

temperature gradients, cooling rates) on the melt pool level. Therefore, a two-color thermography system is suitable for the accurate acquisition of temporally and spatially high-resolution temperature information in a temperature range relevant for low-melting alloys in PBF-LB/M. The spectral sensitivity of the system lies in the NIR range (λ1: 1350–1580 nm, λ2: 1620–1800 nm) in order to be able to record relevant temperature information during the processing of low-melting powder alloys (e.g., along the outgoing melt pool). Figure 8.22 shows a prototypically designed experimental setup. This comprises two camera sensors arranged orthogonally to each other, a lens and filter package, and a glow wire placed in the focal plane as the measurement object. The intensity ratio is calculated pixel by pixel from the synchronously recorded individual images. As an example, a thermogram of a static measurement object is shown in (b) [18].

8.4 Summary and Conclusion

In order to expand the limited material portfolio of laser based powder bed fusion (PBF-LB/M) with high-performance materials, new process strategies have been developed. For lightweight applications and highly stressed components in the aerospace and automotive sectors, for example, Al-Cu or Al-Mg-Si wrought alloys such as EN AW-2xxx, EN AW-5xxx or EN AW-6xxx are required. However, these have a tendency to hot cracking when processed by PBF-LB/M. It is demonstrated that wrought Al-Cu alloys can, in principle, be processed on conventionally available PBF-LB/M equipment if the process is suitably controlled. Furthermore, controlling the solidification conditions for EN AW-2024 by adjusted processing parameters resulting in different grain morphologies leads to the fabrication of dense specimens. The processing window is transferable to EN AW-6082 and can be expanded by adding TiB_2 nanoparticles. The particles act as grain refiners and suppress hot cracks. Combinations of different parameter sets within a cubic structure or changing the part geometry further show that in situ heat treatments are possible. Therefore, tailored microstructures and thus mechanical properties can be achieved. In addition to pre-alloyed powders, the processing of heterogeneous, microscale powder mixtures was also investigated to explore the process of alloy formation and modification. In situ alloy formation based on elemental powders increases flexibility and offers particular advantages in the defined adjustment of component properties. With an appropriate relationship between the particle size, powder layer thickness, and melt pool size, homogeneous alloys can be melted without modification of the system technology. Since the melt pool dynamics present in the PBF-LB/M process have a decisive influence on the processing procedure, the adjustment of the melt pool dynamics by using adapted intensity distributions generated with diffractive optical elements (DOE) or in-house beam shaping offer a promising approach to improve the processability of high-strength aluminum alloys. In addition to achieving lower dynamics in the melt pool, it was also possible to push the limits of the process window based on single weld tracks by improving the surface quality and reducing hot cracking. However, precise temperature control in the PBF-LB/M process is essential for accurate adjustment of the component properties and for the defect-free processing of even larger components. Therefore, a two-color thermography system was developed. However, further research is required to transfer the designed temperature determination method to the PBF-LB/M application. By transferring the measurement system to the PBF-LB/M process, the process understanding for additive manufacturing in general should be extended and, in particular, transferable knowledge for hot cracking in the powder bed should be acquired. If the hot cracking tendency can be correlated with suitable measurement data (e.g., melt pool temperature), a control for the PBF-LB/M process based on two-color thermography is conceivable. This knowledge can be used in the future to process wrought aluminum alloys that are susceptible to hot cracking in series production without defects and to adjust the final component properties in a targeted manner.

References

1. Ahuja, B., Karg, M., Nagulin, K., et al.: Fabrication and characterization of high strength al-cu alloys processed using laser beam melting in metal powder bed. Phys. Procedia **56**, 135–146 (2014)
2. DIN German Institute for Standardization. Din en 573-3:2019-10, aluminium and aluminium alloys - chemical composition and form of wrought products - part 3: Chemical composition and form of products (2019)
3. Karg, M., Ahuja, B., Kuryntsev, S., et al.: Processability of high strength aluminium-copper alloys aw-2022 and 2024 by laser beam melting in powder bed. In: Proceedings of the 25th Solid Freeform Fabrication Symposium, Austin, Texas, USA (2014)
4. Karg, M., Ahuja, B., Wiesenmayer, S., et al.: Effects of process conditions on the mechanical behavior of aluminium wrought alloy en aw-2219 (al-cu6mn) additively manufactured by laser beam melting in powder bed. Micromachines **8**, 23 (2017)
5. Karg, M., Laumer, T., Schmidt, M.: Additive manufacturing of gradient and multimaterial components. In: Proceedings of International Conference on Competitive Manufacturing (2013)
6. Karg, M., Munk, A., Ahuja, B., et al.: Expanding particle size distribution and morphology of aluminium-silicon powders for laser beam melting by dry coating with silica nanoparticles. J. Mater. Process. Technol. **264**, 155–171 (2019)
7. Karg, M., Rasch, M., Schmidt, J., et al.: Laser beam melting in powder bed of non-spherical aluminium microparticles dry coated with metal nanoparticles. In: Proceedings of the 6th International Conference on Additive Technologies iCAT, Nürnberg, Deutschland (2016)
8. Karg, M., Rasch, M., Schmidt, K., et al.: Laser alloying advantages by dry coating metallic powder mixtures with siox nanoparticles. Nanomaterials **8**, 862 (2018)
9. Martin, J., Yahata, B., Hundley, J., et al.: 3d printing of high-strength aluminium alloys. Nature **549**, 365–369 (2017)
10. Mauduit, A., Pillot, S., Gransac, H.: Study of the suitability of aluminium alloys for additive manufacturing by laser powder bed fusion. UPB Sci. Bull. Ser. B: Chem. Mater. Sci. **79**, 219–238 (2017)
11. Nahr, F., Bartels, B., Rothfelder, R., et al.: Influence of novel beam shapes on laser-based processing of high-strength aluminium alloys on the basis of en aw-5083 single weld tracks. J. Manuf. Mater. Proc.
12. Nahr, F., Rasch, M., Burkhardt, C., et al.: Geometrical influence on material properties for ti6al4v parts in powder bed fusion. JMMP **7**, 82 (2023)
13. Rafi, H., Karthik, N., Gong, H., et al.: Microstructures and mechanical properties of ti6al4v parts fabricated by selective laser melting and electron beam melting. J. Mater. Eng. Perform. **22**, 3872–3883 (2013)
14. Rasch, M., Bartels, D., Sun, S., et al.: Alsi10mg in powder bed fusion with laser beam: an old and boring material? Materials **15**, 5651 (2022)
15. Rasch, M., Heberle, J., Dechet, M., et al.: Grain structure evolution of al-cu alloys in powder bed fusion with laser beam for excellent mechanical properties. Materials **13**, 82 (2019)
16. Rasch, M., Roider, C., Kohl, S., et al.: Shaped laser beam profiles for heat conduction welding of aluminium-copper alloys. Opt. Lasers Eng. **115**, 179–189 (2019)
17. Schulze, G.: Metallurgie des Schweißens. Springer, Berlin, Heidelberg (2004)
18. Schwarzkopf, K., Rothfelder, R., Rasch, M., et al.: Two-color-thermography for temperature determination in laser beam welding of low-melting materials. Sensors **23**, 4908 (2023)

Michael Schmidt has headed the Institute of Photonic Technologies since its founding in 2009 at Friedrich-Alexander-Universität Erlangen-Nürnberg. His research interests include laser application from micro- to macroscopic scales within industrial manufacturing, additive manufacturing, and medical engineering. He was the vice spokesperson of the Collaborative Research Center 814 Additive Manufacturing.

Part III
Process Strategies for Multi-material and Multi-functional Parts

Chapter 9
Selective Laser Beam Sintering of Multiphase Systems

Matthias Lindbüchl and Dietmar Drummer

Abstract Multiphase systems are material systems consisting of at least two components. This can be a material system with two polymers (blend production) or a polymer with fillers. The use of multipolymer material systems aims to improve properties through synergistic effects, for example, 'positive' material properties of both polymers mixed together should improve the final component properties. The use of fillers in a polymer matrix, on the other hand, is aimed at functionalization and improvement, especially of the mechanical properties of the components. These two concepts are presented in the following chapter.

9.1 Introduction

Multiphase systems are material systems consisting of at least two components. This can be a material system with two polymers (blend production) or a polymer with fillers. The use of multipolymer material systems is aimed at improving properties through synergistic effects, for example, 'positive' material properties of both polymers mixed together should improve the final component properties. The use of fillers in a polymer matrix, on the other hand, is aimed at functionalization and improvement, especially of the mechanical properties of the components [1]. In blend production, attention must be paid to the process-side compatibility of the polymers. This means that the polymers can be processed in the same process window, whereby the process windows are determined by differential scanning calorimetry (DSC) measurements. The use of fillers presents manufacturers with process-related challenges in terms of mixing strategy and application behavior.

Fillers are generally understood to be inorganic glass- and ceramic-based, metallic, and carbon-based materials. These are added to the polymer in amounts of between 5 and 65% by volume. Fundamentally, there are three reasons for adding

M. Lindbüchl · D. Drummer (✉)
Friedrich-Alexander-Universität Erlangen-Nürnberg, Institute of Polymer Technology,
Am Weichselgarten 10, 91058 Erlangen-Tennenlohe, Germany
e-mail: dietmar.drummer@fau.de

D. Drummer and M. Schmidt (eds.), *Progress in Powder Based Additive Manufacturing*, Springer Tracts in Additive Manufacturing,
https://doi.org/10.1007/978-3-031-78350-0_9

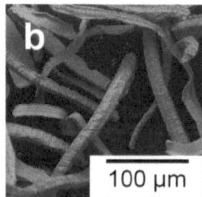

Fig. 9.1 Different filler shapes of copper: **a** plates, **b** fibers, **c** dendrites, and **d** spheres

fillers to polymer systems. To achieve a reinforcing effect, glass or carbon fibers can be added to a polymer matrix to enhance mechanical properties such as tensile strength and stiffness. Equally important is the possibility of functionalization: By incorporating fillers, the thermal, magnetic, or electrical properties can be specifically influenced and adjusted. The last aspect is the possibility of cost reduction by decreasing the use of an expensive polymer matrix material [1, 2].

The fillers can have different shapes, as shown in Fig. 9.1 and, in addition to spherical fillers, there are also platelets, fibers, and dendrites.

When mixed, the fillers have a decisive influence on the material properties. These include powder properties such as bulk density, tap density, and flowability. The powder properties in turn influence thermal properties (heat capacity) or optical properties (diffuse reflection). On the process side, properties refer to the crystallization behavior, which defines the process window and thus the adjustable process parameter ranges, such as the installation temperature. A change in the material or process-side processing properties leads to different characteristics of the component properties such as geometry and morphology, and thermal and mechanical properties. Therefore, a holistic approach from material characterization through the process to the finished component is indispensable for understanding the interactions.

9.2 Material Processing Properties

The material properties comprise the thermal and optical properties, which are significantly influenced by the powder properties. The optical properties of powders and powder mixtures can be characterized by the characteristic value of diffuse reflection. This provides information on the extent to which the radiation introduced during the process is reflected at the powder bed surface. The main factors influencing diffuse reflection are the filler materials and shapes, and also the filler concentrations, the effects of which are shown in Fig. 9.2.

The diffuse reflectance value for unreinforced polyamide 12 (PA12) powder is given as a reference value, which is close to 2%. For copper, there is a strong dependence on the filler form and concentration. While the diffuse reflectance of copper platelets is above the value of polyamide, copper dendrites display roughly the same value as polyamide whereas copper spheres are below the value of polyamide. The

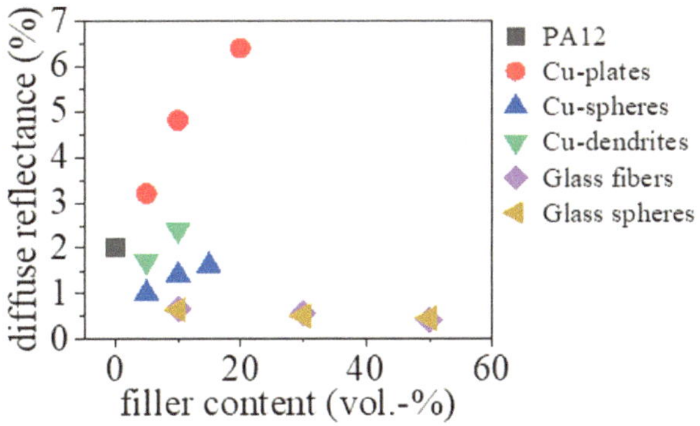

Fig. 9.2 Diffuse reflection dependending on different filler materials for a wavelength of 10.6 μm, shapes and concentrations [7, 9]

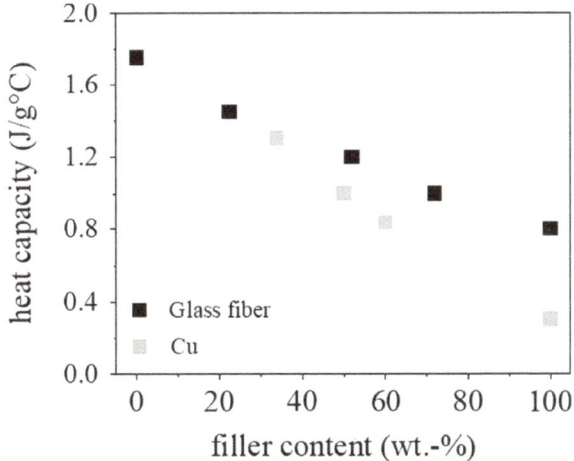

Fig. 9.3 Heat capacity (J/g°C) resulting from different filler materials (Cu, glass) and filler concentrations [5, 9]

size of the copper particles is not decisive here as all three filler forms of the copper particles are in the order of magnitude for $d_{50.3}$ of 28–34 μm. Glass particles show no dependence in the expression of diffuse reflection on filler shape and size in the order of magnitude considered for $d_{50.3}$ of 12 and 27 μm. Additionally, the filler concentration has no significant influence on the characteristic value.

Changes in diffuse reflection can lead to a modified energy input. While glass absorbs a large part of the laser radiation, copper—particularly copper plates with higher filler contents—exhibits higher diffuse reflection due to the increasing interaction areas. As a result, the penetration depth of the laser beam and thus the layer thickness are influenced, which can have effects on the processing behavior. The reason for this can be the different absorptivities of the filler materials compared to polymers [15]. The thermal properties of the powders or powder mixtures can be described by the specific heat capacity and Fig. 9.3 shows the differences between glass and copper fillers.

With increasing filler content, the heat capacity decreases significantly for both material systems. The decrease is independent of the filler form and thus purely concentration-related. The reduction is approximately linear for both materials, with the decrease being greater for metals. On the process side, a lower heat capacity means a higher energy input and modified design in terms of component placement. To avoid heat accumulation, filled components must be placed further apart than components made of pure PA12 [5, 9].

The powder properties, such as the flowability, depend on the type of fillers added to the polymer powders. As a quotient of the tap and bulk densities, the Hausner factor enables a quantitative statement to be made about the flowability and thus the processability with regard to the output or deposition of the powder on the build plate in selective laser sintering. The quotient of both densities is called the Hausner factor or Hausner ratio. In addition to the filler shape and concentration, the mixing strategy plays a decisive role. In Fig. 9.4, the influence of the filler shape and concentration is shown in addition to the effect of the mixing strategy on the flowability for polypropylene (PP).

With increasing filler concentration, the Hausner factor increases, and the flowability thus decreases, whereby this behavior is independent of the filler form and the mixing strategy. Good flowability is achieved with a Hausner factor below 1.25. If the Hausner factor is in the range of 1.25–1.4, the flowability is reduced, although discharge in selective laser sintering is generally still possible. For values above 1.4, cohesive behavior between the powder particles dominates, which means that discharge is generally no longer possible. The superficially non-spherical, fissured glass particles show poorer trickling behavior than the glass fibers. By modified filler incorporation via precipitation, in which the filler particles are introduced directly into the polymer particles, the flowability properties can be significantly improved. The reason for this is the spherical and smoother surface of the polymer particles compared to the glass fillers, which dominates the precipitation process (Fig. 9.5). The fillers enclosed in the polymer particles thus have hardly any negative influence on the Hausner factor. With increasing filler content, and due to the particle size distribution of the fillers, however, longer fillers repeatedly protrude from the polymer particles, which explains the higher Hausner factor compared to pure polymer [6].

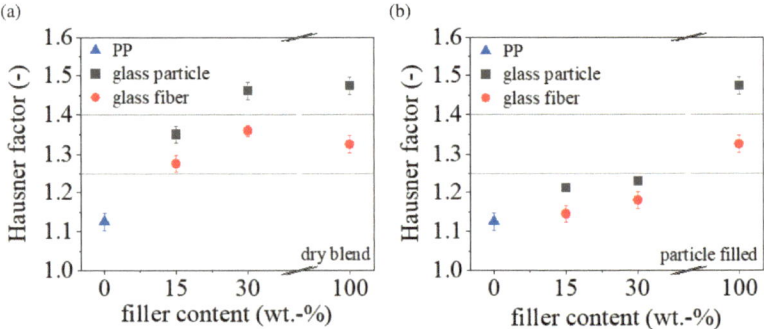

Fig. 9.4 Influence of filler form and concentration on the Hausner factor, **a** dry blend, **b** particle-filled system [6]

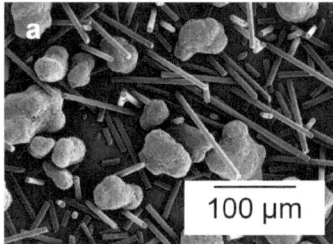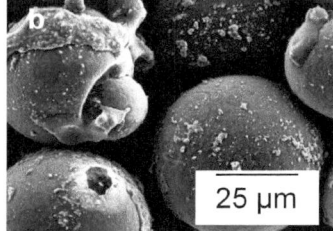

Fig. 9.5 Microscopy images of **a** dry blends with glass fibers and **b** particle-filled systems with glass particles [6]

9.3 Processing Properties

By influencing the material properties (optical/thermal), the incorporation of fillers has an effect on the process-side properties such as the crystallization behavior and thus on the process window. For glass fibers, a nucleating effect can be demonstrated by means of isothermal differential calorimetry. One parameter that can be used to estimate the effect is the half crystallization time $t_{0.5}$, which describes the point in time at which 50% of the crystallization is completed. The higher the filler content, the faster the crystallization process although this behavior does not apply to all polymers to the same extent, as shown in Fig. 9.6.

The described relationship can be observed to a greater extent for glass fiber-reinforced PA12 and to a lesser extent for PP. The difference in the decrease of half the crystallization time, a reduction of up to 64% for PA12 and 35% for PP, may be due to the different macromolecular mobility of the two polymer types. Due to the accelerated crystallization, a reduced process window is present in the real selective laser sintering process, which can negatively influence the process robustness [6, 8].

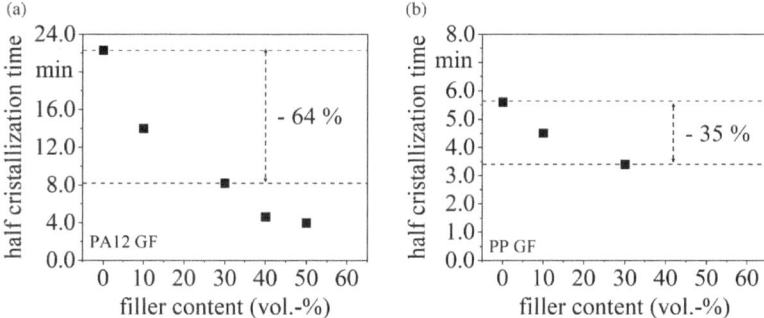

Fig. 9.6 Half crystallization time $t_{0.5}$ of **a** PA12 GF and **b** PP GF depending on the filler concentration [6, 8]

When processing polymer blends, the crystallization behavior and the process window must also be considered and, using the DSC curves, two semi-crystalline polymers can be identified whose process windows overlap. For the production of blends in the selective laser sintering process, it is necessary that the temperature ranges of both materials overlap. If the temperature range of a blend partner is exceeded, it would melt uncontrollably, while if the temperature falls below this range, one phase would already begin to crystallize and residual stress-induced distortion ("curling") of the molten layer would occur [14]. Blend production with a semi-crystalline and an amorphous component offers the possibility of expanding the material variety and setting specific properties via the mixing ratio, whereby one possible material combination is the PBT/PC blend.

However, a mere analysis by means of DSC is not sufficient, since a defined process window can only be determined for the semi-crystalline component; according to DSC, the processing temperature of PBT is 205 °C (Fig. 9.7a). At this temperature, PC is in the molten state, which would prevent processing. Therefore, dynamic mechanical analysis (DMA) is additionally used to determine the storage modulus of both components based on their viscosity state over temperature (Fig. 9.7b). The processing temperature of the blend is then set on the basis of the glass transition temperature of the amorphous component, which in this case is 139 °C for PC. In addition, a sufficiently high energy density must be set in the process that can melt both components [3].

After determining the diffuse reflection, the heat capacity, and the crystallization, respectively the process window, the energy density in the selective laser sintering process has to be adjusted. The influence of the energy density is the subject of current research and selected results are summarized in the compilation of microscopy images in Fig. 9.8. Shown are images of the morphological structure as a function of the energy density and the filler concentration. The filler used is glass fiber, which can be seen in the micrographs.

The polymer-filler blend used here is a dry blend of PA12 with glass fibers, and a homogeneous fiber distribution with no segregation phenomena is observed across all settings. If the energy density is too low (0.3 J/mm^2), the applied energy is not

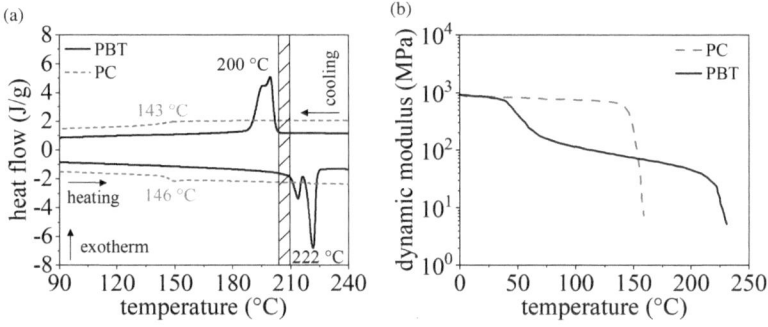

Fig. 9.7 DSC curve for PBT and PC **a** and DMA curve of PBT and PC **b** [3]

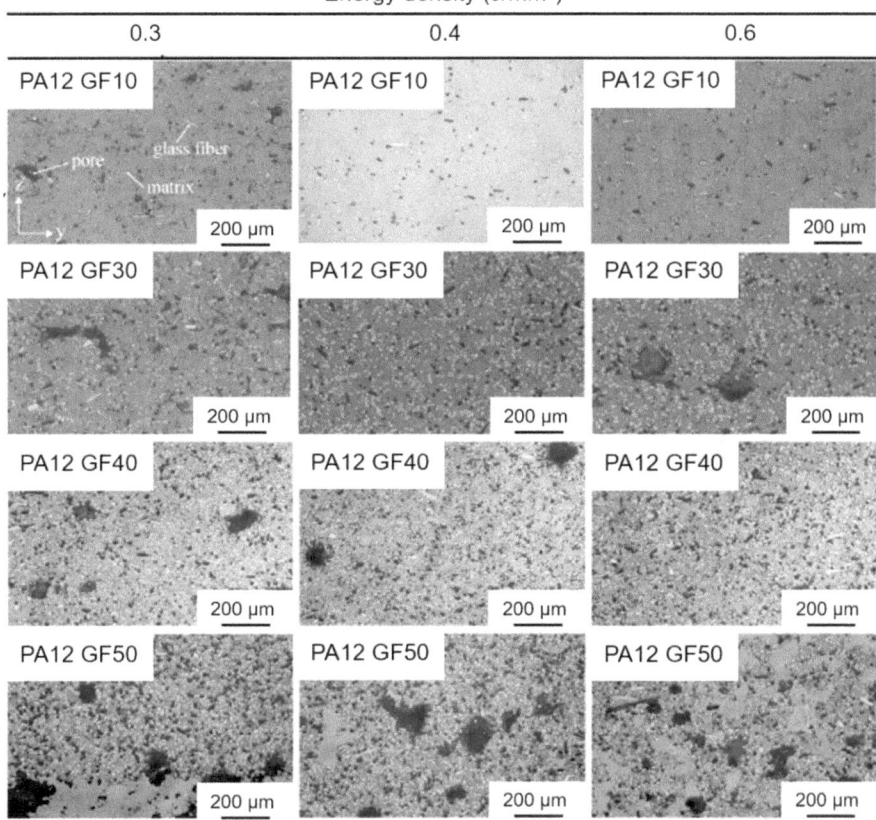

Fig. 9.8 Microscopy images of PA12 with different filler concentrations and various energy densities [7]

sufficient to completely melt all polymer. As a result, the fibers are not completely wetted with the polyamide, and the particles are not fully melted, which promotes the formation of pores. At sufficiently high energy densities, the polymer particles are completely melted, the fibers are sufficiently wetted by the polymer, and thus a sufficiently good bond between polymer and filler is created. This results in significantly fewer pores. With increasing filler concentration, the wetting behavior is negatively influenced and, as a result, more pores appear. At 50% by volume, a critical fiber volume concentration is reached, at which highly porous components are formed [7].

In the effects described so far, the incorporation was carried out on a global scale, as the fillers were mixed with the polymer particles using different mixing strategies and subsequently introduced and processed in a selective laser sintering system [6].

The current research focus is on the local incorporation of fillers into the powder bed. For the realization of the local introduction, a nozzle discharge system is available, which is attached to a gantry structure. The nozzle system is shown in

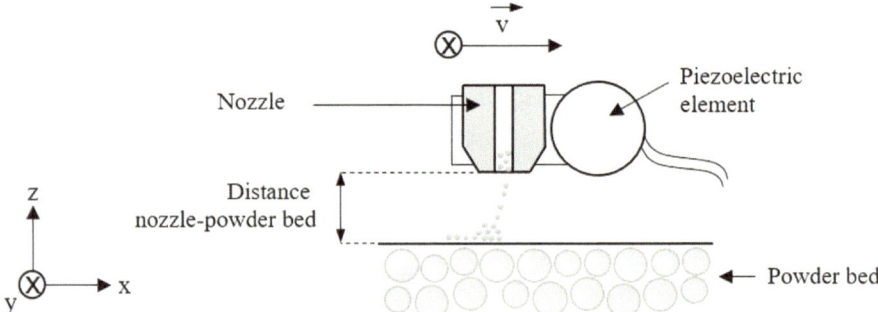

Fig. 9.9 Schematic overview of the principle structure of a nozzle system for local powder injection, according to [11]

Fig. 9.9 as a principle sketch and consists of a glass nozzle with an opening diameter of 0.7 mm, into which the pure fillers or polymer-filler mixtures can be fed.

A piezoelectric element is mounted in the nozzle holder, the vibration of which can be used to ensure discharge from the nozzle. The piezo element can be excited with voltages in the range of 0–1000 V and a frequency of 0–1000 Hz. The discharge speed, which can be set between 0 and 100 mm/s, is controlled via the servo-motors on the x- and y-axes. In addition, the distance of the nozzle to the powder bed surface can be varied. The x-y positioning system is placed in the build chamber of a DTM Sinter Station 2000 above the build platform as shown in Fig. 9.10.

The excitation or frequency pattern influences the discharge quantity. With a sinusoidal excitation pattern, the discharge quantity can be maximized, while the minimum discharge quantity is achieved with a ramp. In Fig. 9.11, the difference between four possible excitation patterns for glass fiber reinforced PP is highlighted.

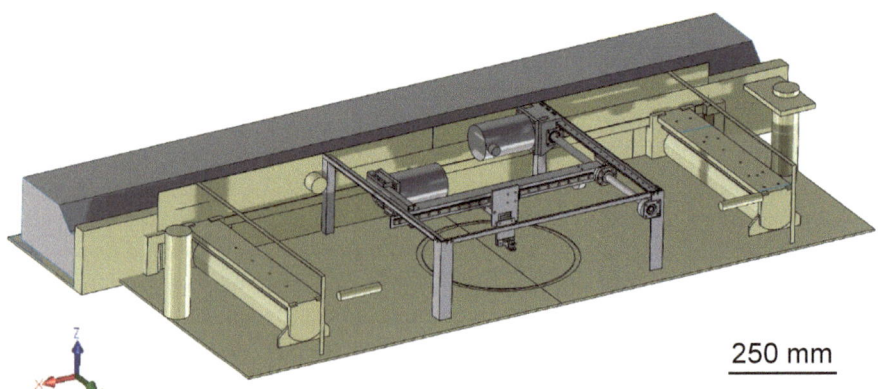

Fig. 9.10 SCAD model illustrating the insertion of the x-y positioning system into a selective laser sintering system (DTM Sinterstation 2000)

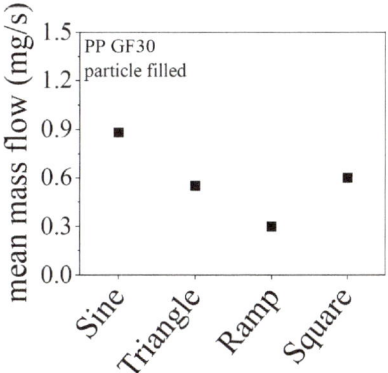

Fig. 9.11 Influence of different excitation patterns of the piezoelectric element of the local nozzle discharge tower on the mass flow using particle filled PP GF30

9.4 Component Properties

The properties of a component manufactured in the selective laser sintering process are the result of the interaction of the material and process-side processing properties. As shown in Chap. 2 concerning the material processing properties, changes in the diffuse reflection can influence the energy input into the powder bed, which can directly result in differences in the formation of the layer thickness for individual layers. Figure 9.12 shows the single layer thicknesses as a function of filler shape, filler type, and filler concentration from previous studies for copper and from current findings from research for glass-filled PA12.

The higher reflection of the copper flakes leads to lower energy coupling and results in lower layer thicknesses compared to the copper spheres, which, when mixed with the PA12 particles, exhibit lower reflection than the pure polyamide. The generally higher absorption of glass compared to copper results in higher energy

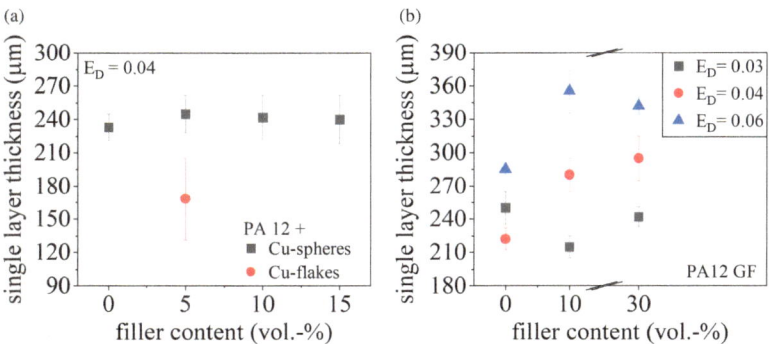

Fig. 9.12 Comparison of single layer thickness for **a** Cu-filled PA12 and **b** glass-filled PA12 as a function of filler form and concentration [5, 9]

coupling into the powder bed, which tends to yield a greater film thickness at comparable energy density (ED $= 0.04$ J/mm^2). In glass fiber reinforcement, the thickness of the individual layers increases when the energy density is increased at constant filler content. An increasing filler concentration at constant energy density also leads to an increase in the thickness of the individual layers due to the absorption behavior and reduced heat capacity [5, 9].

In order to demonstrate the scalability of the described effects to multi-layer components, the geometrical dimensions such as width and thickness of manufactured components can be analyzed. In Fig. 9.13, both parameters are listed as a function of the filler concentration and the energy density.

Analogous to the single layer, an increase in component thickness is observed with higher filler content. This can also be explained by the increased absorption and the increased specific heat capacity, which is responsible for the deviation in geometry. A higher energy density leads to an increased component thickness, since the energy input into the powder bed is increased. It appears that the effect of the energy density becomes smaller at a higher filler content. At a very high filler content of 50% by volume, the energy density is irrelevant for the part thickness and thus the filler content is the more dominant factor [5]. Geometric dimensional tolerances must also be observed when using blends in selective laser sintering. In Fig. 9.14, both the single layer thickness and a sample area are plotted as a function of the mixing ratio of both blend components and the energy density.

With increasing PC content, a slight trend towards lower film thicknesses can be observed. Physical aspects, such as higher viscosity or surface energy, may be causal for this. An increase in surface wettability, which leads to better spreading behavior on the surface of the polymer blend material, may also be a possible reason. Furthermore, chemical influences cannot be ruled out, as it is possible that amorphous phases of PBT are compatible with PC, resulting in a flatter surface geometry. The target size for the exposed area is 625 mm^2 and below 20 wt.% PC, no measurements are possible. PC could affect both PBT phase transformation and surface wetting. In addition, the higher viscosity of molten PC could block droplet formation in the PBT

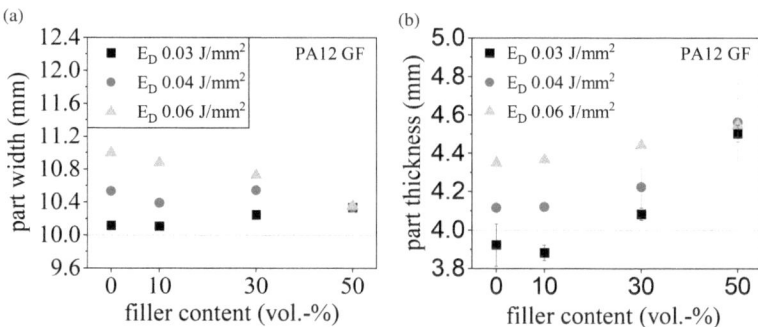

Fig. 9.13 Comparison of the width and thickness of glass fiber-filled PA12 components as a function of the filler content and the energy density [5]

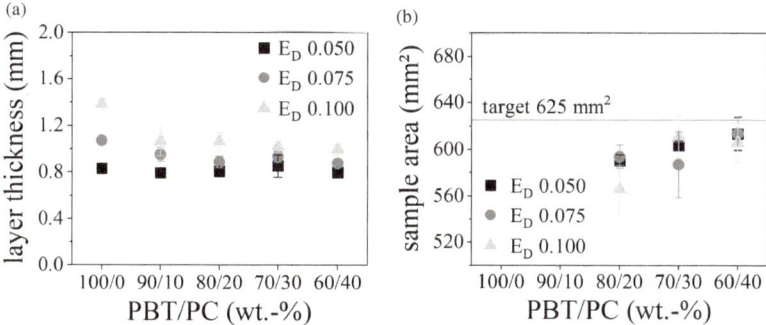

Fig. 9.14 Influence of the energy density and the ratio of the blend components on the single layer thickness and sample area of PBT/PC blends [3]

matrix. The high values for the standard deviations are due to high porosities, and thus no influence of the energy input on the monolayer formation in the horizontal plane can be detected [3]. Another important aspect with regard to the component properties is the mechanical properties, such as the Young's modulus, the tensile strength, and the elongation at break. The following is a summary of current findings from research on the mechanical properties of PP and PA. Figure 9.15 shows the tensile moduli of PP and PA12 as a function of filler concentration, mixing strategy, and energy density.

For PP with different mixing strategies, an increasing tensile modulus with increasing filler content can be observed. The values of the Young's modulus are higher for particle-filled systems than for dry-mixed powder systems. The reason for this is the lower component porosity, which is due to the higher wetting area of the melt at the fillers [4]. For PA12, it is shown that a higher fiber content leads to a significant increase in tensile modulus and a fiber content of 30 vol.% can increase the stiffness up to three times compared to the neat material. At a filler content of 50 vol.% and the lowest energy density of 0.03 J/mm^2, the tensile modulus decreases

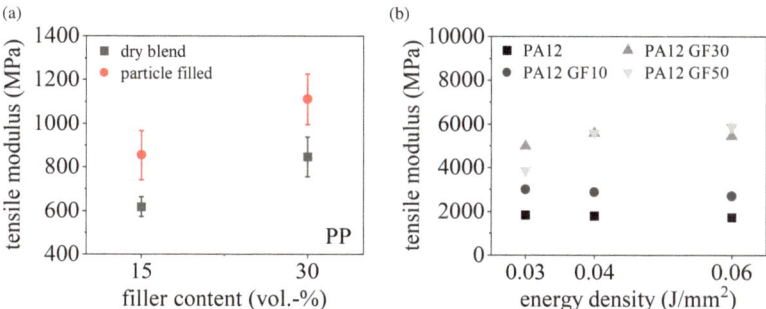

Fig. 9.15 Influence of filler concentration, energy density, and blending strategy on the tensile modulus of PP **a** and PA12 **b** [4, 7]

Fig. 9.16 Comparison of theoretical model and experimental tensile modulus values for glass fiber-reinforced PA12 [7]

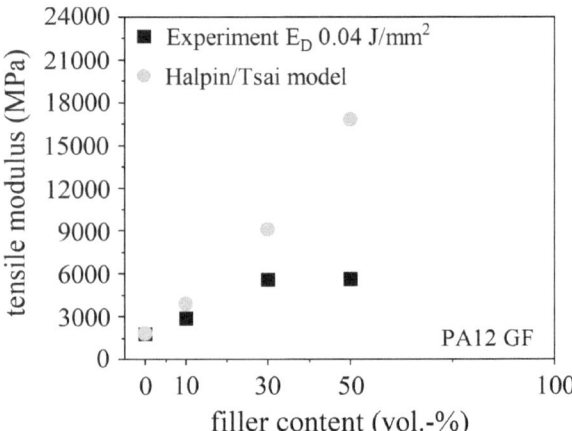

because the energy input is too low to melt the polymer sufficiently. At the higher energy densities, the stiffness remains constant. There is no significant increase in tensile modulus between 30 and 50 vol.% because the partial porosity of the latter is too high. This effect has a stronger impact on the mechanical properties than an increase of the filler content [7]. In Fig. 9.16, the experimentally obtained values of the tensile modulus are compared with the theoretically optimal values that can be achieved by calculation according to the model of Halpin/Tsai.

There is a discrepancy between the theoretically achievable tensile moduli and those of the laser-sintered components ($E_D = 0.04$ J/mm^2). While the tensile moduli of the laser-sintered components plateau at a concentration of 30% by volume, the theoretical moduli still increase. In the SLS process, the porosity of the powder bed and the components, a low fiber-matrix adhesion that does not transfer stresses, and a reduced flowability with increasing fiber content are to be considered as reasons for the differences. At a concentration of 50% by volume, the component porosity is too high and the possibility of coalescence too low, since only a small amount of matrix powder remains, so that no increase in modulus takes place [7]. In contrast to the improvement in stiffness, the tensile strength displays different results, as shown in Fig. 9.17.

For the PP powder filled with glass particles, higher values can be observed than for the dry blends. This can be justified by the poorer flowability and the resulting higher component porosity. In contrast to fiber-reinforced composites, the glass-particle systems have no reinforcing effect with respect to tensile strength. This decreases with increasing filler content. In the case of PA12 reinforced with glass fibers, the tensile strength is constant up to a concentration of 30 vol.%. A marked decrease in tensile strength occurs at a filler content of 50 vol.% because the high volume content of short glass fibers leads to a low powder porosity, and thus a lower flowability of the powder, whereby the remaining polymer matrix content is too low for sufficient coalescence [4, 7]. Corresponding to the increase in stiffness, the general decreasing tendency for the elongation at break is shown in Fig. 9.18.

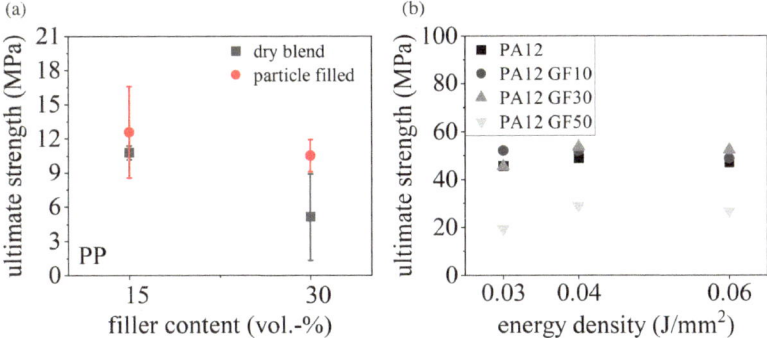

Fig. 9.17 Influence of filler concentration, energy density, and blending strategy on the ultimate strength of PP **a** and PA12 **b** [4, 7]

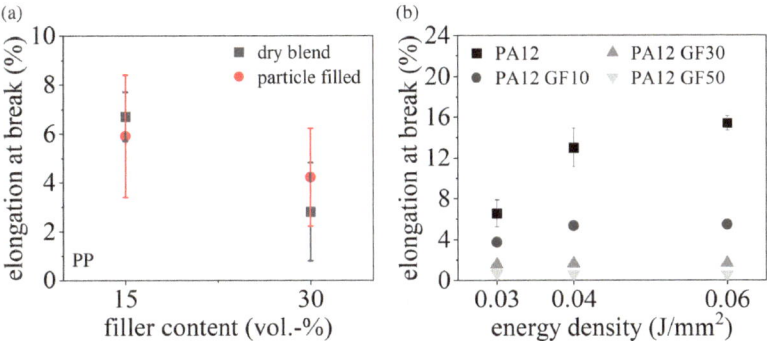

Fig. 9.18 Influence of filler concentration, energy density, and blending strategy on the elongation at break of PP **a** and PA12 **b** [4, 7]

The elongation at break of PP reacts sensitively to material or process changes, which is why a high standard deviation can be observed here. However, a clear statement regarding the mixing strategy cannot be made in this case. For PA12, it is also shown that the elongation at break decreases continuously with increasing filler content. The elongation at break of pure PA12 increases with higher energy density, since more polymer particles are melted. In addition, the brittleness is inversely proportional to the elongation, whereby the brittleness is related to various properties such as the impact strength and thermal expansion of the polymers. With a higher filler content, above 30% by volume, no difference between the energy densities can be observed, since the effect of the high fiber content dominates the effect of the energy density variation [4, 7]. The thermal conductivity can be improved by adding copper particles to the polyamide particles. No change in thermal conductivity with increasing filler content is visible, as shown in Fig. 9.19.

The copper particles are isolated in the polymer matrix, which represents a high resistance to heat flow, and even a variation of the component orientation does not lead

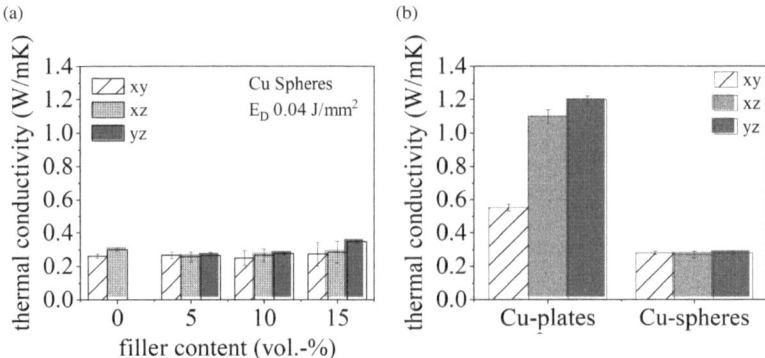

Fig. 9.19 Influence of filler shape and concentration, as well as build direction of copper-filled PA12, on thermal conductivity [9]

to an increase in the thermal conductivity. However, a comparison of the copper shape in Fig. 9.19 shows a significant difference in the achievable thermal conductivity even at a low concentration of 5 vol.%. The copper plates with an anisotropic shape give a value of 1.2 W/mK with a yz partial orientation. The processing direction leads to a predominant direction. The particles partially touch each other and the conductivity resistance of the matrix is minimized. The difference in the achieved thermal conductivity of the xy- and yz-orientation is due to the processing direction during powder application. The filler orientation layers are normal to the z-direction, but parallel to the heat flow. Therefore, the orientation of the parts must be taken into account during virtual placement in the chamber [9].

9.5   Possible Functionalization

Functionalization in terms of thermal conductivity and adaptation of mechanical properties was discussed in the previous chapter. However, further functionalizations are conceivable.

9.5.1   Flame Retardancy

Current findings from research include the increase of the flame retardant effect or the manufacturing of flame retardant components through the addition of flame retardants. For this purpose, polyamide powders with flame retardants can be prepared and processed as dry blends. When reinforcing with flame retardants, the same aspects must be observed as when working with other fillers. Figure 9.20 shows the essential parameters such as the diffuse reflectance and the process window.

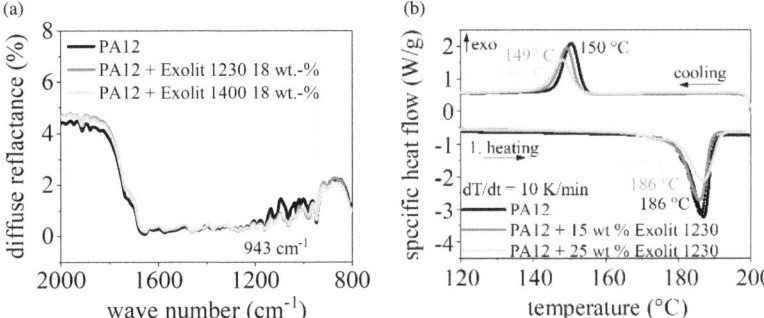

Fig. 9.20 Diffuse reflectance as a function of different flame retardants **a** and influence of flame retardants on the process window **b** for PA12 [13]

The pure PA12 powder shows a low diffuse reflection and thus a high absorption at the CO_2 laser wavelength of 943 cm^{-1}. Both flame retardants (Exolit 1200 and Exolit 1400) show a somewhat higher diffuse reflection at this wavelength. Nevertheless, a high and almost equal absorption can be observed for both blends of PA12 and Exolit. Therefore, the blends should be produced with similar energy densities as typical for PA12 powders and it is not necessary to adjust the process parameters to compensate in case of a change in diffuse reflectance. From the DSC curve, it can be seen that as the flame retardant content increases, the melting temperature does not change, although the enthalpy decreases due to a smaller amount of plastic material that can melt and crystallize. The onset and peak maximum of crystallization decrease slightly with increasing flame retardant content. Thus, the additive does not act as an initial nucleus. Nevertheless, the reduced crystallization temperature leads to a change in processing behavior and the theoretical processing window. In Fig. 9.21, the effects of the addition of the flame retardants at the component level are shown in the form of the LOI (limiting oxygen index) values and the classification according to UL-94 [13].

Fig. 9.21 LOI and UL-94 results of the laser-sintered specimen for different flame retardant concentrations [13]

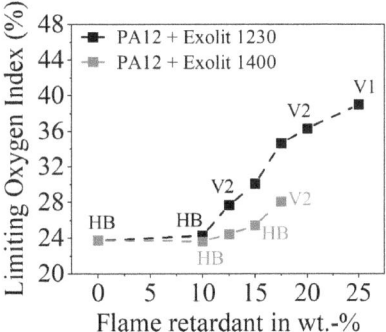

At low filler contents, the LOI remains at a constant level of 24%, regardless of the flame retardant used. At filler contents above the threshold value of 10 wt.%, the LOI increases. This behavior may be due both to a lower plastic content and to the increasing effect of the flame retardant. With the flame retardant Exolit 1230, higher LOI values can be achieved compared to Exolit 1400. Thus, the flame retardant Exolit 1230 exhibits better fire behavior with the same flame retardant content. Increasing filler contents lead to a more brittle material behavior, which manifests itself in a changed mechanical behavior that is decisive in laser sintering. The UL-94 classification indicates an unacceptable fire behavior for aerospace or electrical applications with a UL-94 rating of HB. A concentration of 10 wt.% flame retardant does not improve the fire behavior. As the flame retardant increases, the fire performance can be raised to a UL-94 rating of V-1 for Exolit 1230 flame retardant at a concentration of 25 wt.%. The measured LOI for this specimen is 39%. All tested flame retardant concentrations between 10 and 25 wt.% achieved a V-2 rating with constantly increasing LOI values. The V-1 rating means that the summed burning time per sample does not exceed 30 s, while at least one sample has a combined burning time for both ignitions of between 10 and 30 s. The target UL-94 rating would be V-0 and, to achieve this rating, all samples must burn for less than 10 s without flaming droplets as flaming drops automatically result in a V-2 rating for burn times up to 30 s. The laser sintered specimens with 25 wt.% Exolit 1230 show inhomogeneous firing behavior as three specimens would be rated V-0 and the other two would be rated V-1. Since this could be attributed to inhomogeneous density and local flame retardant concentrations, microscopic images of the tensile specimens were analyzed. When using Exolit 1400, the achievable LOI values and UL ratings are worse for the same amount of flame retardant [13].

9.5.2 Electrical Conductivity

Another functionalization option in current research is the use of the nozzle system for the local introduction of filler systems. By using electrically conductive fillers such as Cu-spheres and Sn63Pb37 solder powder in combination with the possibility of local injection, electrically conductive traces can be deposited and sintered in the process. The use of pure copper fillers is not practical from a process point of view because of the strong reflection behavior, since no energy coupling of the laser into the powder bed can take place outside the deposited copper track. The solder powder used has a similar melting temperature field to the PA 12 powder. The influence of the nozzle system parameters listed in Chap. 3, such as the distance between the nozzle and the powder bed, on the formation of the height profile and the geometric shape of the track is shown in Fig. 9.22 [10].

At a distance of approximately 0 mm, the track is deposited on the powder bed surface, while with increasing distance, the track is introduced into the powder bed surface. In the process, some of the powder bed surface particles are displaced along the track spreading direction and thrown up as a "hill" along the track. At short

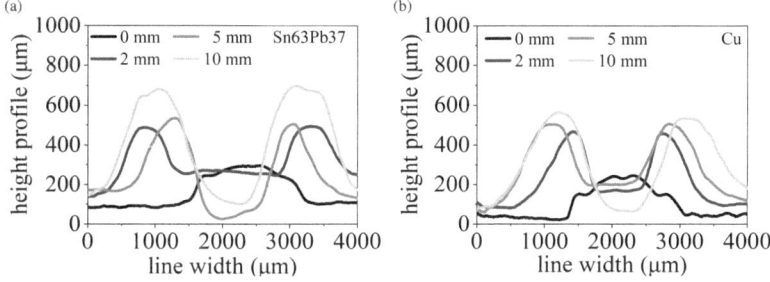

Fig. 9.22 Height profiles resulting from the distance of the nozzle to the powder bed in comparison for **a** solder (Sn63Pb37) and **b** copper [10]

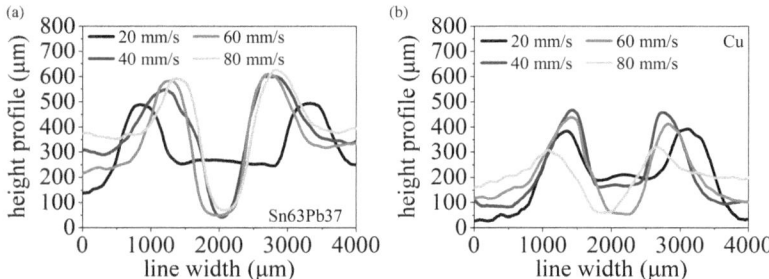

Fig. 9.23 Height profiles resulting from the traversing speed in comparison for **a** solder (Sn63Pb37) and **b** copper [10]

distances, the geometry of the track is U-shaped in cross-section, and the "hills" thrown up are relatively small in proportion. As the spacing increases, the geometry becomes v-shaped and the "hills" thrown up assume a maximum, whereby the deposition behavior of the track is identical for both materials. For the deposition speed parameter, a similar picture emerges, as shown in Fig. 9.23. With increasing speed, the track geometry changes from a U-shape to a V-shape and here, too, the laterally raised "hills" can be seen along the track. Unlike the gap, however, it is not possible to deposit them on the powder bed. With regard to the individual layers produced, a very poor bonding behavior of the track to the individual layer can be seen for the deposit on the powder bed surface [10].

The exposure melts the "hills" and thus flattens them significantly. After exposure, the height differences are in the range of 0.1–0.2 mm and come very close to the 0.1–0.15 mm targeted for the selective laser sintering process. The electrical conductivity of the manufactured, sintered coils can be analyzed by means of four-point electrode measurements in accordance with DIN EN ISO 3915 while the electrical conductivity can be verified for the discharged solder traces. The measured electrical resistance is normalized to the measured distance, and values between 0.02 and 0.17 Ω/mm are obtained. No electrical conductivity can be detected for the removed copper traces due to the greater spacing of the particles seen in Fig. 9.24, which means that no

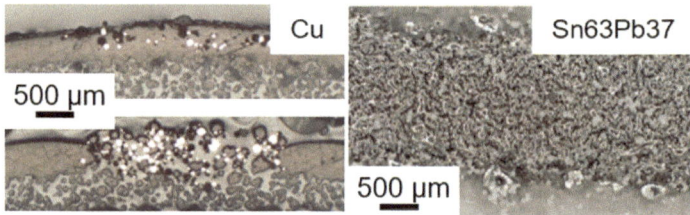

Fig. 9.24 Microscopic image of two copper traces in cross section (top) with 20 vol.-% Cu, (bottom) 50 vol.-% copper (left) and a solder trace in plan view [10]

network can be formed that exceeds the percolation threshold [10]. In contrast, the solder particles that lie close together are partially fused and sintered, resulting in an electrically conductive network.

9.5.3 Extension of Process Limits

The process limits can be extended by the addition of fillers, which represents a further possibility for functionalization. For this purpose, material systems are modified in situ and a powder mixture with aliphatic PA12 and p-aminobenzoic acid (pABA) is prepared. The thermal properties determined using DSC, shown in Fig. 9.25, show a significant effect of the p-aminobenzoic acid content on the melting and crystallization characteristics of the aliphatic PA12. The pABA content correlates with the melting and crystallization peaks, with both characteristic values being

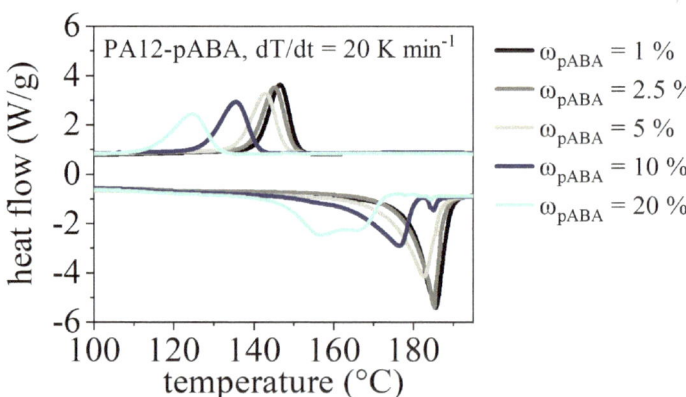

Fig. 9.25 Adapted thermal characteristics of powder blends exhibiting varying mass fractions of pABA [12]

Fig. 9.26 Depiction of the relation of the engineering stress and corresponding engineering strain in dependence of the fraction of pABA [12]

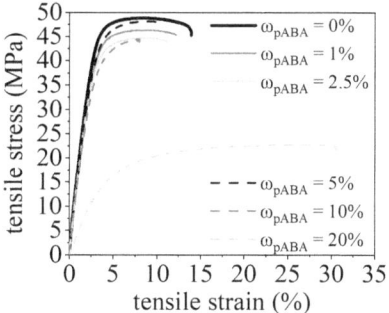

reduced with increasing content, allowing to reduce required isothermal processing temperatures [12].

The required processing temperatures can be reduced to significantly lower values of 141.5 °C. A higher proportion of pABA also leads to changes in the enthalpy of melting and crystallization, associated with an extended thermal melting range.

The mechanical properties are slightly reduced with an addition of less than 20 wt% pABA (Fig. 9.26). With an addition of more than 20 wt.%, there is a clear, significant decrease in tensile strength and modulus of elasticity. In contrast, the elongation at break can be significantly increased to values of 24.65% ± 2.87%, yielding ductile components. Compared to unreinforced polyamide, the characteristic value is thus almost doubled [12].

9.6   Conclusion

The integration of fillers and additives enables the considerable extension of part functionalities in powder bed fusion processes through the global as well as spatially selective deposition. Regarding material processing properties, there is a clear influence of the filler form and concentration on the optical properties such as diffuse reflection or the thermal properties such as heat capacity. Both properties are strongly dependent on the underlying material. The powder properties, such as the flowability, are also negatively influenced by the incorporation of fillers. An improvement of the flow properties can be achieved by other mixing strategies, such as the precipitation of the fillers into the polymer particles. With regard to the process properties, it is shown that, taking into account the compatibility of the material systems, it is possible to process both semi-crystalline blends and blends with a semi-crystalline and amorphous component. A common process window can be defined by evaluating the DMA of both components and the energy density can be used to influence the porosity of filler-reinforced material systems within certain limits. In addition, a critical fiber volume concentration can be defined at which the bonding of the fibers to the matrix is no longer sufficient for a compact component. The general possibility of locally introducing filler systems by means of a nozzle system and an x-y positioning system is given. At the component level, the geometry of blends can be influenced by the mixing ratio of the two components. For fillers, an influence on the geometry can also be observed both with regard to the di-corner of the individual layers and at the component level. The mechanical properties are negatively influenced to a large extent by the addition of fillers, so that both the tensile strength and the elongation at break decrease significantly, while an increase can be achieved in the Young's modulus. Here, too, an improvement can be achieved by different blending strategies. The general tendencies in mechanical behavior occur with both PP and PA12 and indicate matrix- or plastic-independent behavior while the thermal properties are comparable to those of injection molding as they depend on the filler mold. In addition, however, attention must also be paid to the job layout and the placement of the components in the build space. With regard to possible functionalization, the addition of flame retardants shows a significant decrease in flammability with good processability at the same time. By adding p-aminobenzoic acid to PA12, the process limits can be extended and the process window shifted. Additionally, the local introduction of solder particles makes it possible to realize electrically conductive tracks.

References

1. Saechtling Kunststoff Taschenbuch. Hanser, München, 31 edn (2013)
2. Elsner, P.: DOMININGHAUS - Kunststoffe: Eigenschaften und Anwendungen, 8th edn. VDI-Buch. Springer, Dordrecht (2011)
3. Greiner, S., Wudy, K., Lanzl, L., Drummer, D.: Selective laser sintering of polymer blends: bulk properties and process behavior. Polym. Testing **64**, 136–144 (2017)
4. Lanzl, L., Drummer, D.: Gefüllte systeme für die additive fertigung: Erhöhung der reproduzierbarkeit durch innovative mischstrategien. In: 7. Industriekolloquium des SFB 814
5. Lanzl, L., Drummer, D.: Process behavior of short glass fiber filled systems during powder bed fusion and its effect on part dimensions. Polymers **13**(18) (2021)
6. Lanzl, L., Greiner, S., Wudy, K., Drummer, D.: Glass-filled polypropylene-systems für enhancing reproducibility during selective laser beam melting and effect of the mixing strategy on powder properties. In: Proceedings of 7th International Conference on Additive Technologies
7. Lanzl, L., Wudy, K., Drummer, D.: The effect of short glass fibers on the process behavior of polyamide 12 during selective laser beam melting. Polym. Testing **83**, 106313 (2020)
8. Lanzl, L., Wudy, K., Greiner, S., Drummer, D.: Selektives laserstrahlschmelzen von glasfasergefüllten polymersystemen. In: 5. Industriekolloquium des SFB 814, pp. 35–53
9. Lanzl, L., Wudy, K., Greiner, S., Drummer, D.: Selective laser sintering of copper filled polyamide 12: characterization of powder properties and process behavior. Polym. Compos. **40**(5), 1801–1809 (2019)
10. Lindbüchl, M., Drummer, D.: Local integration of electrically conductive paths using an in situ x-y positioning system in selective laser sintering. In: 38th International Conference of the Polymer Processing Society
11. Rothfelder, R., Lanzl, L., Selzam, J., Drummer, D., Schmidt, M.: Vibrational microfeeding of polymer and metal powders for locally graded properties in powder-based additive manufacturing. J. Mater. Eng. Perform. **30**(12), 8798–8809 (2021)

12. Schlicht, S., Drummer, D.: Eutectic in situ modification of polyamide 12 processed through laser-based powder bed fusion. Materials (Basel, Switzerland) **16**(5) (2023)
13. Schneider, K., Wudy, K., Drummer, D.: Flame-retardant polyamide powder for laser sintering: powder characterization, processing behavior and component properties. Polymers **12**(8) (2020)
14. Wudy, K., Drexler, M., Drummer, D.: Selektives laserstrahlschmelzen von polymer-blends: Prozess- und werkstoffanforderungen. RTeJournal-Fachforum für Rapid Technologie
15. Wudy, K., Lanzl, L., Drummer, D.: Selective laser sintering of filled polymer systems: bulk properties and laser beam material interaction. Phys. Procedia **83**, 991–1002 (2016)

Dietmar Drummer succeeded Prof. Ernst Schmachtenberg as Head of the Institute of Polymer Technology at Friedrich-Alexander-Universität Erlangen-Nürnberg on May 1, 2009. He was the spokesperson for the Collaborative Research Center 814 Additive Manufacturing, which focused on process understanding, development, and process-adapted material characterization. Furthermore, he heads the Polymer Group of Neue Materialien Fürth GmbH and 2 Keylabs at the Bavarian Polymer Institute.

Chapter 10
Laser Beam Melting of Plastics with Reactive Liquids

Robert Setter and Katrin Wudy

10.1 Introduction

Novel manufacturing technologies aim to overcome the restrictions of state-of-the-art processing to ignite market-changing products. One example is represented by multi-material additive manufacturing (AM). The potentials of state-of-the-art AM are, for example, complex part production, light-weight design, and the absence of tooling. By extending the material spectrum through multi-material AM, new potentials are added such as strategically altered mechanical properties and/or the introduction of secondary part properties such as thermal and electrical conductivity. In the field of plastics, common applications for multi-material AM are hard-soft structures such as smart structures and 4D-printing. Despite the numerous potentials of multi-material AM, one decisive aspect is yet to be addressed, namely the combination of thermosets and thermoplastics within functional parts [9, 25]. Based on the available commercial systems and the state of research, multi-material AM of polymers can be divided into three categories:

1. Thermoset multi-material AM
2. Thermoplastic multi-material AM
3. Hybrid multi-material AM of thermosets and thermoplastics.

State-of-the-art thermoset-based multi-material AM systems are predominantly located within the fields of material jetting (MJ) and vat photopolymerization. Investigations by Bartlett et al. [3] and Boopathy et al. [4] concentrate on the combination of rigid thermosets and elastomer-like thermosets with MJ to be used for robotic applications and for increased energy absorption. Sakhaei et al. [16] also utilize MJ

R. Setter · K. Wudy (✉)
Technische Universität München, Professorship of Laser-based Additive Manufacturing, Freisinger Landstraße 52, 85748 Garching, Germany
e-mail: katrin.wudy@tum.de

© The Author(s) 2025
D. Drummer and M. Schmidt (eds.), *Progress in Powder Based Additive Manufacturing*, Springer Tracts in Additive Manufacturing,
https://doi.org/10.1007/978-3-031-78350-0_10

modeling for the improvement of the strength and overall performance of solely thermoset parts for ratchet-like mechanisms while Choi et al. [5] introduce a rotating multi-vat system to the vat photopolymerization process for the combination of different thermosets within a single layer.

Thermoplastic multi-material AM is predominantly represented by technologies such as material extrusion (MEX) and laser-based powder bed fusion of plastics (PBF-LB/P). With MEX, different thermoplastics can be combined by using multiple print heads [2] and/or static mixing units [12, 15] within the nozzle. In the case of PBF-LB/P, current investigations and companies focus on electrophotographic powder deposition [13] and pneumatic recoating technologies [1] to selectively introduce variations of thermoplastic powders to the powder bed to alter the mechanical and conductive properties of future parts.

Hybrid multi-material AM of thermosets and thermoplastics represents the subsequent combination of AM with state-of-the-art processes such as molding technologies. Investigations by Szenbenyi et al. [21] and Dorigato et al. [7] demonstrate the combination of additively manufactured thermoplastic structures with an epoxy matrix. The thermoplastic structures are printed first and are subsequently infiltrated by epoxy. However, hybrid multi-material AM is highly restricted by the attainable complexity of the thermoset components since demolding must be considered.

Evaluation of the state of research and technology leads to the conclusion that there is a distinct need for a novel AM technology to simultaneously process thermosets and thermoplastic. Therefore, the Professorship of Laser-based Additive Manufacturing at the Technical University of Munich investigates the combination of thermoplastics with thermosets represented by reactive liquids within a single AM process as part of CRC 814 "Additive Manufacturing" [19–24]. A first concept of the required AM process is based on the integration of a drop-on-demand print head system within PBF-LB/P. The process was named Fusion Jetting (FJ) since it represents the combination of two state-of-the-art AM technologies: Laser-based Powder Bed Fusion of Plastics (PBF-LB/P) and Binder Jetting (BJT). The use of a print head enables selective and precise deposition of the reactive liquid within the thermoplastic powder bed. As a result, the attainable complexity of the thermoset components of multi-material parts is nearly unlimited. A schematic representation of the process steps is depicted in Fig. 10.1. The process starts with recoating of a new powder layer followed by selective deposition of the reactive liquid within the thermoplastic powder bed. After complete infiltration, the curing of the reactive liquid is performed with a UV lamp. Eventually, the powder bed is heated up with infrared lamps and selectively melted with CO_2-laser radiation. Following this, the process steps are repeated until the multi-material part is finished.

The vision of this research project can be summarized by the successful demonstration of AM of multi-material polymer parts which combine thermoplastics and thermosets with subsequent experimental analysis. The approach for reaching this aim can be subdivided into four interconnected sectors:

1. Material characterization and qualification
2. Process design

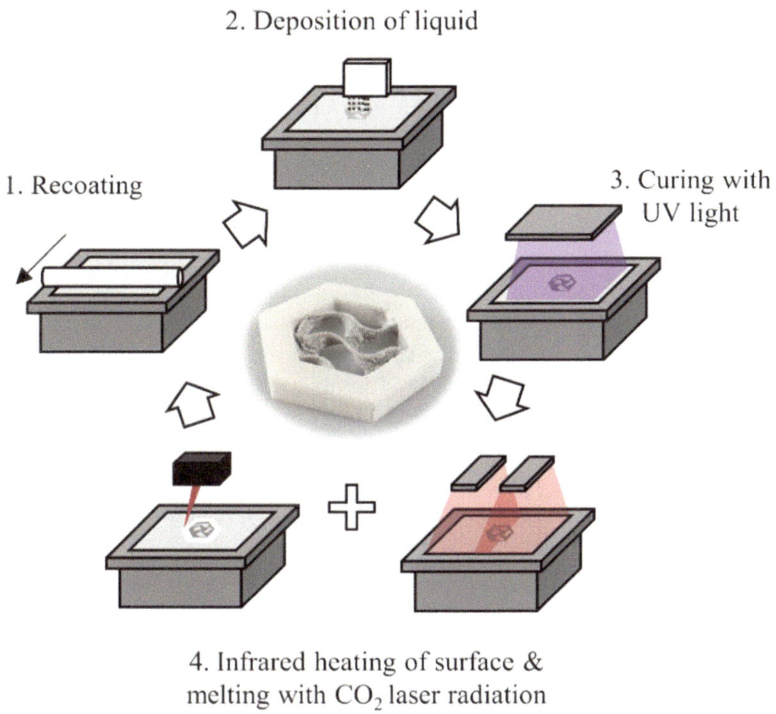

Fig. 10.1 Schematic process steps of the "Fusion Jetting" process

3. Processing of complex multi-material parts
4. Extension of the material spectrum.

The material characterization concentrates on the experimental analysis and understanding of potential materials for future multi-material AM in their individual and combined states. This includes the analysis of the UV-initiated curing behavior of the reactive liquids, as well as the infiltration behavior of the reactive liquids within the powder bed. Parallel to that, the process design concentrates on the utilization of all depicted process steps of Fig. 10.1. The knowledge gathered regarding the materials and the process design is then used to generate complex multi-material parts. These parts are analyzed concerning the interaction zone between the individual materials and the impact of the variation of process parameters on the mechanical properties and failure behavior. In the final step, the material spectrum is broadened with the target to implement secondary material properties within the multi-material parts, for example, to increase the electrical or thermal conductivity. In the following, the scientific gain of knowledge of each sector is summarized, including exemplary results.

10.2   Material Characterization and Qualification

The material characterization concentrates on the evaluation and qualification of different thermoset and thermoplastic systems for use in the FJ process. The process-related requirements for each polymer class are depicted in Table 10.1. Based on the requirements, strategies for material characterization and qualification were derived. The thermoset component requirements are predominantly influenced by the prevailing environmental conditions of PBF-LB/P. To enable the error-free processability of the thermoplastics with PBF-LB/P, the temperature of the process chamber must be elevated, whereby the chamber temperature (usually above 100 °C) is dependent of the crystallization and melting behavior of the thermoplastic. For successful utilization and process stability, the liquids must withstand elevated temperatures and demonstrate sufficient curability in the high-temperature environment. To minimize the risk of unwanted or premature temperature-related curing, the selection of potential thermoset systems was reduced to photopolymers. Photopolymers enable pinpoint control over the starting point of the curing reaction, which is triggered by UV light. To minimize future layer times, the selection was further reduced to acrylate systems. Compared to epoxy systems, acrylates demonstrate higher curing speeds [8, 11]. Accordingly, two commercial acrylate systems from vat photopolymerization were selected for further analysis: UV DLP Hard and UV DLP Firm (Photocentric 3D, UK). The materials contain diacrylates and dimethacrylates as their proprietary building elements and are referred to as Acrylate A (UV DLP Hard) and Acrylate B (UV DLP Firm).

10.2.1   Thermal Stability of the Thermosets

In the first step, the thermal stability of both acrylates was analyzed with thermogravimetric analysis (TGA) coupled with Fourier-transform infrared spectroscopy

Table 10.1 Process-related requirements for the material selection of thermosets and thermoplastics

Thermoset	Thermoplastic
- Liquid	- Powder
- UV-curable	- Low melt viscosity
- Low viscosity for printability with print head and fast infiltration	- Homogeneous packing for uniform infiltration behavior
- No spontaneous thermal curing	- Large temperature interval between melting and crystallization
- Fast curing speeds	
- Reasonable thermal stability to withstand temperatures required for PBF-LB/P processing	- Sufficient flowability and average particle diameter to achieve uniform layers at 100 μm

(FT-IR). The materials are analyzed in a liquid state as well as in a cured solid state. With FT-IR, the characteristic spectrum or "fingerprint" of the evolving gases of each material during temperature increase was analyzed. The spectra were then compared to characteristic database values for the determination of the volatile components. Exemplary results for both materials are depicted in Fig. 10.2 [19].

The TGA results indicate that the liquid samples show significant mass losses above temperatures of 120 °C. Acrylate A shows an increased mass loss compared to Acrylate B, which was correlated to the presence of additional "multi-functional acrylate" within liquid Acrylate B and therefore a lower presence of potentially more unstable diacrylates. The eluted materials showed the highest similarity to the solvent "2-propenoic acid, 2-propenyl ester" (or "Allyl acrylate") for both materials analyzed [6]. The fingerprint was determined at the peak of each spectrum. The TGA results of the cured samples indicated a reduced mass loss, with evolving gases of carbon dioxide and evaporating water. Accordingly, for future processing at temperatures above 120 °C, the curing should be performed as fast as possible to minimize the deterioration of the acrylates in a liquid state [19].

10.2.2 Curing Kinetics of the Thermosets

The analysis of the curing behavior was performed with differential scanning calorimetry (DSC) with the final aim to model the curing kinetics and predict the curing behavior during processing. The modeling of the curing kinetics facilitates two aspects: the ability to draw conclusions concerning the types of chemical reactions present and to predict the curing behavior during the FJ process. For the DSC measurements, UV light guides were used to cure liquid samples of Acrylate A and Acrylate B. UV-intensities of 7.5 mW/cm^2, 15 mW/cm^2, 30 mW/cm^2, and 60 mW/cm^2 at temperatures of 30, 90, and 150 °C were analyzed. The temperatures of 90 and 150 °C were chosen based on the processing temperature ranges of common polymer powders for PBF-LB/P processing such as TPEs like TPU and PEBA. The typical melting temperatures of these materials are located within the temperature interval and are at 136 °C and 150 °C respectively. As described in [19], the DSC results indicate an increase in the curing speed for increasing temperatures and increasing UV-intensities. However, due to thermal decomposition, the beneficial effect of the temperature on the processing speed is restricted. Based on the DSC results, the conversion or degree of cure for each measurement was determined:

$$\alpha = \frac{\Delta H}{H_{\text{total}}} = \frac{\int_{t_{\text{start}}}^{t} \dot{Q}\, dt}{H_{\text{total}}} \tag{10.1}$$

The achieved conversion α is the ratio between the partially emitted exothermic amount of specific energy ΔH and the total amount of reaction enthalpy H_{total}. In the case of DSC or UV-DSC, ΔH is described as the integral of the specific heat flow

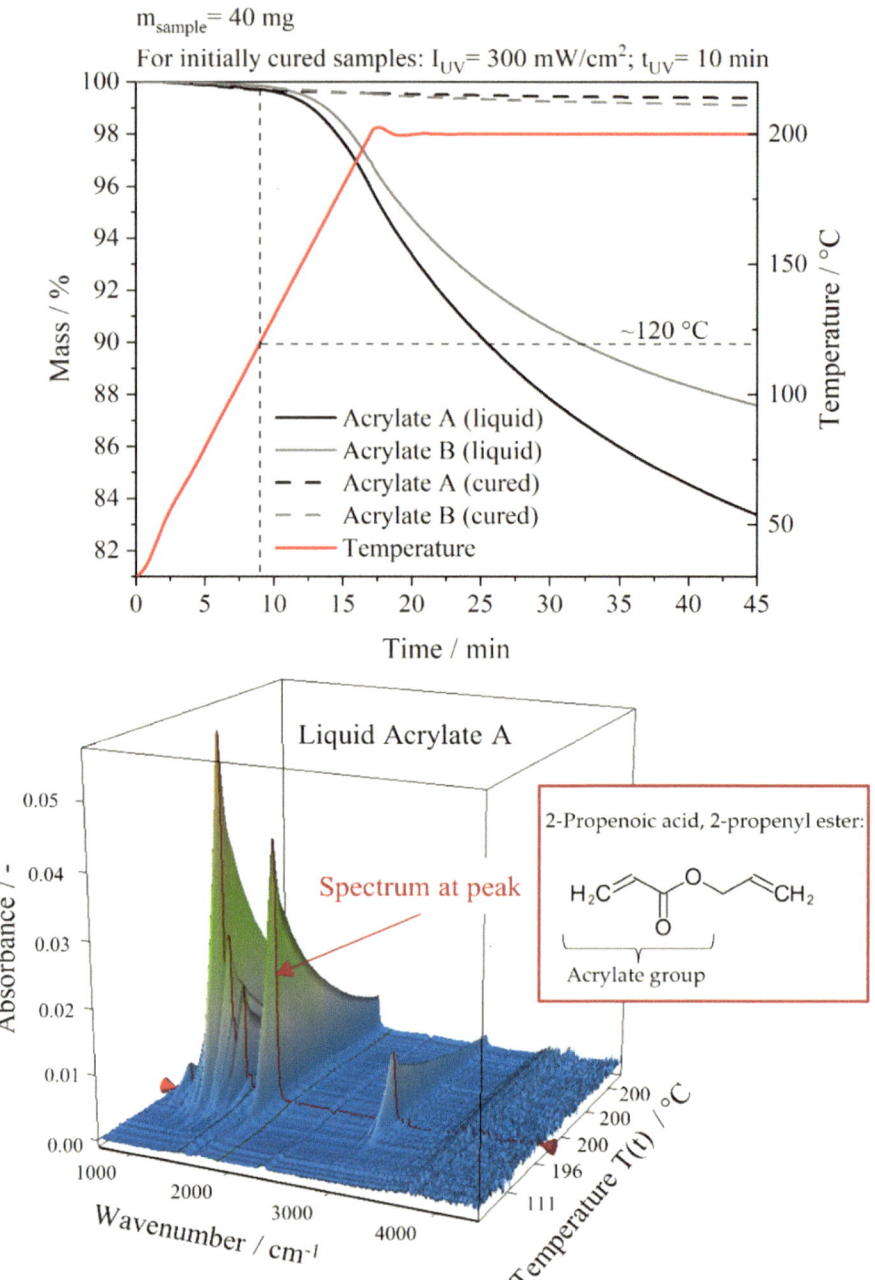

Fig. 10.2 TGA analysis of Acrylate A and Acrylate B with the exemplary in-situ FT-IR results of the evaporating gases from liquid Acrylate A samples at elevated temperatures [19]

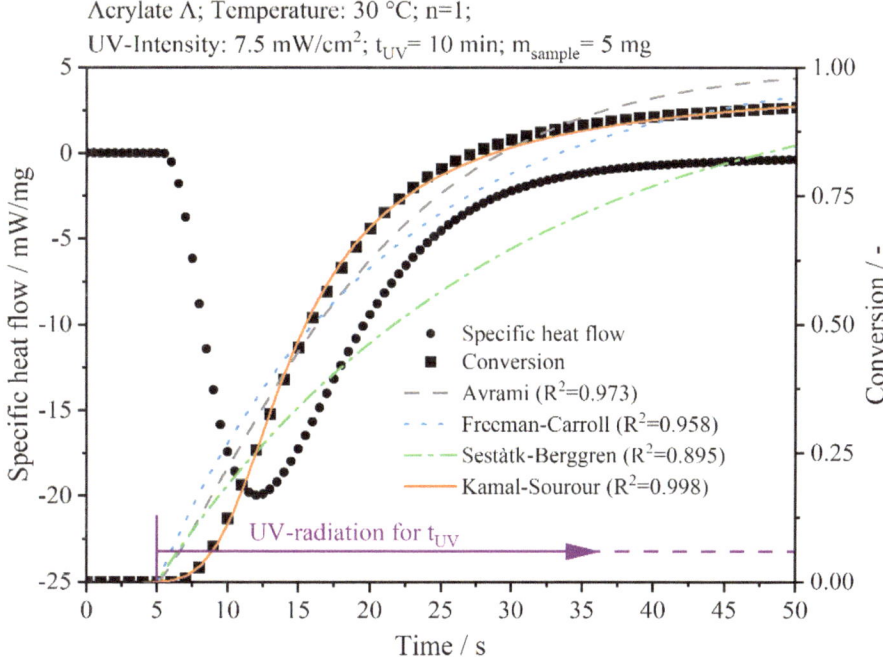

Fig. 10.3 Comparison of different kinetic models to mathematically describe DSC results of Acrylate A at a temperature of 30 °C and a UV-intensity of 7.5 mW/cm2 [18, 19]

$\dot{Q}$ over the time t from the start of the reaction t_{start}. Exemplary results are presented in Fig. 10.3 for a UV-intensity of 7.5 mW/cm^2 and a temperature of 30 °C.

Based on the results, different kinetic models were fitted to the experimental data and evaluated towards a maximum coefficient of determination. First, only single measurements were analyzed to limit the number of models to the most promising candidates. The Kamal-Sourour model and the Avrami model showed the highest coefficients of determination of all models analyzed and were selected for further investigations [18, 19]. After the single measurement analysis, multiple DSC measurements were analyzed simultaneously in clusters of identical temperatures (30, 90, and 150 °C) and identical UV-intensities (7.5 mW/cm^2, 15 mW/cm^2, 30 mW/cm^2, and 60 mW/cm^2). This approach is based on the recommendations developed by the Kinetics Committee of the International Confederation for Thermal Analysis and Calorimetry (ICTAC) [22]. The modeling results identified the Kamal-Sourour method as superior to the Avrami method for the sufficient depiction of the curing kinetics. Furthermore, the best results were achieved by analyzing the measurements in clusters of identical temperatures. Therefore, the present curing mechanism most likely consists of two reaction pathways: an autocatalytic reaction and a n-th order reaction. Introducing an Arrhenius-like exponential decay function to describe the reaction rate constant led to the following expanded Kamal-Sourour equation for ideal depiction of the curing kinetics:

$$\frac{d\alpha}{dt} = A \cdot e^{-\frac{D_1}{I}} \cdot (1-\alpha)^{n_{gen}} + A \cdot K_{cat,gen} \cdot e^{-\frac{D_2}{I}} \cdot \alpha^{m_{gen}} \cdot (1-\alpha)^{n_{gen}} \qquad (10.2)$$

$\frac{d\alpha}{dt}$ is the reaction rate that is dependent on the conversion α. $K_{cat,gen}$ is a weight factor to represent the contribution of the autocatalysis reaction and I is the UV-intensity. m_{gen} and n_{gen} are the temperature-independent reaction orders of the auto-catalytic and n-th order reaction respectively. $D_{1,2}$ are referred to as exponential factors and A as a pre-exponential factor that is dependent on the temperature. With this equation, it was not only possible to successfully model the curing behavior based on experimental results, but further reasonably predict the curing behavior for not measured UV-intensities at isothermal temperatures. Exemplary results are depicted in Fig. 10.4 with predictions for non-measured UV-intensities of 45 mW/cm^2. The ability to predict the curing behavior implies a high relevance for the FJ process. If it is desired to pair the analyzed acrylates with alternative thermoplastic polymer

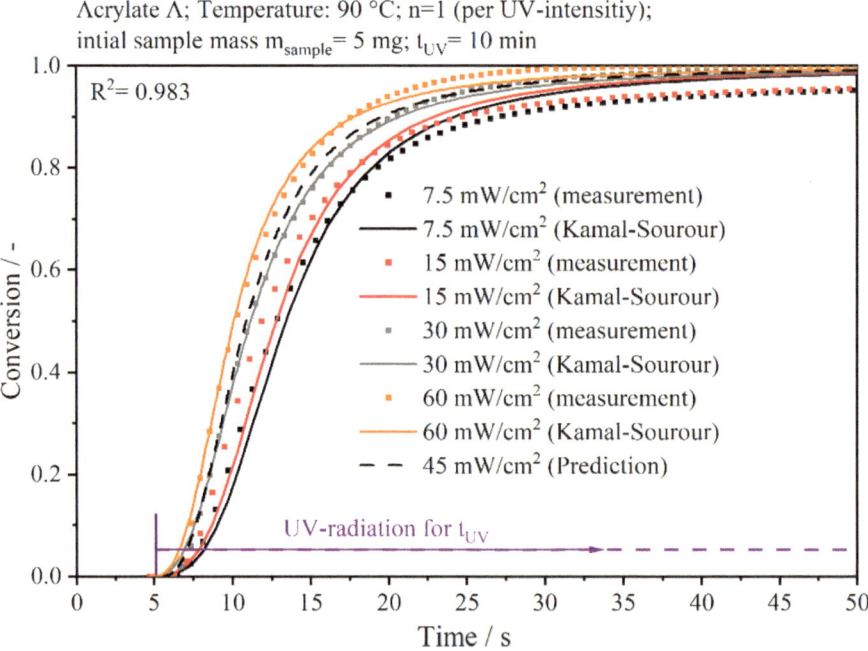

Fig. 10.4 Modeling and prediction of the curing kinetics of Acrylate A at different UV-intensities based on clustered measurements at an isothermal temperature of 90 °C [18]

materials and, therefore, alternative isothermal processing temperatures, it is possible to predict the curing behavior of the acrylate without any additional experimental effort [18].

10.2.3   Infiltration of the Thermoset Within the Powder

The investigation of the infiltration behavior of Acrylate A and Acrylate B within the thermoplastic powder bed is based on the experimental analysis and mathematical modeling of the droplet height over time at different isothermal temperatures. Both are crucial parameters for the right timing of the FJ process steps and the potential necessity of delay times to guarantee a uniform print. Incomplete infiltration would lead to potential geometric deviation and process errors due to collisions with the recoating unit. For experimental determination of the droplet height, as well as the contact angle, an OCA 20 optical tensiometer along with an oven chamber and a high-speed camera (DataPhysics Instruments GmbH, Filderstadt, Germany) was utilized. To mathematically describe the droplet height over time, a model based on the Washburn equation [23] was used. The model describes the droplet height over time $h(t)$ dependent on the penetration length L with the initial droplet height of a liquid h_0 on a powder bed for the boundary condition of $L(t) \leq h_0$:

$$h(t) = h_0 - |L| = h_0 - \sqrt{\frac{\alpha \cdot D_h \cdot \cos(\phi) \cdot \sigma_l \cdot t}{4\eta}} \qquad (10.3)$$

D_h is the pore diameter dependent on the Sauter mean diameter, which is an average particle size defined as the diameter of idealized identical spheres with a volume-to-surface area ratio identical to the volume-to-surface area ratio of a particle collective. ϕ is the contact angle between tube wall and liquid. σ_l is the surface tension of the fluid and η the dynamic viscosity. α represents a material-specific form factor, which most likely correlates to particle distribution-related packing. Exemplary experimental results along with the mathematical modeling based on Eq. (10.2) are depicted in Fig. 10.5 for Acrylate A.

The powders analyzed for this investigation were two different PA11 powders (provided by subproject A1) and commercial thermoplastic elastomer (TPE) powder (type: PEBA 2301, EOS, Germany). The TPE represents an ideal partner to be combined with the rigid acrylate systems for future AM of hard/soft structures. The PA11 powders were produced through chemical precipitation and are referred to as PA11-5 and PA11-20, depending on the initial mass contents. The mean diameters of the PA11 powders were specifically tailored to achieve approximate mean diameter values of half and double the values of the TPE powder (PA11-5: $d_{50.3} = 35\,\mu\text{m}$; PA11-20: $d_{50.3} = 77\,\mu\text{m}$; PEBA 2301: $d_{50.3} = 140\,\mu\text{m}$). For an isothermal temperature of 60 °C, the infiltration speed indicates a strong dependency on the type of material used. Compared to the PA11 powders, the acrylate shows slowed infiltration

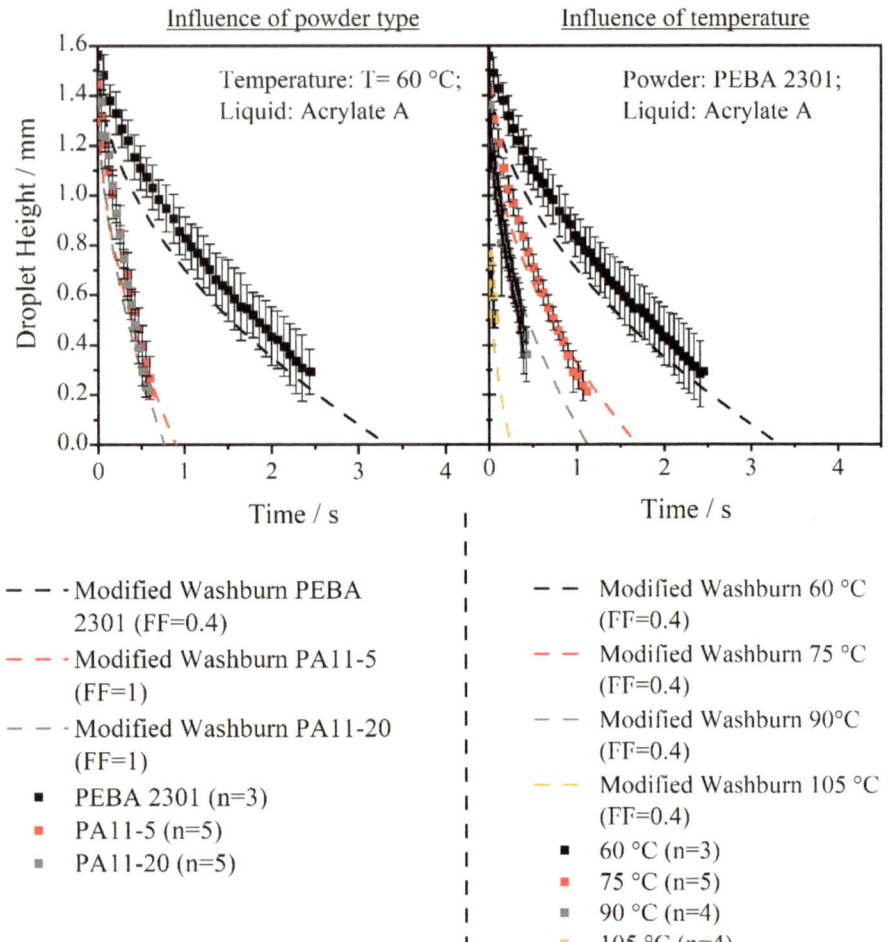

Fig. 10.5 Influence of the powder type and the temperature on the droplet height over time of Acrylate A [17]

behavior within the TPE powder. A hypothesis for this behavior is based on the particle shape as the milled TPE particles have sharper edges and irregularities compared to the cauliflower-like chemically precipitated PA11 particles. Therefore, the PA11 powders have a more homogeneous powder distribution with random close packing, while the TPE powder has a more heterogeneous distribution with irregular packing. As previously stated by Zhou et al. [26] and Hapgood et al. [10], irregular packing contains more macro-voids which hinder drainage and cause extended infiltration pathways. Another decisive aspect is represented by the temperature, which significantly increases the infiltration speed due to the lowering of the viscosity. Figure 10.5

shows the droplet height over time of Acrylate A within the TPE powder at isothermal temperatures of 60, 75, 90, and 105 °C. The starting temperature of 60 °C was gradually increased until the maximum achievable temperature of the experimental setup was reached, which is close to the approximate PBF-LB/P processing temperatures of TPEs such as PEBA and TPU (melting temperatures of 136 °C and 150 °C respectively). The results show that for maximum temperatures, the infiltration time was significantly reduced from more than 3 s to lower than 0.5 s, which represents an important takeaway for the calculation and prediction of future machine movements and minimum layer times. Based on the results, the temperature and the particle shape-dependent powder distribution were declared as primary factors for the infiltration speed. In comparison, the mean particle diameter was declared as a secondary aspect. For mathematical modeling of the infiltration of acrylates within TPE powder, a universal form factor of 0.4 was determined whereas for the description of the infiltration of the acrylates within PA11 powders, no form factor was necessary ($\alpha = 1$). This supports the hypothesis of the necessity of a form factor for powders with a heterogeneous powder distribution with irregular packing. In summary, the results of the infiltration behavior show high potential for pairing TPEs with acrylate systems in FJ processing. It could be shown that for temperatures close to the processing window of potential TPE materials (> 100 °C), the infiltration of the acrylate is rapid with infiltration times lower than 1 s and no spontaneous or unwanted curing due to the temperature exposition occurs. This further indicates that additional delay steps to guarantee complete infiltration are most likely not necessary. Further, the mathematical description of the droplet height over time facilitates the prediction of the infiltration speed of alternative temperature profiles once a material-specific form factor is found [17].

The key findings from this section can be summarized as follows:

- Acrylate photopolymers are selected as the liquid component for FJ processing due to their rapid curing speeds and sufficient thermal stability for surrounding temperatures up to 120 °C.
- Increased UV intensities and temperatures up to 90 °C result in increased curing speeds of the acrylates. This is beneficial for fulfilling the requirement of fast curing speeds to minimize the layer time of the FJ process.
- A modified version of the Kamal-Sourour equation demonstrates promising coefficients of determination (> 0.99) to describe the curing kinetics of the acrylates. The model was further used for plausible predictions of the curing behavior for not-measured temperatures and UV intensities, which is highly relevant for the FJ process. For processing alternative materials at material-specific isothermal processing temperatures, it is possible to predict the curing behavior of the acrylate without any additional experimental effort and plan the processing sequence accordingly.
- Thermoplastic elastomers are selected as the powder component for the FJ process due to their difference in mechanical properties compared to acrylates and therefore represent an ideal partner for future AM of hard soft structures.

- The acrylates demonstrated sufficient infiltration behavior within the TPE powder with infiltration speeds below 0.5 s at temperatures near the TPE processing temperature.
- The infiltration behavior was mathematically modeled with a modified version of the Washburn equation.

10.3 The "Fusion Jetting" Process

With the determination of the curing speeds, the demonstration of sufficient thermal stability of the acrylates at TPE process temperatures, and profound insights concerning the temperature-related infiltration speeds, it was possible to conceptualize the FJ process. To incorporate all process steps schematically depicted in Fig. 10.1, an experimental process chamber was conceptualized and designed. An entirely self-built print system was developed with temperature-sensitive components located outside the build chamber. The single process steps are depicted in Fig. 10.6. The machine is equipped with a CO_2-laser with a maximum power of 55 W (Coherent,

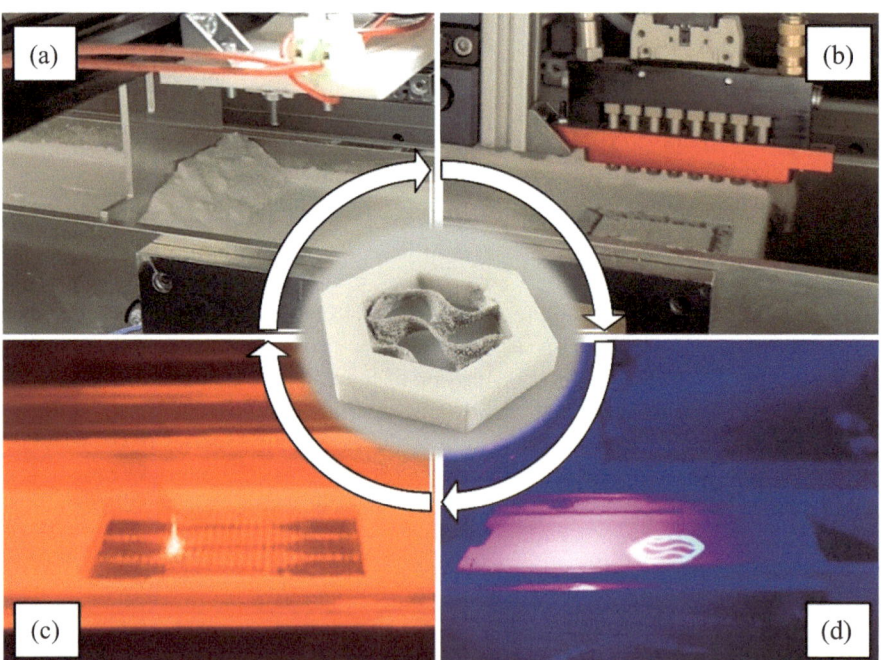

Fig. 10.6 Single process steps to generate hybrid multi-material polymer parts: **a** Recoating; **b** Material deposition; **c** UV-curing; **d** Laser exposure

US), a scanning unit with an f-theta lens (Scanlab, Germany), IR-heaters, chamber heating, a movable UV lamp, and a drop-on-demand print head system (Gyger, Switzerland). For controlling the machine movements and process parameters, the PLC hardware (Beckhoff, Germany) is orchestrated by a holistic machine control framework (Autodesk, US). The framework enables the unrestricted composition of sequential process steps in any desired sequence, for example, the sequential execution of multiple laser exposure, repeated deposition of liquids within a single layer, or the variation of the laser power in selected layers [20].

10.4   Fusion Jetting Process Design

The lessons learned from the material characterization enabled the printing of multi-material parts combining acrylate photopolymers and TPE. The acrylate component is represented by the previously-mentioned Acrylate A, which was selected due to its generally lower viscosity and improved infiltration behavior with high reproducibility compared to Acrylate B. As a representative of TPE, thermoplastic polyurethane powder (TPU) (type: TPU 1301; EOS, Germany) was selected to be combined with the acrylate. Preliminary single-layer experiments showed sufficient infiltration and curing behavior of Acrylate A within the TPU powder. The single-layer experiments further demonstrated promising attachment between the melted TPU regions and the incorporated Acrylate A to be qualified for further investigations. After the single-layer experiments, multi-layer experiments were first performed to understand the process design. Selected results are presented in Fig. 10.7 (left). The micrograph image shows the resulting material transitions after alternating execution of BJT and

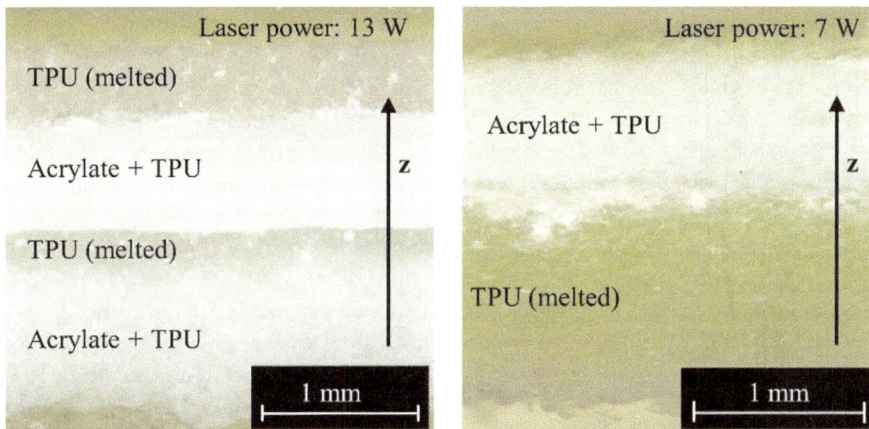

Fig. 10.7   (Left) Influence of the sequence of the process steps on the transition areas between TPU and acrylate regions (color of acrylate: white); (Right) Increased infiltration of acrylate within TPU regions with low laser powers (color of acrylate: white) [20]

PBF-LB/P, starting with BJT. PBF-LB/P was performed at a constant laser power of 13 W (Hatch distance: 250 μm; Laser speed: 2500 mm/s; Resulting energy density: 0.021 J/mm^2). The layer height of BJT was selected at 1 mm and for PBF-LB/P at 0.1 mm, whereby the latter represents a common layer height for commercial PBF-LB/P processing. The number of layers of BJT was restricted to a single layer while each PBF-LB/P procedure consists of five subsequent layers [20].

The first BJT layer displays a rough bottom surface due to unhindered infiltration of the acrylate within the powder bed (Fig. 10.7). Between the first BJT layer and the first PBF-LB/P layers, a nearly graded material transition is visible with a smooth transition between both materials. Due to the higher density of the already melted material within the first PBF-LB/P layers located beneath, the second BJT procedure demonstrates a distinct interface with decreased infiltration. The second BJT layer showcases higher saturation levels and increased acrylate density compared to the previous BJT step. Accordingly, between the second BJT layer and the second PBF-LB/P layers the material interaction is restricted. Further experiments indicated increased infiltration of acrylate within the melted TPU regions by using lower laser powers of 7 W (hatch distance: 250 μm; scan speed: 2500 mm/s; area energy density: 0.011 J/mm^2). An exemplary result is presented in Fig. 10.7 (right). This can be seen as a potential process strategy to counter delamination behavior, which was visible during the experimental analysis of the mechanical properties and the failure behavior described below [20].

To evaluate the impact of the introduction of acrylate reinforcements (Acrylate A) within TPU parts on the mechanical properties, three different tensile specimens were prepared. The designs are shown in Fig. 10.8. To compare the impact of the sequence of the process steps on the mechanical properties, seven layers of BJT were introduced in an alternating fashion (Type: Zebra7L) and an interconnected fashion (Type: Block7L) to the tensile specimen geometry. All specimens were processed with the same laser power of 15 W (hatch distance: 250 μm; scan speed: 2500 mm/s; area energy density: 0.024 J/mm^2). All specimen types are compared to TPU specimens without reinforcements. The evaluation of the stress-strain behavior shows that the introduction of acrylate reinforcements increases the Young's modulus depending on the process sequence and the filler content. The results for all specimen types are visualized in Fig. 10.8. The Block7L and Zebra7L specimens achieve increased average values of approximately 120 MPa and 100 MPa respectively compared to TPU specimens. Since both specimen types contain the same acrylate content, it can be concluded that the interlaminar connection of the acrylate-reinforced layers benefits stiffer material properties at small deformations and therefore higher Young's moduli [20]. The progressions of the tensile test results are visualized in Fig. 10.9 and the maximum stress and strain values of the TPU specimens exceed the maximum values of all reinforced configurations. In comparison to the Zebra7L specimens, the Block7L type reaches the highest stress values of all reinforced specimens. Consequently, interlaminar connections between the reinforced layers lead to the improvement of the mechanical properties and increased stress values. The TPU specimens and selected Zebra7L specimens show similar continuous curve progressions before specimen failure (Type 4 after DIN EN ISO 527-1). Block7L and

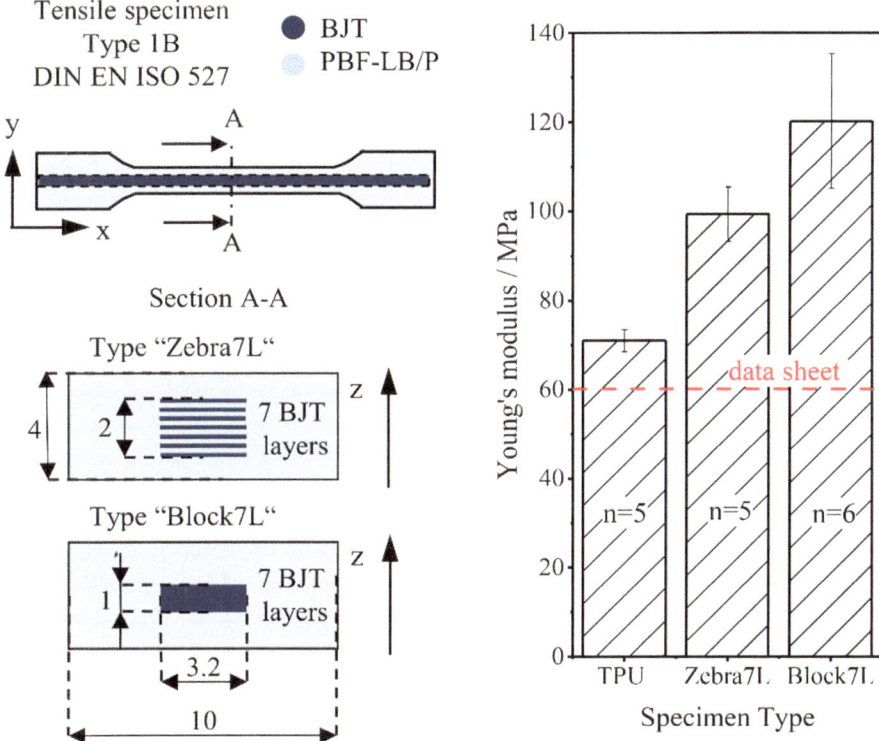

Fig. 10.8 (Left) Tensile specimen design with variation of the processing sequence and the acrylate content; (Right) Young's modulus of an acrylate-reinforced TPU specimen compared to a TPU-only specimen [20]

Zebra7L show segmented curve progressions which can be divided into four steps according to the failure mode depicted in Fig. 10.9. A characteristic sawtooth-shaped failure behavior is visible, which can be correlated to the initial failure of the acrylate (Fig. 10.9, image 1). The acrylate failure is followed by necking and partial failure of the TPU at regions between the remaining acrylate segments (Fig. 10.9, image 2). Following this, a significant drop in the stress is visible which is correlated to delamination between acrylate-reinforced layers and sole TPU layers (Fig. 10.9, image 3). For Zebra7L specimens, the final delamination is significantly reduced, representing a potential processing strategy for the reduction of delamination in consideration of the inherent reduction of the maximum stress [20].

The key findings from this section can be summarized as follows:

- Acrylate reinforcements in TPU tensile specimens lead to a significant increase of the Young's modulus (factor 2 compared to TPU). However, the maximum achievable stress and strain is reduced compared to that observed in TPU samples alone.

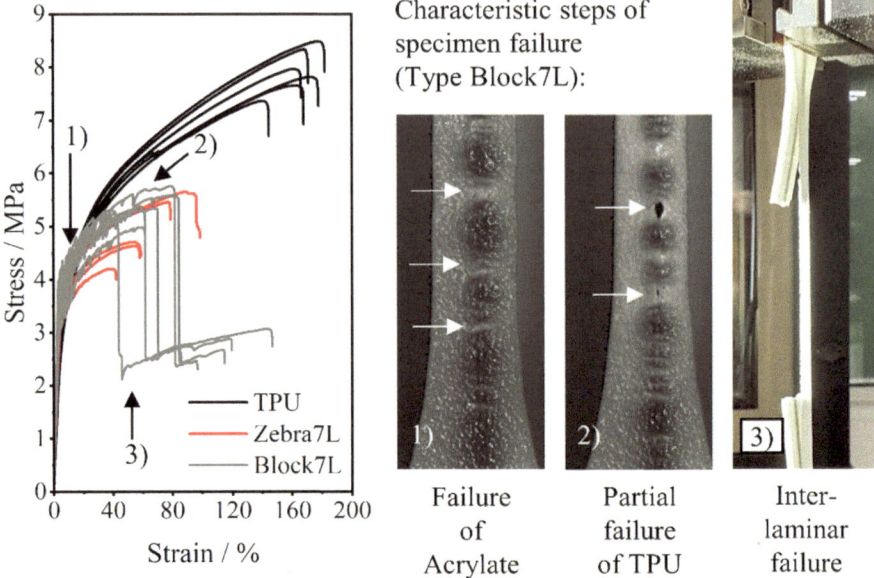

Fig. 10.9 Comparison of the stress-strain behavior between acrylate-reinforced TPU specimens and TPU-only specimens with characteristic steps of the failure behavior of acrylate-reinforced TPU Block7L specimens: **1** Failure of Acrylate; **2** Partial failure of TPU; **3** Interlaminar failure (color of acrylate: gray) [20]

- The multi-material tensile specimens show a characteristic failure behavior divided into three steps: Acrylate failure, selective TPU failure and necking, and delamination between reinforced and non-reinforced layers.
- The delamination behavior is visible for acrylate-reinforced specimens, which can be countered by different stacking patterns of the reinforced layers (for example alternating stacking instead of a block reinforcement).
- Microscopic analysis of the transition area indicates that small laser powers (<15 W) can potentially be utilized to counter delamination, by promoting infiltration of the acrylate in already melted but porous regions of TPU.

10.5 Complex Multi-material Parts

Based on the results and optimized process parameters from the section, the first complex multi-material parts were processed, and the outcomes are presented in Fig. 10.10. Figure 10.10a shows an acrylate-reinforced TPU sandwich structure in which the acrylate is included for increased bending stiffness in the face plates of the sandwich structure. The core structure is TPU, designed as a lattice structure.

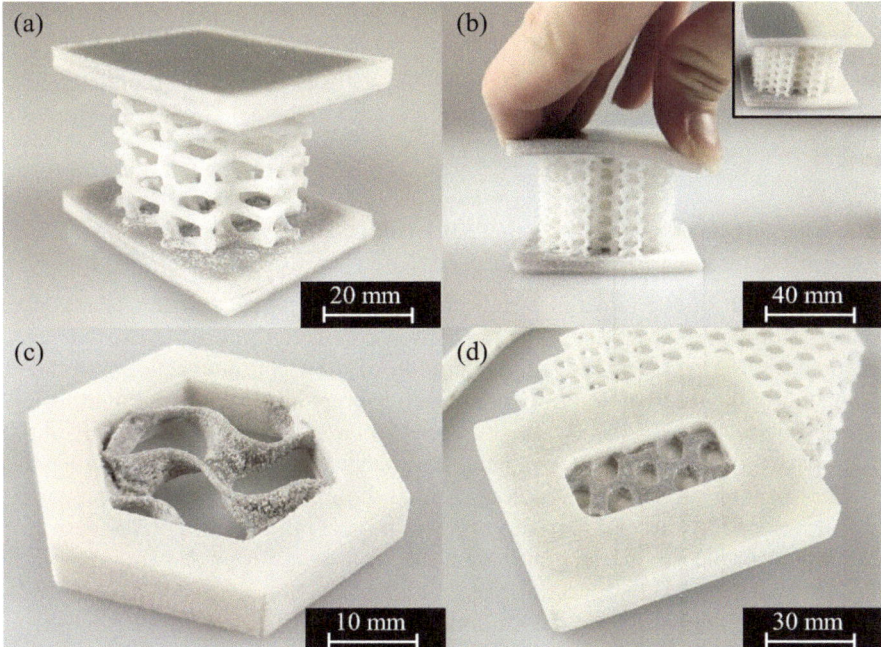

Fig. 10.10 a and **b** TPU sandwich structure with acrylate-reinforced face plates and TPU-lattice structures; **c** multi-material part with TPU outer ring and acrylate-reinforced core structure; **d** TPU block with embedded acrylate honeycomb structure [20]

To emphasize the increase of the bending stiffness within the face plates, a second partially reinforced structure was designed with increased compression stiffness of the core structure (Fig. 10.10b). The compression stiffness was increased by a reduction of the sizes of the unit cells of the lattice structure. The structures in Fig. 10.10a and b showcase the achievable complexity of TPU regions and the successful scalability of the acrylate-reinforced regions. To highlight the achievable complexity and the realization of more filigree acrylate-reinforced regions, a stator-like structure (Fig. 10.10c) and a sandwich structure with acrylate-reinforced honeycomb core elements (Fig. 10.10d) were generated. Especially the stator-like structure highlights the achievable complexity of the acrylate-reinforced regions as the acrylate can be incorporated into both the melted and non-melted regions of TPU. The achievable wall thickness of the acrylate regions was approximated to lie between 0.7 and 1 mm. The print resolution was restricted by the print head technology used, and alternative print heads with reduced minimal material output could potentially achieve even more filigree structures [20].

The key findings from this section can be summarized as follows:

- It is possible to manufacture highly complex multi-material parts combining thermosets and thermoplastics.

- High structural complexities can be achieved for both the thermoset and thermoplastic components.
- Thermoset-based reinforcements within the shells of sandwich structures prove to be a promising demonstrator for future research in the field of FJ.

10.6 Extension of the Material Spectrum

The extension of the material spectrum targets the implementation of secondary material properties such electrical or thermal conductivity to FJ parts. One example analyzed is represented by the alteration of the electromagnetic properties by substituting the reactive liquid with conductive graphite-based inks. Possible applications are wide-band microwave absorbers for communication technology and the aerospace industry. For these applications, especially graded transitions of the material are of interest. The methodology behind the ink composition and preparation was developed as part of a previous research project [14]. Instead of UV-curing, the liquid part of the ink is evaporated with the IR-heaters for pre-heating the powder bed for later PBF-LB/P processing steps. After evaporation of the liquid, the graphite nanoparticles remain within the thermoplastic powder bed and, during subsequent PBF-LB/P steps, the nanoparticles are enclosed within the melted thermoplastic. Figure 10.11 shows one of the first attempts to incorporate a graded material transition in the build direction (z-direction) within a TPU cube.

By continuously increasing the print head speed with every new layer, the material output was also lowered. For analysis of the graphite content, thin cuts were taken from the center of the sample and analyzed with a microscope. The microscopic

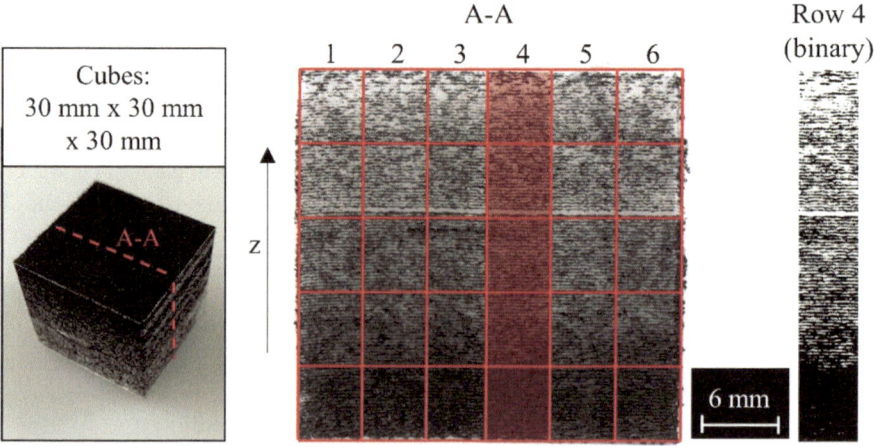

Fig. 10.11 Isometric and microscopic image of a microtome slice originating from a TPU cube with continuously decreasing graphite content in the build direction

images were turned into binary black-and-white images and segmented into grids. The segments were subsequently analyzed regarding their black-and-white pixel distribution. Apart from knowing the initial graphite content of the ink (approximate mass-percentage of 10%), the results just represent qualitative distributions since the actual graphite content after the IR-heating and the laser treatment is not known. For the accurate determination of the graphite content within each segment, TGA analysis is necessary.

The key findings from this section can be summarized as follows:

- The substitution of reactive liquids with conductive inks represents a promising extension of the FJ technology to implement secondary material properties within thermoplastic parts.
- Conductive graphite nanoparticles were successfully implemented within TPU parts and graded transitions of the graphite particles were successfully realized.

10.7   Summary and Outlook

In addition to generating valuable results in the field of material characterization and the process understanding of BJT and PBF-LB/P, an entirely new AM process named FJ was designed and investigated. The overall aim to reproducibly generate hybrid multi-material polymer parts combining thermosets and thermoplastics was fulfilled. In preliminary material qualifications, the curing behavior of acrylate-based photopolymers was analyzed regarding the thermal stability and curing behavior at different UV-intensities and elevated temperatures. Kinetic models were created based on the experimental results from DSC measurements. The output was a model based on the Kamal-Sourour equation to describe and predict the curing behavior of acrylate systems at different UV-intensities and isothermal temperatures. This reduces the experimental effort for material qualification in FJ. Besides being relevant for FJ processing, the lessons learned can further be transferred and utilized for the material qualification of other AM processes that make use of acrylate-based photopolymers at elevated temperatures. Another important scientific output is the successful characterization and modeling of the time- and temperature-dependent infiltration behavior of reactive liquids within thermoplastic powders. The results also represent transferable and universally applicable insights on the processability and the process understanding of generic BJT systems. Based on the fundamental insights from material characterization, multi-material parts were first generated and analyzed. The microscopic analysis of the boundary layer between the different materials identified selectively lowered laser powers within single layers as a potential process strategy to counter delamination. The analysis of the mechanical properties based on quasi-static tensile tests showed a significant increase in the Young's modulus for acrylate-reinforced TPU specimens (approximately factor 2). The processing sequence (alternating vs. interconnected acrylate reinforced layers) proved influential for the visible delamination behavior between non-reinforced and reinforced layers and for the maximum values of the Young's modulus. Finally, the material spectrum was successfully extended by substituting the reactive liquids with conductive inks containing graphite nanoparticles. With this setup, it was possible to generate TPU parts containing graphite particles with graded transitions. In future experiments, the reproducibility of single layers will be analyzed and evaluated regarding its impact on the delamination behavior. Further research will target the utilization of increased processing speeds, as well as the extension of the material spectrum with an increased focus on conductive inks.

References

1. Aerosint. https://aerosint.com/ (2022)
2. Ali, M., Mir-Nasiri, N., Ko, W.: Multi-nozzle extrusion system for 3d printer and its control mechanism. Int. J. Adv. Manuf. Technol. **86**(1), 999–1010 (2016)
3. Bartlett, N., Tolley, M., Overvelde, J., Weaver, J., Mosadegh, B., Bertoldi, K., et al.: A 3d-printed, functionally graded soft robot powered by combustion. Science **349**(6244), 161–5 (2015)
4. Boopathy, V., Sriraman, A., Arumaikkannu, G.: Energy absorbing capability of additive manufactured multi-material honeycomb structure. Rapid Prototyp. J. **25**(3), 623–9 (2019)
5. Choi, J.-W., Kim, H.-C., Wicker, R.: Multi-material stereolithography. J. Mater. Process. Technol. **211**(3), 318–28 (2011)
6. Craver, C., Society, C.: The Coblentz Society Desk Book of Infrared Spectra. The Society (1982)
7. Dorigato, A., Rigotti, D., Pegoretti, A.: Novel poly(caprolactone)/epoxy blends by additive manufacturing. Materials (2020)
8. Esposito Corcione, C., Frigione, M., Maffezzoli, A., Malucelli, G.: Photo - dsc and real time - ft-ir kinetic study of a uv curable epoxy resin containing o-boehmites. Europ. Polym. J. **44**(7), 2010–23 (2008)
9. GarcÃÃƒâ£ŁšÃ,Âa-Collado, A., Blanco, J., Gupta, M., Dorado-Vicente, R.: Advances in polymers based multi-material additive-manufacturing techniques: state-of-art review on properties and applications. Add. Manufact. **50**, 102577 (2022)
10. Hapgood, K., Litster, J., Biggs, S., Howes, T.: Drop penetration into porous powder beds. J. Colloid Interface Sci. **253**(2), 353–66 (2002)
11. Jiang, F., Drummer, D.: Curing kinetic analysis of acrylate photopolymer for additive manufacturing by photo-dsc. Polymers (2020)
12. Khondoker, M., Asad, A., Sameoto, D.: Printing with mechanically interlocked extrudates using a custom bi-extruder for fused deposition modelling. Rapid Prototyp. J. **24** (2018)
13. Kopp, S.-P., Stichel, T., Roth, S., Schmidt, M.: Investigation of the electrophotographic powder deposition through a transfer grid for efficient additive manufacturing. Procedia CIRP **94**, 122–7 (2020)
14. Lehmann, M., Kolb, C., Klinger, F., Zaeh, M.: Preparation, characterization, and monitoring of an aqueous graphite ink for use in binder jetting. Mater. & Des. **207**, 109871 (2021)
15. Ren, L., Song, Z., Liu, H., Han, Q., Zhao, C., Derby, B., et al.: 3d printing of materials with spatially non-linearly varying properties. Mater. & Des. **156**, 470–9 (2018)
16. Sakhaei, A., Kaijima, S., Lee, T., Tan, Y., Dunn, M.: Design and investigation of a multi-material compliant ratchet-like mechanism. Mech. Mach. Theory **121**, 184–97 (2018)

17. Setter, R., Riedel, F., Peukert, W., Schmidt, J., Wudy, K.: Infiltration behavior of liquid thermosets in thermoplastic powders for additive manufacturing of polymer composite parts in a combined powder bed fusion process. Polym. Compos. **42**(10), 5265–79 (2021)
18. Setter, R., Schmölzer, S., Rudolph, N., Moukhina, E., Wudy, K.: Modeling of the curing kinetics of acrylate photopolymers for additive manufacturing. Polym. Eng. & Sci. **63**(7), 2149–68 (2023)
19. Setter, R., Schmölzer, S., Rudolph, N., Moukhina, E., Wudy, K.: Thermal stability and curing behavior of acrylate photopolymers for additive manufacturing. Polym. Eng. & Sci. **63**(7), 2180–92 (2023)
20. Setter, R., Wudy, K.: Simultaneous processing of thermosets and thermoplastics in additive manufacturing of multi-material polymer parts. In: Annual Technical Conference - ANTEC, Conference Proceedings (2023)
21. Szebenyi, G., Czigany, T., Magyar, B., Karger-Kocsis, J.: 3d printing-assisted interphase engineering of polymer composites: concept and feasibility. Express Polym. Lett. **11**(7), 525–30 (2017)
22. Vyazovkin, S., Burnham, A., Favergeon, L., Koga, N., Moukhina, E., Pérez-Maqueda, L., et al.: Ictac kinetics committee recommendations for analysis of multi-step kinetics. Thermochim. Acta **689**, 178597 (2020)
23. Washburn, E.: The dynamics of capillary flow. Phys. Rev. **17**(3), 273–83 (1921)
24. Wudy, K., Drummer, D.: Infiltration behavior of thermosets for use in a combined selective laser sintering process of polymers. JOM **71**(3), 920–7 (2019)
25. Zheng, Y., Zhang, W., Baca Lopez, D., Ahmad, R.: Scientometric analysis and systematic review of multi-material additive manufacturing of polymers. Polymers (2021)
26. Zhou, Z., Buchanan, F., Mitchell, C., Dunne, N.: Printability of calcium phosphate: calcium sulfate powders for the application of tissue engineered bone scaffolds using the 3d printing technique. Mater. Sci. Eng. C **38**, 1–10 (2014)

Katrin Wudy is Professor for Laser-Based Additive Manufacturing at the Technical University of Munich. Her areas of expertise include laser-based additive manufacturing techniques with polymers and metals. The focus of her research is on innovative process strategies, novel monitoring techniques, and new fields of application for the transformation of additive manufacturing into a green manufacturing technology.

Chapter 11
Additive and Formative Manufacturing of Hybrid Parts with Locally Adapted, Tailored Properties

Jan Hafenecker, Richard Rothfelder, Michael Schmidt, and Marion Merklein

Abstract Hybrid parts offer a suitable solution for coping with the increasing demands of mass customization and open up new fields in industry. The combination of forming and additive manufacturing (AM) processes confer the benefits of both disciplines while simultaneously overcoming their drawbacks. However, with the combination of the processes, interactions emerge that have to be considered. In detail, additively manufactured structures deteriorate the formability of the sheets during the forming operation due to stress concentrations and resulting necking. Consequently, these interactions will be addressed in the following chapter to fully exploit the potential of hybrid components.

11.1 Introduction

Hybrid parts offer a suitable solution for coping with the increasing demands of mass customization [29] and open up new fields in industry [21]. The combination of forming and additive manufacturing (AM) processes confer the benefits of both disciplines while simultaneously overcoming their drawbacks [36]. However, with the combination of the processes, interactions emerge that have to be considered. In detail, additively manufactured structures deteriorate the formability of the sheets during the forming operation due to stress concentrations and resulting necking [27]. Consequently, these interactions have to be investigated to fully exploit the potential of hybrid parts. Each component of hybrid parts, which consists of sheet metal and additively manufactured structures, has its own microstructure and thus mechanical

J. Hafenecker · M. Merklein (✉)
Friedrich-Alexander-Universität Erlangen-Nürnberg, Institute of Manufacturing Technology, Egerlandstraße 13, 91058 Erlangen, Germany
e-mail: marion.merklein@fau.de

R. Rothfelder · M. Schmidt
Friedrich-Alexander-Universität Erlangen-Nürnberg, Institute of Photonic Technologies, Konrad-Zuse-Straße 3/5, 91052 Erlangen, Germany

© The Author(s) 2025
D. Drummer and M. Schmidt (eds.), *Progress in Powder Based Additive Manufacturing*, Springer Tracts in Additive Manufacturing,
https://doi.org/10.1007/978-3-031-78350-0_11

properties. With a forming operation in mind, it is important to know the differences, since they have a significant influence on the material's behavior. Therefore, the properties of additively manufactured specimens are investigated. Derived from this investigation, local in-situ alloying by vibrational microfeeding proves to be a crucial step for the combination of AM and forming technologies when using Ti64 (Ti64). Without it, the decreased formability of the hybrid parts outweighs the benefits of the process combination. Thus, the necessary steps to create a controlled local in-situ alloying device are elaborated on. As an essential requirement for local in-situ alloying, a target alloy must be defined. Since the base alloy Ti64 has no high deformation capabilities, the increase of β-phase content in the new alloy is a viable option. The first requirement for the target alloy is a high β-phase content to improve formability. The target alloy must also be processable with the same process parameters as the base alloy. In addition, the added material needed to create the target alloy must not disturb the coating process. Subsequently, the amount of alloying elements must be able to be set precisely. These four requirements are the basis for the process development conducted. With a target alloy defined, which can be processed using the same parameters as for Ti64, the next step is to develop a system for transporting the required amount of alloying elements into the process zone. For this task, a vibration-based powder feeding concept developed in subproject B6 in the second period of the CRC 814 was adapted [45]. The original concept was designed to supply the whole powder layer without the need for an additional coating process. By applying vibration to a powder-filled funnel and moving the setup via an axis system, the powder was distributed across the powder bed surface. However, this coating process is time-consuming and since the forming simulations show that only a small area of the part needs to be modified, the system was downscaled for dispensing small amounts of powder with high precision. With the frame for the laser-based powder bed fusion of metals (PBF-LB/M) local in situ alloying set, the adaption of the scanning strategy and the suppression of feedstock contamination by additional powder are initiated. The necessary adaption of laser power, scanning speed hatch distance, and layer thickness in the PBF-LB/M process for the target alloy and the base alloy to create a dense volume for both is conducted. With feasible parameters, the last and final step of the process development could be addressed, namely the production of locally in-situ alloyed hybrid parts.

11.2   Objectives and Methodology

The research outlined in this chapter aims to address the challenges that arise in the manufacturing and processing of hybrid parts. In order to do so, it is important to understand the individual processes, as they are the basis for the process interactions. Each process influences the subsequent process, which has to be considered. With the knowledge gained on the cause-and-effect relationships, process characteristics can be used to improve the properties of hybrid parts. To increase the potential of hybrid components even further, methods for local material property adjustments are

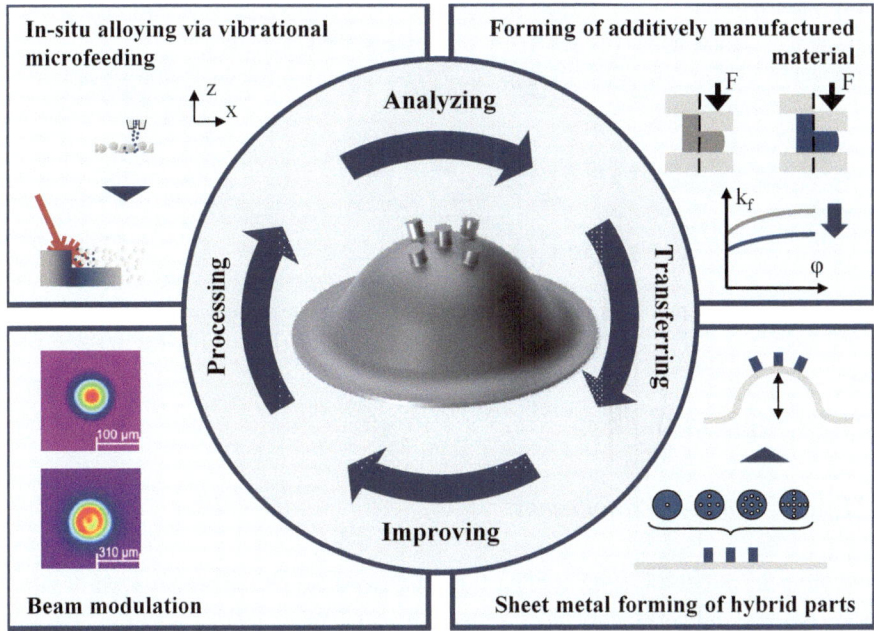

Fig. 11.1 Methodology used to investigate the interaction of additive manufacturing and forming to manufacture parts with locally adjusted properties

the subject of further research. Henceforth, the research gap concerning hybrid parts can thus be closed and the methodology used is shown in Fig. 11.1.

To understand the process sequence as a whole, each process is analyzed and by performing characterization tests on substrate and additive material, differences can be identified. Based on these results, the influence on the sheet-forming process of hybrid parts is determined combining experimental and numerical approaches, the sheet-forming process can be investigated in even more detail, highlighting the critical regions. To expand the process boundaries, the chemical composition of the investigated titanium alloy is locally adjusted via vibrational microfeeding. Alloying elements are precisely added and molten during the PBF-LB/M process. The challenges that arise when using this approach are compensated for by beam shaping.

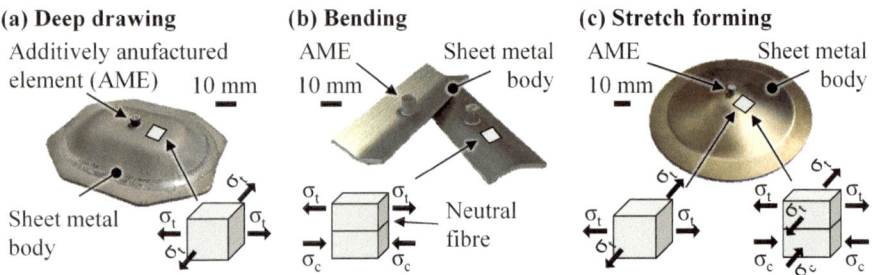

Fig. 11.2 Forming processes applied to hybrid parts with the resulting stress states according to [14]

11.3 State of the Art

11.3.1 Forming Processes

Forming describes the shaping of parts through mechanical deformation while simultaneously retaining the mass of the semi-finished product [22]. The plastic deformation, which permanently manipulates the shape of the part, also leads to changes in the mechanical properties i.e., an increase in strength [22], and can be applied to either bulk or sheet metal [20]. Nevertheless, the characteristic benefits of forming processes are its high material usage, improved part properties, and the possibility to produce large lot sizes with low cost and time investments [20]. However, the flexibility and freedom of geometrical design are limited by the high costs of the tools and machinery used in these processes [22]. To overcome these limitations, the combination of AM processes to produce hybrid parts offers a viable solution [29]. According to standards, forming processes can be also divided by the present stress state [1]. A selection of sheet metal forming processes, in which hybrids are investigated, is shown in Fig. 11.2.

11.3.2 Titanium Alloys

Titanium (Ti) is a light metal, with a density of $4500\,kg/m^3$ [26], high mechanical strength, and good corrosion resistance, leading to the use of this element in aerospace and medical applications [2]. Additional distinctive features of Ti when compared to other materials are a high electrical resistance and low heat conductivity [30]. The microstructure of Ti at room temperature is the hexagonal dense packed α-phase. With increasing temperatures exceeding the so-called β-trans us temperature (for pure Ti 882.5 °C), the microstructure transforms to the cubic body centered β-phase [23]. Ti has allotropic material properties [49] and both phases have advantages concerning their mechanical properties. To make use of this circumstance, the $\alpha + \beta$

alloy Ti64 was developed in the 1950s for aerospace applications [35]. It is the most commonly used Ti alloy with a market share of approximately 50% [2]. Ti64 is challenging for conventional processing such as forming [35]. In AM, it can be processed with relative ease for beam-based powder bed fusion processes such as PBF-EB/M [39] and PBF-LB/M [25]. The combination of forming operations and the PBF-LB/M process to produce Ti64 hybrid parts is the fundamental idea behind the experiments that were conducted.

11.3.3 Local Alloy Adjustment via Vibrational Microfeeding

Parts produced by means of AM can reach higher mechanical strengths than cast parts and can also match forged parts [47]. As mentioned in the previous section, Ti64 is challenging to forge but shows good processability for the PBF-LB/M process. The absorptivity of Ti64 is very high for the commonly used near-Infrared (NIR) laser systems of powder bed machines [5]. In combination with the high energy incoupling for small powder particles [48], Ti64 allows for the efficient use of laser power in the AM process. Attempting to combine AM and forming the high stiffness of AM-Ti64 components on sheet substrate leads to boundary conditions that limit the potential of the hybrid process [15]. An option to overcome these obstacles is the local in-situ alloying via microfeeding additional alloy elements. The cubic body centered β-phase reaches higher degrees of deformation before material failure in comparison to the hexagonal dense packed α-phase. By adding Aluminum (Al) and Vanadium (V) to Ti, the β-phase of Ti64 can be kept at room temperature [35]. By locally stabilizing the β-phase even further, the process limits can be expanded. Since this increased formability is only needed in the area of material flow during deep drawing, and a globally mixed powder feedstock would destroy the beneficial material properties of Ti64 in the complete additively manufactured volume, a new approach to graded material properties is needed, namely microfeeding. By adding extremely small fractions of additional powder locally, the material properties can be changed with high spatial resolution [38]. By evaluating the precise influence of the effect of in-situ alloying elements on Ti64 in the PBF-LB/M process, beneficial target alloys can be identified [17]. To diminish AM-related process restrictions, the exposure strategies have to be adapted [16], and to reduce powder feedstock contamination by additional alloying elements, beam shaping can be used [37]. With the knowledge gathered in this context, a robotic powder dispensing system for local in-situ alloying was designed.

11.3.4 Hybrid Parts

The term "hybrid" refers to the combination or fusion of different entities with the aim of making improvements to the single components [43]. Consequently, hybrid parts can either be the combination of processes [10] or materials [7]. In terms of the research presented, the hybrid nature of the parts results from the combination of forming technology and AM, seen in Fig. 11.3.

The manufacturing of hybrid parts can be performed with different processes and process sequences respectively. The additive structure can be manufactured either by PBF-EB/M [42] or PBF-LB/M [41]. The sheet forming operations comprise, among others, deep drawing [4], bending [27], and stretch forming [15]. With regard to the degree of plastic deformation, the forming operation can be performed before or after the additive manufacturing step [8]. Each component has a different granular structure [42] and thus mechanical properties as the result of the respective production process. This offers the possibility to cover different functions in the part but also leads to challenges during manufacturing.

11.4   Forming of AM Parts

To understand the differences between AM and sheet material, characterization tests must be performed. On the one hand, tensile and compressive tests are performed to investigate the general formability of the material. On the other hand, cup-backwards-extrusion experiments are used to investigate the influence of a forming operation on the additively manufactured material. The material investigated is the titanium alloy Ti64, which is manufactured by PBF-LB/M and by PBF-EB/M. Due to the high cooling rates during the PBF-LB/M process, the granular structure of the specimens is martensitic [3] and thus has a high hardness and low ductility [24]. To cope with this, a heat treatment according to [46] is used, which consists of a holding time of 2 h and a temperature of 850 °C to increase the ductility. As-built and heat-treated parts are then investigated in tensile and compressive tests, as well as cup-backwards-extrusion experiments. For the cup-backwards-extrusion experiments, the tool setup described in [11] is used. The roughness of the parts is investigated using tactile measuring while the geometry of the backwards-extruded specimens is analyzed optically. Hardness measurements are performed on a micro-hardness scale. The investigations focused on the differences between the mechanical properties of additively manufactured

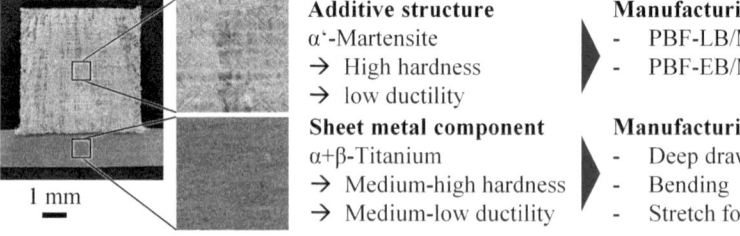

Fig. 11.3 Micro-section of hybrid part made with sheet metal and additive structure

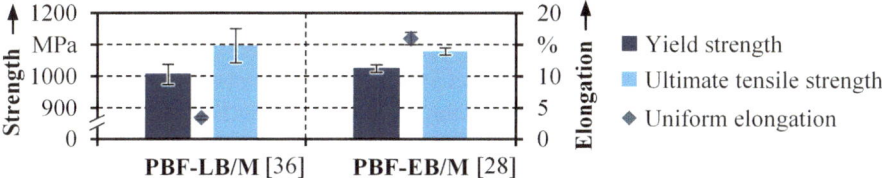

Fig. 11.4 Experimental results from tensile tests with specimens manufactured via PBF-LB/M [40] and PBF-EB/M [42]

materials with regard to the process used. Apart from the beam source used, the high building chamber temperature of 650 °C during PBF-EB/M is a characteristic that leads to changes in mechanical properties. The resulting yield stress, ultimate tensile strength, and uniform elongation from tensile tests performed at room temperature are shown in Fig. 11.4.

The yield and ultimate tensile strength show no distinct differences regarding the process used and only the standard deviation for the PBF-EB/M parts is reduced. However, the uniform elongation of PBF-LB/M parts is approximately 75% lower compared to the parts produced with an electron beam. This can be attributed to the lower cooling rates as a result of the higher building chamber temperature. Due to the high strength and partially low ductility, further processing of Ti64 is performed at elevated temperatures.

11.4.1   Results

The current research findings are separated into the results from the tensile/compressive testing and the cup-backwards-extrusion experiments. To gain an enhanced impression of the additively manufactured Ti64, tensile and compressive tests are performed in different states. A preliminary approach to increase the ductility is to apply a heat treatment according to [46], which corresponds to a holding time of two hours at a temperature of 850 °C and thus below the β-temperature of Ti64. A second approach is the variation of the forming temperature, which is why the range between 200 and 600 °C is investigated. Due to the tensile-compressive-asymmetry of titanium, compressive tests are performed in addition to the tensile tests, and the results are presented in Fig. 11.5.

Figure 11.5a shows the true stress-true strain curves of Ti64 in different states. The as-built material has the highest yield strength and, at the same time, the lowest ductility. Concerning the forming operation of hybrid parts with additive components, these characteristics are the least suitable, and thus different forming temperatures are investigated (Fig. 11.5b), whereby 400 °C offered the best combination of increased ductility and reduced strength. The heat treatment that was mentioned leads to an increase in ductility and a reduction of strength, which is also found within compressive tests (Fig. 11.5c). With this change, the difference in mechanical properties between the additively manufactured material and sheet metal becomes smaller. However, the sheet component has a lower strength and in a forming operation, the plastic deformation of the sheet material would start before the deformation of the additively manufactured component. With regard to this, an important aspect of these investigations is the local adaption of the chemical composition during the PBF-LB/M process to reduce the strength and increase the ductility without the need for heat treatments. The resulting true stress-true strain curve of the material with adapted chemical composition shows the success of this adaption (Fig. 11.5a). The low yield strength, high ductility, and distinct strain-hardening comprise suitable conditions for forming operations. The results from the investigations concerning the influence of a forming operation i.e., a cup-backwards-extrusion process on additively manufactured material, are shown in Fig. 11.6. The forming operation is examined regarding the influence of the surface roughness, geometry, and hardness of additively manufactured semi-finished products.

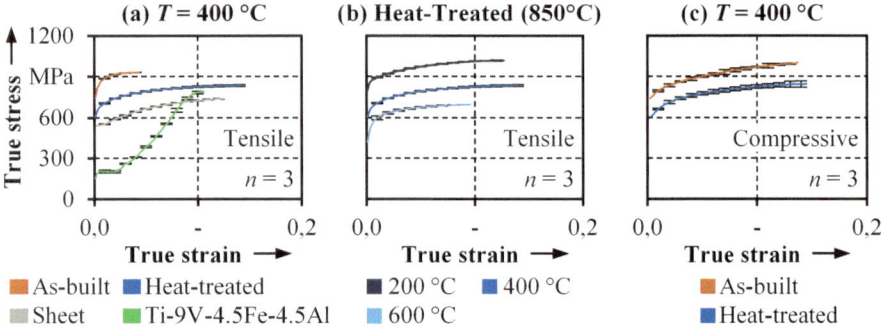

Fig. 11.5 Results of the tensile and compressive tests for additively manufactured material: **a** comparison of true stress-true strain curves of different material states, **b** influence of the forming temperature, and **c** influence of the heat treatment for compressive testing

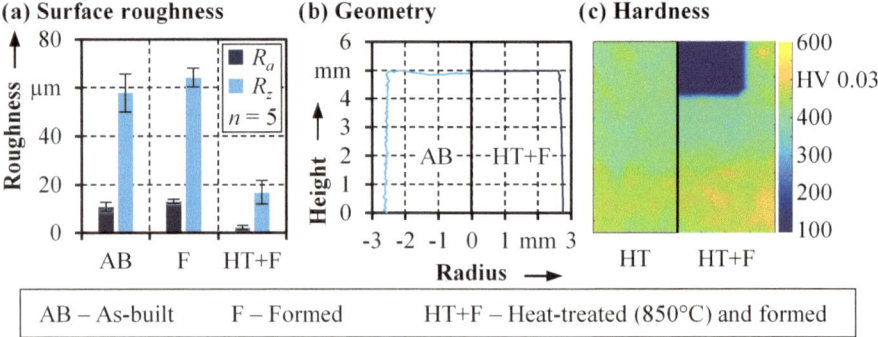

Fig. 11.6 Results of the cup-backwards-extrusion experiments: **a** influence of the forming operation on the surface roughness, **b** influence on the geometry, and **c** influence on the hardness

Due to the outlined material properties of Ti64 in the as-built state, the achievable strains during forming operations without material failure in terms of cracking are rather low [11] although, with an upstream heat treatment, these limits can be widened. Regarding surface roughness, the material failure occurs before a visible improvement of the material without heat treatment. However, with a heat treatment, the average roughness (R_a) and peak roughness (R_z) can be decreased significantly by forming (Fig. 11.6a). In detail, the average roughness is decreased by 80% and the peak roughness by 70%. Additionally, the geometry is more defined after the forming operation (Fig. 11.6b) as the corners are sharpened and the surfaces are flattened and even. The strains of the plastic deformation also lead to an increase in hardness (Fig. 11.6c). This is most pronounced for the bottom and the wall of the cup, at which point contact stresses and strains induce dislocations. These results are important, as they help to elucidate the material behavior of hybrid parts, which consist of sheet and additively manufactured components.

11.5   Forming of Hybrid Parts

The investigated hybrid parts consist of Ti64 sheet metal substrates with additively manufactured functional elements on them. The fields of research in this context comprise the behavior of hybrid parts in forming processes such as superplastic forming [40], bending [8], stretch forming [15], and deep drawing [14]. For each process, a tool is specifically designed and manufactured to enable the forming of hybrid parts at elevated temperatures. Additionally, the interaction between the additive and sheet components of the hybrid parts is investigated concerning the joint strength. To do so, tool setups for shearing and tensile loads are used [42]. The results from previous investigations comprise the investigations regarding the joint strength of hybrid parts in different conditions and the use of hybrid parts in different forming operations. As previously mentioned, the beam source and heat treatment can influence the properties of the additively manufactured material and therefore the joint strength of hybrid parts. Moreover, the stress state and a previous forming operation also influence the joint strength. The resulting strengths under different conditions are shown in Fig. 11.7.

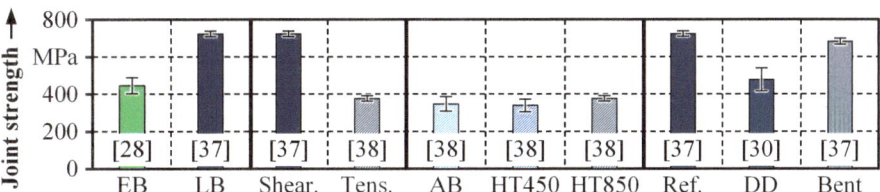

Fig. 11.7   Joint strength of Ti64 hybrid parts in different conditions

The first investigations on the joint strength of Ti64 hybrid parts report a shearing strength of 447 ± 44 MPa [42] for manufacturing with electron beam (EB). Parts made with a laser beam (LB) are found to have shearing strengths up to 724 ± 14 MPa [33]. The tensile strength, however, is measured to be almost 50% lower (376 ± 15 MPa) [34]. Investigations concerning the heat-treatment temperature show no significant improvement in the tensile joint strength [34]. For the joint strength under shearing loads, an upstream forming operation reduces the strength, regardless of whether this is a deep drawing process (DD) [4] or a bending operation [33]. Due to the lack of fusion defects in the joining zone caused by uneven surfaces, the joint strength of hybrid parts with deep drawn sheets is approximately one-third lower (478 ± 62 MPa) compared to the non-formed counterpart. However, the influence of the bending operation (683 ± 17 MPa) is almost comparable to that of the non-formed part. With this in mind, the forming operation of hybrid parts can be investigated more profoundly. Additional insight is provided by using numerical simulations of the investigated forming processes. To do so, the sheet and additively manufactured materials are characterized in tensile tests at different forming temperatures, strain rates, and stress states. The comparison of different material modeling approaches resulted in the yield surface modeling according to [9] as being the best fitting [40] and the approach of [31] for the extrapolation of the flow curve at elevated temperatures [28] for Ti64. To combine the additive and sheet material within a numerical forming simulation, a local material modeling approach is needed. Combining the numerical simulation and physical experiments, a comprehensive investigation of the bending of hybrid parts is found in [32]. According to [32], a transition radius between the additively manufactured element and the sheet substrate reduces the stress-concentrating effect of the geometry and thus lowers the strains. However, increasing the diameter of the functional element promotes stress concentrations and thickness reductions at the transition area, as well as an increasing height [32]. A comparison of different sheet thicknesses results in increased strains for higher thicknesses [32]. Regarding the tool dimensions, a smaller die width and punch radius increase the stresses at the transition area, thereby promoting earlier failure of the parts [32]. Based on these results, recent studies are focused on the forming of hybrid parts with more than one functional element.

11.5.1 Results

The combination of the multi-material modeling approach developed for the numerical simulation [13] and the deep drawing tool in which the sheet is loaded with different stress states [14] offers the possibility to holistically investigate the forming of hybrid parts. The findings derived for the bending of hybrid parts are also tested for deep drawing, as shown in Fig. 11.8. The diagrams show the maximum drawing depth at which no failure in terms of cracking arises.

In the deep drawing experiments at 400 °C, the same effect is visible as for the bending experiments, whereby an increasing height and diameter reduce the achievable drawing depth. Generally, the presence of the AME leads to reduced formability since the sheet without an element can be formed further without failure. The cracking of the parts arises in two regions for the hybrid parts: the transition area and the area of contact loss between sheet and punch. While the latter poses a common area for failures in stretch forming, the former is only found for hybrid parts.

Further research examined the influence of varying diameters on 5 AMEs, as well as the distance towards the center and the general number. The corresponding results are shown in Fig. 11.9.

Derived from Fig. 11.9a, the effect of increasing diameter is the same for one as for five AMEs, whereby an increase leads to the decrease of the achievable drawing depth without material failure. Since a more detailed view of material behavior is gained through numerical simulation, different distances of the outer AMEs towards the center of the hybrid part are investigated by simulation (Fig. 11.9b). The resulting thickness reductions can be used as an equivalent for material failure since high reductions correspond to a probable failure by cracking. Thus, placing the outer AMEs at a distance of 10 mm is most critical for the investigated levels, while the least critical distance is 15 mm. As shown in [12], the stress state that the AMEs are placed in depends on the distance to the center and the drawing depth. At the investigated drawing depth of 15 mm, the highest loads are still found in the center of the part in terms of biaxial tensile stresses. Therefore, placing AMEs there leads to an increase in the stresses and thereby an increased thickness reduction. Comparing different quantities and distribution of AMEs (Fig. 11.9b), it can be concluded that the mere increase in number is not critical since the resulting thickness reduction for one, five, and nine circularly arranged elements is comparable. However, the circular arrangement of nine elements is less critical in terms of thinning than the counterpart with nine elements arranged in a cross. This effect is also found in [13] and attributed to the evenness of distribution. To summarize, increasing the height or diameter of AMEs on hybrid parts designated for forming operations leads to a reduction of formability in terms of greater thickness reductions and, consequently, earlier failure. The location of the AMEs depends on the stress state and loads of

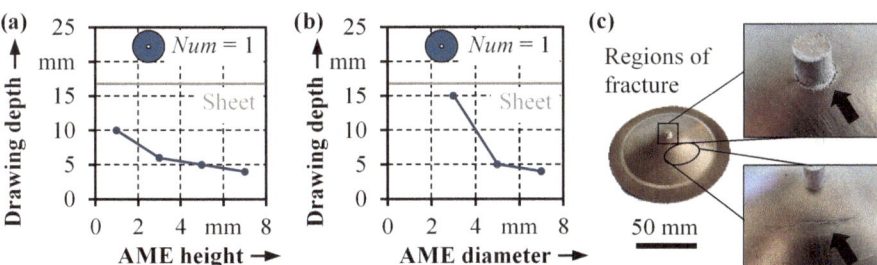

Fig. 11.8 a Achievable drawing depth for Ti64 hybrid parts in dependence of the **a** height and **b** diameter of the additively manufactured element (AME), and **c** regions of fracture

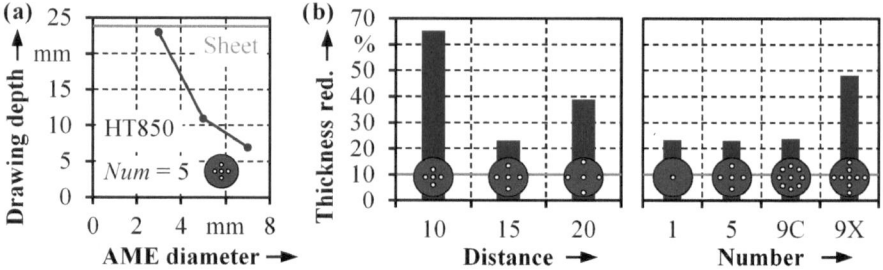

Fig. 11.9 a The maximum drawing depth without material failure from a physical deep drawing experiment with Ti64 hybrid parts, and **b** the maximum thickness reduction from the numerical simulation of Ti64 hybrid parts

the hybrid part during forming. An increased number of AMEs does not generally reduce the formability of the hybrid part, as long as their distribution is even.

11.6 Target Alloy for Local In-Situ Alloying by Vibrational Microfeeding

To define a suitable target alloy as a basis for the local in-situ alloying, the possible options for β-stabilization must be considered and tested in the PBF-LB/M process. The aim is to stabilize the β-phase during solidification to the point of acceptable formability. The theoretical basis for the experiments conducted is the molybdenum equivalent according to [6]:

$$\text{Mo Eq.} = 1.0(\text{wt.\% Mo}) + 0.67(\text{wt.\% V}) + 0.44(\text{wt.\% W}) + 0.28(\text{wt. \% Nb})$$
$$+ 0.22(\text{wt.\% Ta}) + 2.9(\text{wt.\% Fe}) + 1.6(\text{wt. \% Cr}) - 1.0(\text{wt. \% Al})$$

The "moly equivalent Mo Eq." describes the effect of a certain element on the β-stabilization standardized to the impact of molybdenum on the Ti-base system. Because Al stabilizes the α-phase [6], the first set of experiments was conducted with pure Ti. The first potential alloying element is V since Ti64 already has a V content. Iron (Fe) has the highest impact of all the presented elements, with a high lever to the β-phase content. Fe will be the second potential alloying element to be investigated. To determine the content of the β-phase and the resulting mechanical properties, a series of experiments with globally alloyed feedstock with different Mo Eq. are conducted. The resulting specimens are tested for β-phase content, density, and mechanical properties [17]. Based on the knowledge gained, possible target alloys based on Ti64 are defined and investigated.

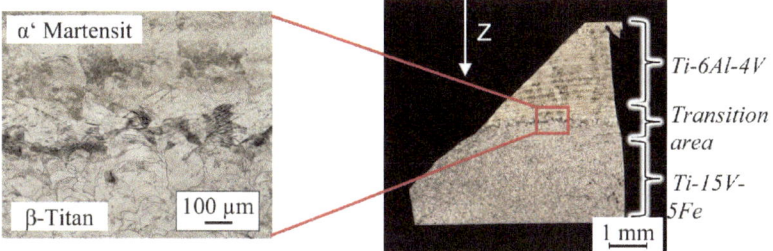

Fig. 11.10 Graded compression specimen with a gradual transition from Ti64 to Ti-15V-5Fe. The magnified cross section shows the transition from α'-martensite to the β-phase

11.6.1   Results

To evaluate the effect of the chosen alloying elements Fe and V on the β-phase stabilization during solidification, six powder mixtures with different Mo Eq. were produced and processed by means of PBF-LB/M. The conclusions according to [17] deliver crucial information on the Ti64-base target alloy. For instance, the Mo Eq. is a suitable tool for predicting the amount of β-phase in the resulting alloy. On the other hand, the Mo Eq. is not suited for predicting the mechanical properties due to more complex phase transformations. The formation of the ω-phase leads to brittle material behavior despite a high β-content. The addition of even small amounts of Fe and V strongly suppresses the formation of α'-martensite, a typical phase for Ti64 parts produced by PBF-LB/M, and increases the formation of the β-phase. Exceeding a Mo Eq. of 10% is sufficient to receive more than 90% β-crystals in the resulting specimens. For 15–20% Mo Eq. in the feedstock, the ω-phase precipitation leads to cracks in the solidified specimens. By exceeding these values, with a Mo Eq. of 25%, a highly ductile material with high fracture strain, almost completely consisting of the β-phase, can be created. Additionally, the addition of these elements does not show a negative impact on the PBF-LB/M processing parameters if the formation of the ω-phase was avoided. A hypothesis derived from these results is the suppression of the ω-phase formation by adding Al. To show that an alloy can be processed with similar parameters is a necessary but not sufficient condition for the creation of locally graded material properties. The materials must form a crack and pore-free transition zone. To test whether Ti-15V-5Fe also satisfies this requirement, specimens with a local transition from Ti-15V-5Fe to Ti64 are produced. The graded compression specimen is visible in Fig. 11.10. The result shows no cracks in the transition zone and displays adapted mechanical properties in comparison to the pure alloys in compression testing.

Considering the experimentally derived information, a Mo Eq. of 25% is chosen for the target alloy based on Ti64. By adding V to a mass fraction of 24% and iron to a mass fraction of 5%, a Ti64-base alloy Ti-24V-5Fe-4Al with a Mo Eq. of ca. 25% is created and processed in the powder bed. The metallurgical analysis of the resulting specimens reveals cracks in the manufactured parts. This leads to the adaption of the hypothesis of Al avoiding the precipitation of brittle phases and rather shifting them to higher Mo Eq. levels. Driven by these new insights, a new possible target alloy is defined: Ti-9V-4.5Fe-4.5Al with a Mo Eq. of 15%. The specimens produced with this material composition show neither cracks nor pores in the additively manufactured volume. The Al content definitively shifts the process window for the in-situ alloyed Ti-base alloys, since this Mo Eq. would not be feasible with a pure Ti-base alloy. The mechanical testing of specimens from Ti-9V-4.5Fe-4.5Al reveals more favorable properties of the new target alloy, which has excellent mechanical properties considering its formability. It reaches higher strengths in comparison to Ti64 parts with a low yield strength and high ductility. This promising composition leads to the use of Ti-9V-4.5Fe-4.5Al as a defined target alloy and to the next step in the process development chain: vibrational microfeeding.

11.7 Vibrational Microfeeding for In-Situ Alloy Adaption

With the knowledge concerning which material composition must be achieved, the dispensing system used in combination with the alloying elements must be characterized. The adapted powder dispensing device used consists of a small powder-filled glass funnel with an opening diameter of a few 100 μm connected to a piezo actor applying vibration to the tube. The input values for material flow control are the funnel diameter and, for the piezo actor, the frequency and amplitude. A crucial requirement for the funnel-material combination is a start-stop function, which can be controlled by the input values. As a result, the material flowability plays an important role [38]. In the first step, the Hall-flow test is used to characterize the flowability of the Ti64, Fe, and V powders used. The second step involves qualitative tests of the material flows through the funnel, albeit only when applying vibration and stopping after the end of the vibration. The third step comprises quantitative tests on the influence of the funnel diameter, frequency, and amplitude on mass flow over time. The mass flow is determined by weighing glass containers before and after dispensing with a high-resolution scale. With the information concerning the reachable control over the mass flow, dispensing tests with a moving dispenser on a powder bed surface are conducted. This fourth step described the connection between the mass flow, dispensing strategy, and resulting layer thickness. For the dispensing strategies, two options are plausible. The first is to start the dispensing process by applying vibration to the funnel and moving it over the powder bed surface to create the local alloy composition, while the second is to move the dispensing unit to the process zone and then start the vibration and stop it again by moving it to the next point. This solution is similar to the early scanning methods for powder bed fusion and vat polymerization, in which the laser was moved from spot to spot and the time the local exposure lasted defined the scanning speed. Considering the powder, several options for the dispensing strategy also exist, whereby one option is to premix the target alloy and place it in the funnel. Although the precision needed for matching the target alloy is reduced significantly, the resulting layer thickness increases accordingly. The second option is to integrate several funnels with the pure alloying elements and sequentially dispense them on the powder bed. This requires the highest precision considering the mass flow since very small amounts of both elements must also be applied sequentially. The third option is to only mix the alloying elements. This option is a compromise between the first two, and it creates a relatively small layer thickness with moderate accuracy. However, as discussed in the following results section, the term "moderate" is relative in this context. With the information concerning the mass flow over time, the dispensing area's dependence on the material or material combination, and the input values of the funnel diameter, frequency, and amplitude, a suitable strategy for dispensing and alloy feedstock can be considered.

11.7.1 Results

The first characterization of powder flowability with the Hall flow tests shows that no material apart from the Ti64 powder is considered to be flowable and that Ti64 shows good flowability. To correlate this information with the mass flow over time of the microfeeding system, further tests are required, whereby the qualitative dispensing tests with the elemental powders lead to surprising results. The most important of these is that for every material combination, a diameter-vibration combination could be found that fulfills the requirement of a start-stop function, which is realized by bridge building. This is a known phenomenon in process technology [44] and is commonly considered to be a problem. However, in this instance, it is a crucial factor for operating the dispensing device. These preliminary results obtained in this context are promising and present a proof-of-concept for the general idea of local in-situ alloying by vibrational microfeeding. Figure 11.11 shows a detailed overview of the dispensing system.

However, due to the excessively large amount of powder applied with elemental powders, a mix of Fe and V is used. Calculating the composition needed for the target alloy, a powder with 82% V and 18% Fe is produced. The input values for the material flow control are the funnel diameter, frequency, and amplitude of the piezo actor. A crucial requirement for the funnel-material combination is a start-stop function that can be controlled by the input values. Therefore, the material flowability plays an important role [38]. The information obtained from the Hall flow test cannot be correlated to the qualitative mass flow tests, and hence the aim of the dispensing system used is to derive precision in the mg area. As a result, quantitative tests are conducted in relation to the qualitative ones. The mass flow over time as a function of the funnel opening diameter, vibration frequency, and amplitude are determined. An important result is that the most significant factor for mass flow is the funnel opening diameter. While this finding may be unsurprising, unfortunately, the effect of frequency and amplitude shows no linear correlation with the mass flow.

A funnel diameter of 630 µm and a frequency of 4000 Hz with an amplitude of 10 V leads to a repeatable mass flow of 0.24 mg/s of V82Fe18 powder. Another important finding is that the mass flow is not stable over time, and while a stable repeatable mass flow could be measured after 10 s, no mass flow could be measured after one second of excitation. The mass flow increases rapidly from the first to the third second, reaching saturation after 10 s. As a result, the point dispensing strategy is chosen to be more suitable than the moving line dispensing strategy. Another benefit of the point dispensing strategy is the increased spatial resolution as even small moving correlated vibrations caused the area of powder application to increase. Moving the system into position and dispensing without further movement increases the process control and mass flow precision. The next step is to test the chosen dispensing strategy on a powder bed surface to determine the resulting layer thickness. The dispensed powder spots are scanned to determine their height, resulting in an additional height of approximately 50–60 μm above the powder bed surface. This must be considered for the PBF-LB/M process parameters. The diameter of the powder spots varies around 0.5 mm. By calculating the mass of Ti64 under the powder layer, the required dispensed mass can be calculated. This leads to the insight that the required precision is much higher than it is precise to the mg. For a common layer thickness of 50 μm for PBF-LB/M, the required mass deposited to match the target alloy composition equals 0.014 mg, which cannot be achieved with the determined parameters. By doubling the layer thickness to 100 μm, this value is increased to 0.027 mg. For a vibration time of two seconds, it is mathematically possible to achieve the required precision. Moreover, it is not necessary to match the exact value at every spot since recent studies in the scope of the CRC 184 showed that in-situ alloyed powders mix homogeneously over the course of several layers [19]. To match the right mass flow, an average value is thus considered acceptable. With the defined dispensing strategy matching the requirements for the target alloy, the next step is to adapt the scanning strategy to the resulting boundary conditions.

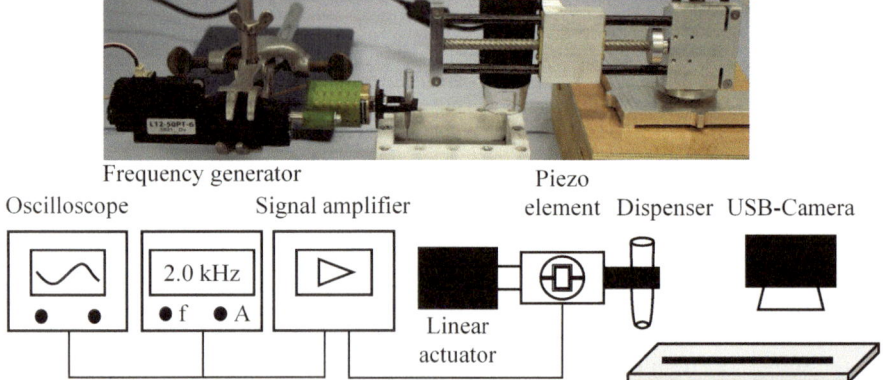

Fig. 11.11 Overview of the dispensing unit for vibrational microfeeding

11.8 Scanning Strategy and Local In-Situ Alloying

Ti64 is processable over a broad range of parameters [25], which is beneficial for addressing the challenges of local in-situ alloying. To increase the acceptable precision for the mass flow, a layer thickness of 100 µm is chosen. Furthermore, to compensate for the additional local layer thickness due to microfeeding, the parameters must create dense specimens for a layer thickness of 150 µm, while the energy input must not be high enough to create keyhole porosity after the area for locally graded material properties ends and only the base alloy is processed with the standard 100 µm layer thickness. The increase in the energy input was shown to be beneficial for melting the in-situ alloyed powders during the process [18]. Hence, the application of the same strategy to the in-situ alloyed powder with more mass is plausible. This could be done with the regular Gaussian beam profile of the standard NIR single mode lasers used in PBF-LB/M machines, although this leads to spatter formation which could move Fe and V particles out of the process zone and lead to Ti64 feedstock contamination over time. To limit this occurrence as far as possible, the use of beam shaping was investigated. To quantitatively characterize spatter formation, tungsten particles were used as tracer particles in Ti64 specimens. One set of specimens was produced using a standard Gaussian intensity distribution while the second set was created using beam shaping with a ring-spot beam profile. The analysis of the distribution of the tracer particles showed a possible reduction of spatter formation of up to 75% [37]. For the subsequent parameter adaptation, the ring-spot beam profile was used accordingly. With the spatter formation decreased, a suitable parameter combination of laser power, scanning speed, hatch distance, and layer thickness is experimentally determined. After suitable parameters with a high build rate to compensate for the increase in build time due to microfeeding were identified, locally in-situ alloyed structures on Ti64 substrate plates were manufactured and then deep-drawn to create locally in-situ alloyed hybrid parts. Those were compared to parts with the same geometry without in-situ alloyed areas regarding the material flow during the forming operation.

11.8.1 Results

The parameter development for Ti64 with beam shaping and high layer thickness leads to a set of parameters suited for the production of high-density specimens with a layer thickness of 150 µm for a laser power of 525 W, a scanning speed of 1375 mm/s, and a hatch distance of 100 µm. Those parameters also have a high build rate to compensate for the extra time required for dispensing additional alloy elements. With a target alloy, a dispensing system for controlled mass flow, and a suitable set of parameters that compensate for the height of the additional powder spots, the first step of manufacturing hybrid parts could start. The simulation displayed a material flow in the area of two mm above the substrate plate. Accordingly, the first twenty layers were mixed with the additional 82V18Fe powder. The substrate plates used in the manufacturing of the hybrid parts were stripped of additional screw drillings to create a sufficient sheet metal area for the specimen geometry. This leads to problematic side effects in combination with the preheating of the Ti64 substrate. At 200 °C preheating temperature, the plates start to bend at the edges, leading to a bulge in the middle of the plate. This again leads to an inhomogeneous powder layer thickness after the initial coating with a very thin layer in the middle of the powder bed and a high powder layer in the surrounding area. As a result, the layer thickness in the bottom area of the pin is too high, leading to a flawed connection between the pin and the plate. In a normal situation, this would be compensated over time. However, in this scenario, it directly affects the locally graded zone. To determine to what extent this is problematic, the created specimens are cut out from the substrate plate, heat treated at 850 °C for two hours, and deep drawn to 20 mm. The comparison to the specimens of the same geometry without the locally alloyed zone shows no measurable impact on the force required. The connection zone of the pins to the locally alloyed zone is damaged but is still connected to the plate, which shows no cracks whereas the plate with the pins without the in-situ alloyed zone shows cracks. With the unknown influence of the damaged connection zone for the graded pins, the comparison still does not show a significant disadvantage of the local in-situ alloying. The existing connection of the additive part and substrate plate after the forming operation can be considered a success in the process combination. Another important aspect of the introduction of a forming step is the time advantage in comparison to producing the whole volume of the part in one step by PBF-LB/M.

The increase in the inter-layer time was measured to be an average of three minutes per layer. The increase in additive production time is significant as the deep drawing that creates the hybrid part significantly increases the overall contour volume of the final part, whereby the 20 mm would mean 200 additional layers for the additive part in this specific case. The potential of the hybrid parts to decrease production time can be further enhanced by increasing the deep drawing distance.

11.9 Conclusion and Outlook

The aim of the research outlined in this chapter was the development of a new process route for the manufacturing of hybrid parts by combining PBF-LB/M and sheet forming to use the geometrical design freedom of AM with the rapid production cycles of forming technologies. The material combination chosen for this effort is the Ti-base alloy Ti64 which complicated the task since Ti64 has limited forming capabilities. To compensate for this drawback of the otherwise excellent mechanical properties of Ti64, a system for local in-situ alloying was developed to create locally graded part properties in the additively manufactured part. By stabilizing the better formable β-phase only in the area of material flow during forming, the process limitations can be reduced. The knowledge of the relevant cause-and-effect relationships gained in the course of the investigation covers the complete process chain of hybrid part manufacturing. The effect of material distribution on the substrate surface and its effects on the formability of the resulting geometry was investigated by a validated numerical process model with local material properties and experiments. A comprehensive understanding of the additively manufactured elements' influences on sheet metal forming operations was developed. The use of forming operations as a post-processing method for additive parts was introduced and successfully applied. Due to the sustainable nature of the process, forming is considered a viable alternative for machining operations. On the AM side, the use of beam shaping to reduce spatter formation yielded novel insights concerning the powder bed fusion process. The adaption of process parameters for high build rates by using ring-point beam profiles instead of Gaussian beam profiles is an addition to the recent topic of beam shaping in AM in general. The development of a device for local in-situ alloying for the creation of locally graded part properties in PBF-LB/M, which can be cost-effectively integrated into a commercially available machine, shows the potential for even further specialized products designed for the AM process. The knowledge derived from the results concerning vibrational microfeeding can drastically increase the speed of future investigations for various material combinations with different goals by providing the tools for the rapid calculation of the required dispensing precision and defining possible target alloys. The results show that the initially specified high accuracies are not sufficiently precise for this process. However, powder mass flows in the sub-milligram range could still be realized in a controlled fashion. By experimentally investigating the effect of β-stabilizing and high melting elements introduced to the PBF-LB/M process by in-situ alloying, a valuable understanding regarding the tailored material properties could be gained. The created target alloy has excellent mechanical properties that are beneficial for forming. The novel insights concerning the process combination of AM and forming technologies create a valuable progression for a broad new field of manufacturing technologies regarding current industrial development.

References

1. Manufacturing processes forming—classification, subdivision, terms and definitions, alphabetical index (2003)
2. Materials Properties Handbook: Titanium Alloys, 4th edn. ASM International, Materials Park, Ohio (2007)
3. Ahuja, B., Schaub, A., Karg, M., Lechner, M., Merklein, M., Schmidt, M.: Developing LBM process parameters for Ti-6AL-4V thin wall structures and determining the corresponding mechanical characteristics. Phys. Procedia **56**, 90–98 (2014)
4. Ahuja, B., Schaub, A., Karg, M., Lechner, M., Merklein, M., Schmidt, M.: High power laser beam melting of Ti-6AL-4V on formed sheet metal to achieve hybrid structures. In: Laser 3D manufacturing II, p 93530X (2015)
5. Alsaddah, M., Khan, A., Groom, K., Mumtaz, K.: Use of 450–808 nm diode lasers for efficient energy absorption during powder bed fusion of Ti6AL4V. Int. J. Adv. Manuf. Technol. **113**(9–10), 2461–2480 (2021)

6. Bania, P.J.: Beta titanium alloys and their role in the titanium industry. JOM **46**(7), 16–19 (1994)
7. Behrens, B.-A., Kosch, K.-G.: Development of the heating and forming strategy in compound forging of hybrid steel-aluminum parts. Materialwissenschaft und Werkstofftechnik (Mat.-wiss. u. Werkstofftech.) **42**(11), 973–978 (2011)
8. Butzhammer, L., et al.: Experimental investigation of a process chain combining sheet metal bending and laser beam melting of Ti-6AL-4V. In: Lev, W.G. (ed.) Proceedings of the Lasers in Manufacturing LIM (2017)
9. Cazacu, O., Barlat, F.: A criterion for description of anisotropy and yield differential effects in pressure-insensitive metals. Int. J. Plast **20**(11), 2027–2045 (2004)
10. Choi, K.S., Liu, W.N., Sun, X., Khaleel, M.A., Fekete, J.R.: Influence of manufacturing processes and microstructures on the performance and manufacturability of advanced high strength steels. Int. J. Plast **131**(4), 151 (2009)
11. Hafenecker, J., Kuball, C.-M., Rothfelder, R., Schmidt, M., Merklein, M.: Surface modification of additively manufactured parts by forming (2021)
12. Hafenecker, J., Merklein, M.: Investigations on sheet metal forming of hybrid parts in different stress states. Prod. Eng. Res. Dev. **38**, 684 (2022)
13. Hafenecker, J., Papke, T., Huber, F., Schmidt, M., Merklein, M.: Modelling of Hybrid Parts Made of Ti-6Al-4V sheets and additive manufactured structures, pp. 13–22. Springer, Berlin Heidelberg (2020)
14. Hafenecker, J., Papke, T., Merklein, M.: Influence of stress states on forming hybrid parts with sheet metal and additively manufactured element. J. Mater. Eng. Perform. **207**(6), 1176 (2021)
15. Hafenecker, J., Rothfelder, R., Schmidt, M., Merklein, M.: Stretch forming of Ti-6AL-4V hybrid parts at elevated temperatures. Key Eng. Mater. **883**, 135–142 (2021)
16. Huber, F., et al.: Customized exposure strategies for manufacturing hybrid parts by combining laser beam melting and sheet metal forming. J. Laser Appl. **31**(2), 22318 (2019)
17. Huber, F., et al.: Systematic exploration of the L-PBF processing behavior and resulting properties of -stabilized Ti-alloys prepared by in-situ alloy formation. Mater. Sci. Eng., A **818**, 141374 (2021)
18. Huber, F., Rasch, M., Schmidt, M.: Laser powder bed fusion (PBF-LB/M) process strategies for in-situ alloy formation with high-melting elements. Metals **11**(2), 336 (2021)
19. Karg, M., et al.: Laser alloying advantages by dry coating metallic powder mixtures with SIOX nanoparticles. Nanomaterials **8**(10), 862 (2018)
20. Klocke, F.: Manufacturing Processes: 4: Forming. Springer, Berlin, New York (2013)
21. Klocke, F., Roderburg, A., Zeppenfeld, C.: Design methodology for hybrid production processes. Procedia Eng. **9**, 417–430 (2011)
22. Lange, K. (ed.): Handbook of Metal Forming. McGraw-Hill, New York (1985)
23. Lepkowski, W.J., Holladay, J.W.: The physical properties of titanium and titanium alloys (1957)
24. Leyens, C., Peters, M.: Titanium and titanium alloys: fundamentals and applications. Wiley-VCH, Weinheim, Chichester (2003)
25. Liu, S., Shin, Y.C.: Additive manufacturing of ti6al4v alloy: a review. Mater. Des. **164**, 107552 (2019)
26. Lütjering, G., Williams, J.C.: Titanium: With 51 Tables, 2nd edn. Springer, Berlin, Heidelberg (2007)
27. Merklein, M., Dubjella, P., Schaub, A., Butzhammer, L., Schmidt, M.: Interaction of additive manufacturing and forming. In: Drstvenšek, I., Drummer, D., Schmidt, M. (eds.) 6th International Conference on Additive Technologies - iCAT 2016: Proceedings, pp 309–316. Interesansa - zavod (2016)
28. Merklein, M., Hagenah, H., Kaupper, M., Schaub, A.: Mechanical response of Ti-6AL-4V alloy on deformation at moderate temperatures. Key Eng. Mater. **549**, 311–316 (2013)
29. Merklein, M., Junker, D., Schaub, A., Neubauer, F.: Hybrid additive manufacturing technologies—an analysis regarding potentials and applications. Phys. Procedia **83**, 549–559 (2016)

30. Mills, K.C.: Recommended Values of Termophysical Properties for Selected Commercial Alloys, 1st edn. Woodhead, Cambridge (2002)
31. Nemat-Nasser, S., Guo, W.-G., Nesterenko, V.F., Indrakanti, S.S., Gu, Y.-B.: Dynamic response of conventional and hot isostatically pressed Ti-6AL-4V alloys: experiments and modeling. Mech. Mater. **33**, 425–439 (2001)
32. Papke, T.: Untersuchungen zur Umformbarkeit hybrider Bauteile aus Blechgrundkörper und additiv gefertigter Struktur. FAU University Press (2022)
33. Papke, T., et al.: Influence of a bending operation on the bonding strength for hybrid parts made of Ti-6AL-4V. Procedia CIRP **74**, 290–294 (2018)
34. Papke, T., Huber, F., Geyer, G., Schmidt, M., Merklein, M.: Characterisation of the tensile bonding strength of Ti-6AL-4V hybrid parts made by sheet metal forming and laser beam melting. In: Schmitt, R., Schuh, G. (eds.) Advances in Production Research, pp. 361–370. Springer International Publishing (2019)
35. Peters, M. (eds.): Titan und Titanlegierungen. Wiley-VCH, Weinheim, 3rd edn. (2010)
36. Pragana, J., Sampaio, R., Bragança, I., Silva, C., Martins, P.: Hybrid metal additive manufacturing: a state-of-the-art review. Adv. Ind. Manuf. Eng. **2**(5), 1–21 (2021) Silva, C., Martins, P.: Hybrid metal additive manufacturing: A state-of-the-art review. Advances in Industrial and Manufacturing Engineering **2**(5), 1–21 (2021)
37. Rothfelder, R., Huber, F., Schmidt, M.: Influence of beam shape on spatter formation during PBF-LB/M of Ti-6AL 4V and tungsten powder. Procedia CIRP **111**, 14–17 (2022)
38. Rothfelder, R., Lanzl, L., Selzam, J., Drummer, D., Schmidt, M.: Vibrational microfeeding of polymer and metal powders for locally graded properties in powder-based additive manufacturing. J. Mater. Eng. Perform. (2021)
39. Scharowsky, T.: Grundlagenuntersuchungen zum selektiven Elektronenstrahlschmelzen von TiAl6V4. Ph.D. Thesis, Friedrich-Alexander-Universität Erlangen-Nürnberg, Thesis year not provided
40. Schaub, A.: Grundlagenwissenschaftliche untersuchung der kombinierten prozesskette aus umformen und additive fertigung (2019)
41. Schaub, A., Ahuja, B., Karg, M., Schmidt, M., Merklein, M.: Fabrication and characterization of laser beam melted Ti-6AL-4V geometries on sheet metal. In Demmer, A. (ed.), Proceedings/DDMC 2014, Fraunhofer Direct Digital Manufacturing Conference. Fraunhofer Verlag (2014)
42. Schaub, A., Juechter, V., Singer, R.F., Merklein, M.: Characterization of hybrid components consisting of SEBM additive structures and sheet metal of alloy Ti-6AL-4V. Key Eng. Mater. **611–612**, 609–614 (2014)
43. Schuh, G., Kreysa, J., Orilski, S.: Roadmap-hybride produktion. Zeitschrift für wirtschaftlichen Fabrikbetrieb (ZWF) **104**(5), 385–391 (2009) -hybride produktion". Zeitschrift für wirtschaftlichen Fabrikbetrieb (ZWF) **104**(5), 385-391 (2009)
44. Schulze, D.: Pulver und Schüttgüter: Fließeigenschaften und Handhabung. Springer Vieweg, Berlin, 3rd edn. (2014). http://gbv.eblib.com/patron/FullRecord.aspx?p=1967825
45. Stichel, T., Laumer, T., Linnenweber, T., Amend, P., Roth, S.: Mass flow characterization of selective deposition of polymer powders with vibrating nozzles for laser beam melting of multi-material components. Phys. Procedia **83**, 947–953 (2016)
46. Vrancken, B., Thijs, L., Kruth, J.-P., van Humbeeck, J.: Heat treatment of Ti-6AL-4V produced by selective laser melting: microstructure and mechanical properties. J. Alloy. Compd. **541**, 177–185 (2012)
47. Wohlers, T., Campbell, R.I., Diegel, O., Kowen, J., Mostow, N., Fidan, I.: Wohlers report 2022: 3D printing and additive manufacturing: global state of the industry (2022). https://researchspace.auckland.ac.nz/handle/2292/62273
48. Zhang, J., et al.: Influence of particle size on laser absorption and scanning track formation mechanisms of pure tungsten powder during selective laser melting. Engineering **5**(4), 736–745 (2019)
49. Zwicker, U.: Titan und Titanlegierungen, 1st edn. Springer, Berlin (2014)

Michael Schmidt has headed the Institute of Photonic Technologies since its founding in 2009 at Friedrich-Alexander-Universität Erlangen-Nürnberg. His research interests include laser application from micro- to macroscopic scales within industrial manufacturing, additive manufacturing, and medical engineering. He was the vice spokesperson of the Collaborative Research Center 814 Additive Manufacturing.

Marion Merklein has been the Head of the Institute of Manufacturing Technology at Friedrich-Alexander-Universität Erlangen-Nürnberg since 2008. Merklein's research topics lie at the interface between materials science and manufacturing technology. She uses numerical methods and process analyses to shorten the development time for components and processes and to make manufacturing processes more robust. In her work, she incorporates issues of large-scale industrial production and searches for resource- and energy-saving solutions.

Part IV
Simulation Techniques

Chapter 12
Robust Structure-Process-Optimization

Daniel Huebner, Jannis Greifenstein, Fabian Wein, and Michael Stingl

12.1 Topics in Structural Optimization for Additive Manufacturing

The combination of structural optimization and additive manufacturing presents an abundance of interesting and fruitful research opportunities, which resulted in numerous research publications on the topic in recent years (see Fig. 12.1). The overview presented here does not comprise a comprehensive list of research in the field but rather highlights the main areas of research and classifies the contributions made in this work. For a more comprehensive introduction to the interplay between additive manufacturing and the structural optimization of mechanical components, we recommend the excellent work [6] and for a literature overview, the review article [20]. The topics in the research area can be structured around the following background themes:

- **support structures**—consider overhang constraints, reduce or avoid the need for support structures, reduce powder use and post-machining cost, and avoid implementing support structures in inaccessible places
- **material aspects**—consider material anisotropy, whether microstructure or multi material
- **porous infill**—use of homogeneous or spatially variable lattice structures, meta material optimization
- **robustness**—consideration of uncertain material properties or manufacturing precision
- **post-treatment**—reduce post-machining efforts, and facilitate conversion of optimization results to printer compatible file formats

D. Huebner · J. Greifenstein · F. Wein · M. Stingl (✉)
Friedrich-Alexander-Universität Erlangen-Nürnberg, Chair of Applied Mathematics (Continuous Optimization), Cauerstraße 11, 91058 Erlangen, Germany
e-mail: stingl@math.fau.de

© The Author(s) 2025
D. Drummer and M. Schmidt (eds.), *Progress in Powder Based Additive Manufacturing*, Springer Tracts in Additive Manufacturing,
https://doi.org/10.1007/978-3-031-78350-0_12

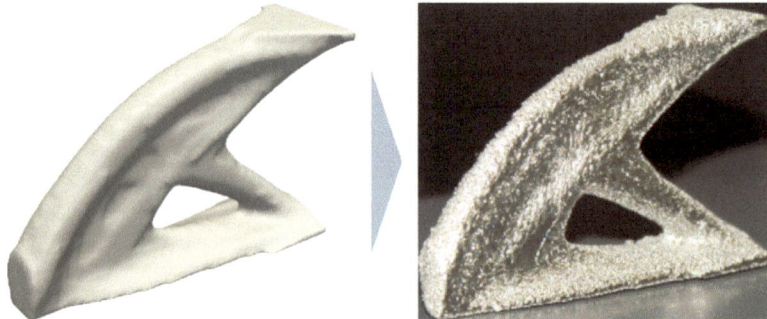

Fig. 12.1 A component that was optimized by topology optimization and its manufactured counterpart

Over the course of the CRC814, we developed novel optimization techniques for several different problems from these topics. Firstly, we will outline studies on the influence of uncontrolled variations in the effective material properties achieved in the additive manufacturing process, which are commonly caused by variations in the temperature field. Next, we will present our results on multi material optimization. Later, we will explore additional design capabilities enabled by additive manufacturing. As the process oftentimes results in slightly anisotropic material properties, we will highlight the advantages of optimizing the printing direction of optimized parts. Further, we will examine the potential performance increase when the grain structure of certain alloys can be influenced locally. Lastly, we will report on our current findings on additively manufactured lattice structures.

12.2 Robust Topology Optimization with Worst-Case Handling of Material Uncertainties

One challenge in additive manufacturing, despite recent progress, is a fully reproducible building process. Among others, temperature differences in the build chamber or microscopic pores can lead to local variations in the effective material properties achieved. For this section, the potential influence of these uncontrolled variations on component performance was methodically investigated. Additionally, a design process incorporating these uncertainties in topology optimization was developed, which guarantees a baseline performance even for the worst-case local distribution of the material variations.

In lieu of probabilistic approaches, as is common in the literature, a robust approach based on a worst-case model was developed which acts as a safeguard against the worst possible material configuration from a specified uncertainty set. The new method is related to works in topology optimization with respect to degradation [1]. The natural type of a worst-case optimization model is a so-called minimax problem, in which an objective function is minimized with respect to design variables ρ on an outer level and maximized with respect to uncertainty parameters δ on an inner level:

$$\min_{\rho} \max_{\delta} J(\rho, \delta) \tag{12.1}$$

As an example of this problem, for this section it was aimed to minimize the compliance of a component, where the outer variables ρ describe the topology of the component, while the inner variables δ characterize uncertainties in the effective material properties. This problem is neither convex nor concave per se, nor can it be approximated or reformulated in such a way. Consequently, minimax formulations have been investigated in detail [10, 12], in particular concerning how to handle the problem by the introduction of an optimal value function.

$$\psi(\rho) = \max_{\delta} J(\rho, \delta). \tag{12.2}$$

Several mathematical challenges had to be tackled in this context as the inner problem must be solved for globally optimal solutions since local optimality would result in discontinuity of the optimal value function. Additionally, the numerical error must be kept low, as numerical artifacts would also lead to discontinuity. Furthermore, Ψ is, in general, not continuously differentiable with respect to the design variable, i.e., pseudo density. To handle this, multiple mathematical methods were employed.

First, to be able to evaluate the optimal value function, an inverse parameterization was employed, which results in a concave inner problem and continuity of Ψ [12]. It was also proven that the inverse parameterization leads to an outer approximation of the original admissible set and thus effectively safeguards against the actual worst case.

To ensure the differentiability of Ψ, two regularization techniques were employed. First, using a Tikhonov type regularization, an explicit formula for the gradient with respect to the outer design variables was derived [12]. Thus, the regularized problem can be solved with first-order optimization methods such as the method of moving asymptotes (MMA) [30]. The optimal value of Ψ was determined by an interior-point method, where a barrier approach was used to treat the emerging inequalities to reduce computational complexity. The resulting inner problem corresponds to a central path problem, where the barrier parameter describes the parameter of the central path.

The optimized designs are shown in Fig. 12.2. Although there are no noticeable differences in the topologies, the redistribution of effective material properties is clearly visible in the design obtained by the worst-case approach (right), when compared to the one obtained without uncertainty protection (left). For significant local material property fluctuations (e.g., by degradation) the newly developed algorithm yields a significant improvement in the objective value [12].

The findings over the course of the project showed that in additive manufacturing the influence of the uncertain material parameters on the compliance of topology optimized parts is only minor. This is quite encouraging with respect to additive man-

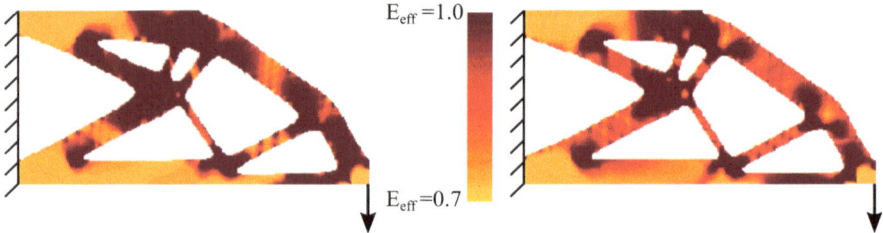

Fig. 12.2 An optimized 2D cantilever beam with bearing on the left and a force in the lower right corner pulling downwards is considered. The color coding represents the local, scaled Young's Modulus to indicate the worst-case manifestation of the material degradation, that is, an uncertainty of 30%. The worst-case compliance of the "nominal" design (without protection against material uncertainty, depicted on the left) is 4.7% higher than the compliance for a constant average Young's Modulus. For the robust design (which offers the best possible protection against material uncertainty, shown on the right), the loss in compliance is, in the worst case, 3.9%

ufacturing of parts as robust optimization methods can consistently further decrease the influence of the worst-case material distribution.

12.3 A Novel Multi-material Optimization Approach

In recent years, the Sequential Global Programming (SGP) algorithm has been developed, which allows researchers—for the first time—to combine discretely and continuously parameterized materials in multi material optimization. In comparison to classical Discrete Material Optimization (DMO) methods, SGP is a flexible approach, that stands out due to several features: any number of materials are treatable at linear complexity, it applies to isotropic and anisotropic materials, arbitrary continuously differentiable interpolation is possible (e.g., modeling of composite materials), it applies to constant and parameterized materials (e.g., local material orientation), and real and complex valued properties (e.g., for viscoelastic material behavior).

A graph-based parameterization of materials has been introduced, which allows for considering both isotropic and anisotropic materials (see Fig. 12.3). Continuously parameterized materials can be mapped by interpolation functions along edges of the material graph and can be combined with discrete parameters. To solve the resulting non-smooth optimization problem, it is replaced by a sequence of easier problems, like in the classical method of moving asymptotes. As a new contribution, the state-dependent functions are not approximated with respect to the interpolation parameters but rather with respect to the effective material tensors, which is similar to methods in FMO (Free Material Optimization) [26, 29].

In this context, the approximations are separable with respect to the design cells. This allows for a cell-wise decomposition of each subproblem, which in turn entails separability with respect to the individual edges of the material graph and yields the desired linear complexity. The resulting subproblems are generally non-linear but

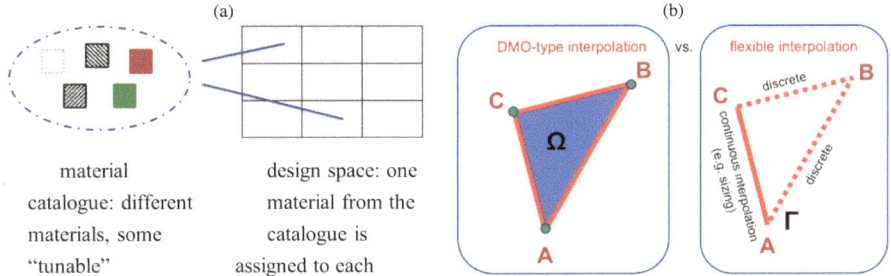

Fig. 12.3 Principle of multi material optimization. **a** The design domain is discretized through, e.g., finite elements. One material taken from a material catalog is assigned to each element. **b** Parameterization of classical DMO methods and SGP. In DMO methods, a continuous interpolation of the properties of material nodes is considered that uniformly penalizes deviations from the nodes. SGP takes a parametrization along edges of the material graph as a basis. This allows for individual interpolation schemes along each edge

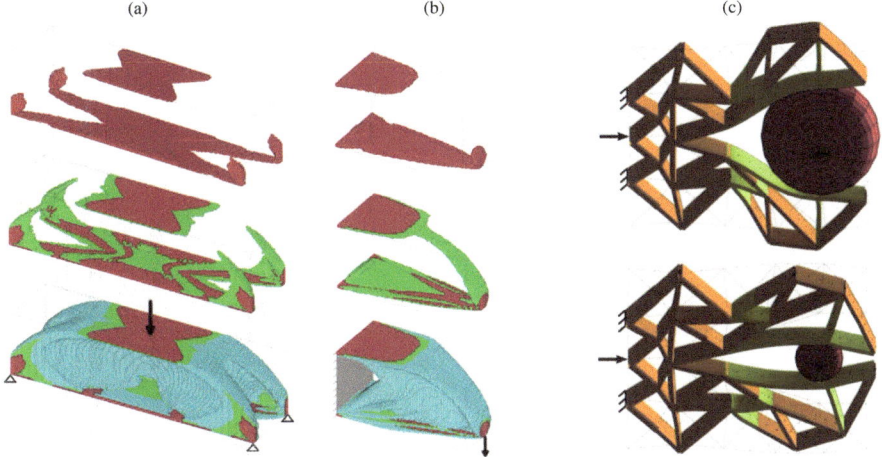

Fig. 12.4 Multi material designs optimized by means of the SGP method. **a** 3D MBB beam and **b** 3D cantilever beam. Both designs were optimized for minimal weight with prescribed mechanical compliance. The materials used are Mg (blue), Al (green), and TiAl6V4 (red). **c** Design of a flexible gripper optimized with a special version of SGP for Timoshenko beam networks. The materials are PA12 (brown) and a ten times softer material (green). The objective was to maximize the amplitude of the gripper movement relative to an input displacement, which is applied at the left center node. The resulting force at the contact surface was modeled by a spring model

continuously differentiable and can, at best, be solved analytically to global optimality [27]. If this is not the case, the low dimensionality allows for (approximately) globally optimal solutions by means of simple multi grid methods [16]. Exemplary optimized designs are shown in Fig. 12.4a and b.

An important aspect in the context of structural design for additive manufacturing is the consideration of manufacturability, especially in terms of minimal structural sizes. These are quite simple to implement in multi material optimization based on continuum models for mechanical compliance minimization, but pose a challenge for other objectives, e.g., in compliant mechanism design [20]. For this reason, a second approach was investigated in which components are represented as networks of linear Timoshenko beams. The description of the individual beams using geometrical parameters allows for a direct limitation of individual structural sizes. Additionally, a method based on the analytic function was developed that does not require discretization along the individual beams under linear elastic assumptions [18]. Furthermore, a specialization of the SGP method was developed for multi material optimization of frame structures [16]. To allow for a design parameterization that is as free as possible, the multi material optimization approach (which already allows for simultaneous optimization of topology, material, and size by itself) was combined by an alternating direction method with a geometry optimization, which yields optimized positions of the nodes in the beam network. With this, it is possible to design a flexible multi-material gripper that adheres to sizing restrictions (see Fig. 12.4c).

12.4 Simultaneous Optimization of Topology and Anisotropic Material

Additive manufacturing with polymers often results in anisotropic material properties. The orientation of a component in the build chamber thus has an impact on its resulting (mechanical) behavior. Using the results from Chap. 5, average anisotropic material coefficients for SLS of PA2200 were determined. The Young's Modulus in the build layer was, on average, 5.5% higher than vertical to it. Using this data, a study was conducted in which designs with minimal mechanical compliance under a volume constraint were optimized for different pre-selected component orientations. The result is shown in Fig. 12.5. Based on this, a topology optimization problem was set up, which considers the component design and the build direction of the part [9]. It was shown that the algorithm reliably reproduces the optimal rotation and design from different starting parameters and is more efficient in terms of computational time than sampling a fixed grid for the orientation as in the study.

For IN718, it is possible to manufacture equiaxed grains with approximately isotropic properties, as well as columnar grains with anisotropic behavior (see Chap. 7) with locally different orientations. The algorithm was thus extended to include both types of grain structure and handle a different orientation of a crystalline microstructure from point to point in addition to the global component orientation in the build chamber. To achieve this simultaneous optimization of grain structure and topology, a preliminary version of the SGP method based on topological derivatives was developed [8, 19, 25]. Later, a method was investigated that also takes the continuity of the grain structure into account [11, 13]. Figure 12.6 shows an exemplary optimized structure.

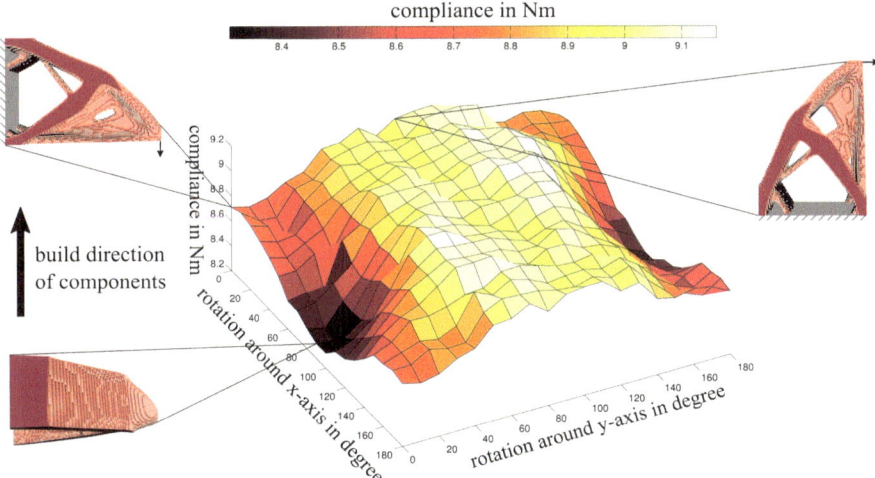

Fig. 12.5 Optimized designs of a cantilever beam for different component rotations in the build space. For a Young's Modulus that is 5.5% stiffer in the plane of isotropy than vertical to it, the mechanical compliance obtained differs by up to 9.3%

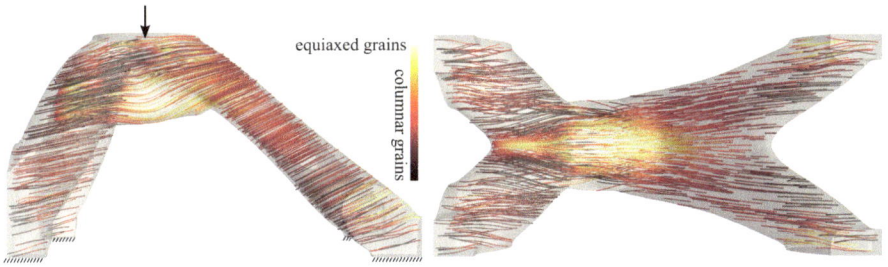

Fig. 12.6 Topology optimization of an IN718 component with optimized crystalline microstructure and build orientation. The colors denote the interpolation value between the isotropic material properties of equiaxed grains (white) and anisotropic material properties of fully columnar grains (black). The streamlines show the direction of the columnar grains with the highest stiffness that are approximately 45° tilted w.r.t. the grain columns. With this design, the compliance could be improved by 13.5% compared to a design obtained with established topology optimization methods and equiaxed grains

12.5 Current Findings from the Research: Lattice Optimization

In the third phase of the CRC814, truss lattice structures have come into focus. Several challenges arise in this context, which are addressed in this section. To optimize components built from cellular materials, two-scale topology optimization methods are often applied, see [24, 28, 37]. These methods enable simultaneous

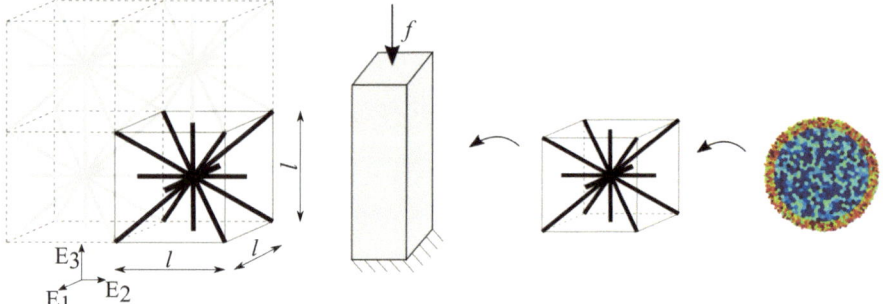

Fig. 12.7 Left: Unit cell for a lattice structure in three dimensions. Right: The properties of individual lattice beams are determined by the grain structure

optimization of the component's geometry and the cellular material's topology based on homogenized properties.

12.5.1 Impact of Grain Structure

Truss lattices are typically composed of thin struts. Here, the influence of the grain structure in metallic components is more noticeable, especially the impact of boundary effects (see Chap. 15). By applying the results from other chapters, for this chapter, the difference between a material model that disregards grain structure and a sophisticated model that takes grain structure into account was investigated [21]. This was done for both a pure compliance minimization problem and a multi-objective optimization problem [7], simultaneously considering mechanical compliance and macroscopic buckling.

As an example, we optimize a cuboid component with an aspect ratio of 4:4:10 made of a cellular material, which is built by unit cells shown in Fig. 12.7. We fix the bottom surface in all three dimensions and apply pressure to the top surface. In the first optimization problem (P1), we seek to minimize the mechanical compliance, while in the second one (P2) we want to maximize the minimal buckling load factor at the same time. The design variable is the local relative lattice volume, which can directly be translated into diameters for the struts of the cellular material at any given point in the macroscopic design domain. It is bounded by 0.018 from below and 0.15 from above. We apply a global volume bound of 5%, use a density filter, and employ a variant of the optimality condition method [3]. The optimized designs are presented in Fig. 12.8 while the corresponding function values are shown in Table 12.1.

For the pure compliance minimization, the compliance of the design respecting grain structure is roughly 19% larger than that of the design that disregards the grain structure, whereby only minor differences can be seen in the optimized designs themselves. This is confirmed by a cross-evaluation of the two designs as the evaluation of the first design with the sophisticated model yields a compliance of 4.5787 (+0.13% compared to the second design), while the second design has a compliance of 3.8421 (+0.055% compared to the first design), when evaluated with the first model.

For simultaneous optimization of compliance and buckling load factor, again, the designs do not differ much, which is confirmed by a cross-evaluation showing a difference of less than 1% in both compliance and load factor. For both material models, the compliance is 1% worse compared to pure compliance minimization, while the load factor is increased by 33% or 35%, respectively. Again, the compliance of the optimization with the sophisticated model is 19% larger than with the other one. Most importantly, however, the resulting load factor is 15% smaller.

This shows that the material model, which is based on the isotropic material assumption on the microscopic scale, overestimates the critical load factor on the macroscopic scale. To clarify, suppose one were to use the model that disregards the grain structure to optimize a component with respect to compliance, while applying a constraint on the load factor with a 10% safety margin to prevent buckling. According to the model, which does account for grain structure, the component would be at risk of collapsing due to buckling. Therefore, to ensure safety, the sophisticated model that considers the grain structure should be used for optimization.

12.5.2 Optimization with Respect to Macro and Microscopic Buckling

A workflow for the simultaneous treatment of macroscopic and microscopic buckling was developed for the results in this chapter for two- and three-dimensional settings [14, 17]. In the following, we will explain important aspects of two-scale optimization with respect to buckling on both scales, starting with a discussion in two dimensions.

First, we present two aspects that are important for the upscaling process: the design of joints of lattice struts and the number of unit cells in a representative volume element (RVE). As an example, we choose a lattice consisting of equilateral triangles that is parameterized by its porosity. A unit cell design with sharp corners where the lattice struts meet leads to local stress concentrations at these corners. To circumvent this issue, we round the corners with circular arcs, while keeping the overall volume of the structure constant. A lattice with rounded unit cells is shown in Fig. 12.9.

It turns out that the choice of the radius parameterizing the arcs only has a minor effect on the Poisson's ratio. Young's Modulus is mainly affected by a small relative lattice volume; for a small volume, a large radius leads to very thin lattice struts due to volume preservation and the lattice loses stiffness. The impact on the homogenized buckling load factor is more significant. For small radii, the buckling resistance increases with increasing radius because stress concentrations are avoided, the lattice struts are better supported, and their relative length-to-width ratio becomes smaller.

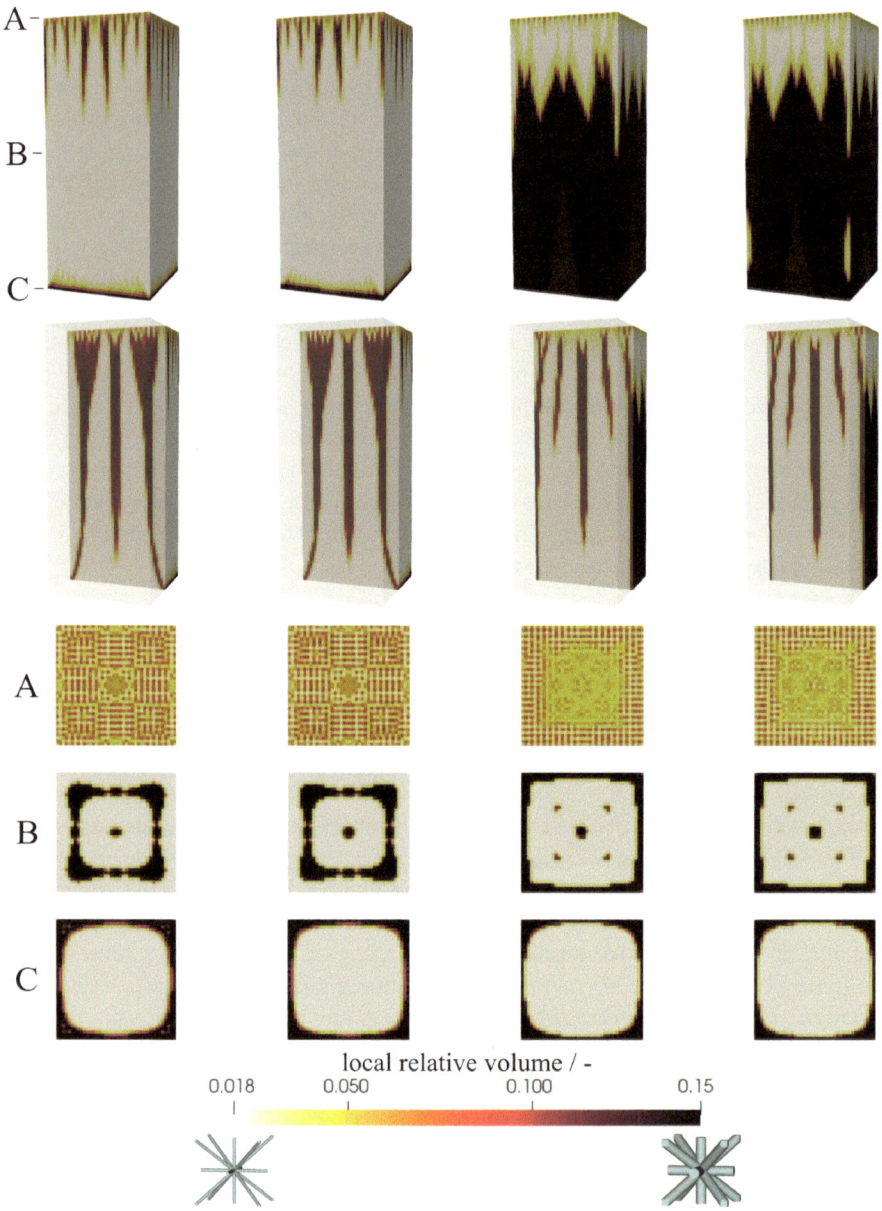

Fig. 12.8 From left to right: Optimized designs for compliance minimization without considering grain structure; with grain structure considered; for simultaneous buckling and compliance optimization without considering grain structure and with grain structure considered. From top to bottom, the side view, a longitudinal section, the top view (**A**), a transverse section (**B**) and the bottom view (**C**) are displayed

Table 12.1 Resulting compliance and buckling load factor for the material model without grain structure and with grain structure for pure compliance minimization (P1) and simultaneous optimization of compliance and load factor (P2). Italic buckling load factors have not been optimized and are only provided as a reference

	Problem	Compliance	Buckling load factor
w/o grains	P1	3.84	8.25
	P2	3.88	*11.00*
With grains	P1	4.57	6.93
	P2	4.64	9.37

Fig. 12.9 Different RVEs containing 1×1, 2×2, 3×3, and 4×4 unit cells and the coordinate system for homogenization

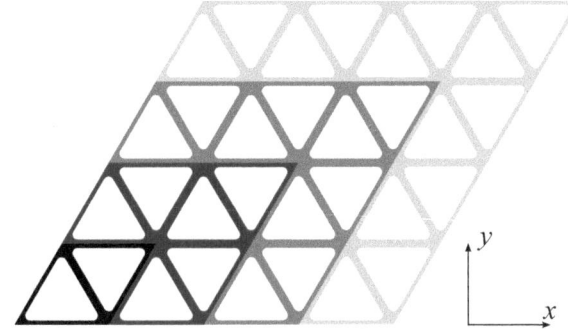

It is noted that more sophisticated geometries and parameterizations are possible with higher computational effort without changing the general method and cell designs can be derived using the more sophisticated inverse homogenization approach. To facilitate manufacturing constraints such as a condition on the smallest feature size, the use of parametrized design methods such as the feature mapping approach [34, 35] or a spline box approach [15] is recommended.

Asymptotic homogenization can only detect high-frequency modes, i.e., modes with a wavelength that is smaller than the size of the representative volume element. However, buckling modes might span over more than one cell. To identify these modes, we encompass $\kappa \times \kappa$ unit cells in the RVE. RVEs containing various unit cells are shown in Fig. 12.9. We note that alternatively the Floquet-Bloch [23] approach along with a sufficiently fine discretization of the Brillouin zone can be used. As explained in [17], the presented method of repeating the unit cell within the RVE is analogous to applying the Floquet-Bloch theory with a special discretization of the Brillouin zone.

We observe that especially an RVE with only one unit cell results in rather high homogenized load factors. This is because the joint of the struts gets locked in rotation due to the periodic boundary conditions. In an RVE with more than one cell, and in the analysis of dehomogenized designs, joints of lattice structures are free to rotate (see Fig. 12.12b).

To incorporate buckling on the microscale, we construct a worst-case model from homogenized stability data, computed offline for every possible macroscopic stress situation. Under the assumption of linear pre-buckling, the homogenized load factor depends linearly on the macroscopic stress:

$$\lambda(\sigma, \rho) = \frac{1}{\|\sigma\|} \lambda\left(\frac{\sigma}{\|\sigma\|}, \rho\right) \tag{12.3}$$

Hence, homogenization can be conducted with macroscopic unit stress (in some given norm) and later, on the macroscopic scale, the homogenized load factor must be divided by the norm of the applied macro-stress. It is thus sufficient to examine stresses on the unit sphere surface S^2 instead of the whole three-dimensional stress space $(\sigma_{xx}, \sigma_{yy}, \sigma_{xy})$.

We use a spherical coordinate system to characterize unit stresses on S^2: The zenith reference (z-axis) is the axis from the origin through the biaxial compression stress $\sigma = (-1, -1, 0)$ and the azimuth reference (x-axis) is the axis from the origin through $\sigma = (1, -1, 0)$ (Fig. 12.10a). In this coordinate system, biaxial compression and tension stress conform to the north and south poles, while all pure shear stresses rest on the equator. The inclination (or latitude if thinking of geographical coordinates) characterizes the type of stress, whereby the rotation invariant biaxial compression and tension stresses conform to poles, and other special types, e.g.,

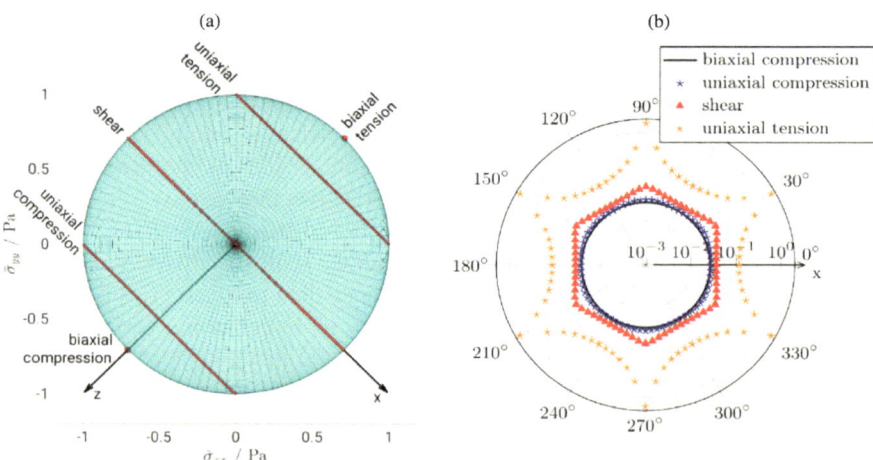

Fig. 12.10 **a** Representation of unit stresses as a unit sphere surface (σ_{xy} out of plane axis). Special stress types form circles (uniaxial/shear stress) or single points (biaxial stress) on the surface. Unit stresses are parameterized by spherical coordinates with zenith reference z and azimuthal reference x. **b** Homogenized buckling load factors for an RVE with 3×3 unit cell repetitions and a relative volume of 30%. Note the logarithmic scaling of the radial axis. The displayed angle is the azimuthal angle from the spherical reference coordinate system in **a** and corresponds to the rotation of the stress around the RVE. The symmetry of the unit cell is reflected in the load factors

uniaxial and shear stresses form circles on the unit sphere surface (circle of latitude). The azimuthal angle (longitude) describes the rotation of the applied macroscopic stress σ relative to the RVE.

The homogenized load factors for uniaxial compression, biaxial compression, and shear stress for an RVE with a relative volume of 30% and three cell repetitions are shown in Fig. 12.10b. We can clearly see that the symmetry of the unit lattice cell is reflected in the load factors. The shape of the uniaxial loading case closely matches the results obtained by [5].

Next, we coalesce all data in a worst-case model. Having evaluated homogenized load factors for different cell repetitions and all stresses on the unit stress sphere $\sigma \in S^2$, i.e., for all stress types and directions, we select the smallest buckling load factor with respect to all unit stresses and number of cell repetitions:

$$\lambda_{wc}(\rho) = \min_{\sigma \in S^2, \kappa \in \mathbb{N}} \lambda(\sigma, \rho) \tag{12.4}$$

This worst-case model depends only on the local volume fraction and no longer on the local stress type or direction. We note that in practice the unit sphere is discretized and the number of cell repetitions bounded from above. The worst case is thus only a worst case with respect to the discretization resolution and maximal cell repetitions. We found, however, that the homogenized buckling load factors for our exemplary lattice cell have high regularity with respect to the stress variable.

The parameterization of the lattice allows us to decouple the two scales like Bendsøe and Kikuchi did in [2]: Prior to any optimization procedure, a discrete subset of the parameter space is chosen, and homogenization is conducted for each of the microstructures gained from this parameter set. The properties obtained are then interpolated in the continuous parameter space and the interpolated material model can later be reused to solve various optimization problems.

For this, we employ a piecewise cubic Hermite interpolation approach [4], which results in a continuously differentiable model that allows for conducting gradient-based optimization later. We obtain the worst-case microscopic buckling load factor associated with macroscopic stress σ by dividing by the norm of this stress (see (Eq. 12.3)):

$$\lambda^I(\sigma, \rho) = \frac{1}{\|\sigma\|} \lambda_{wc}(\rho) \tag{12.5}$$

where λ^I_{wc} is the interpolated version of Eq. 12.4 and is shown in Fig. 12.10a.

The error of the approximated microscopic buckling load factor λ^I comprises several individual errors: the error introduced by finite element discretization on the microscopic level to solve the homogenization problems, the error arising from restriction to the worst case, and the error resulting from the interpolation. The discretization and interpolation errors can easily be controlled by using finer finite element meshes on the microscopic scale and finer interpolation grids. Thus, the worst-case error has the highest significance.

The worst-case error can be avoided if the worst-case model (12.4) and univariate interpolation (12.5) are replaced by a discretization and trivariate interpolation in the combined stress and density space $S^2 \times (0, 1]$. For gradient-based optimization, a continuously differentiable interpolation scheme is essential. Possible techniques to realize that for this worst-case free approach include piecewise cubic Hermite interpolation [4] or interpolation on sparse grids based on cubic B-splines [31].

However, as we will see later, in a three-dimensional setting, stress has six independent entries, which result in five parameters to represent unit stress, requiring interpolation to be carried out in a six-dimensional space. There, due to the curse of dimensionality, the differentiable sparse grid approach [31] is preferable for the worst-case free approach.

The worst-case approach that was described was implemented in the open-source software openCFS [32] and numerical experiments were conducted. In a two-dimensional setting, we choose a rectangular design domain with an aspect ratio of 1:5.2. Movement at the lower edge is restricted in the vertical direction and pressure is applied at the upper edge.

We want to minimize the mechanical compliance $c(\rho)$ and maximize the buckling load factor on both scales under a global volume constraint (12.10). We treat this multiobjective optimization problem with the ϵ-constraint method [22] and move the compliance into a constraint (12.9). A combination of the load factors of both scales in a minimum function would lead to a non-smooth character. Instead, we apply a bound formulation using the slack variable s, as suggested by Bendsøe [3]. We include the first few $L < M$ macroscopic load factors in the objective to handle potential mode switching, and end up with the following optimization problem:

$$\max_{s,\sigma \in \mathbb{R}, \rho \in \mathcal{U}_{ad}} \quad s \tag{12.6}$$

$$\text{s.t.} \quad \Lambda_\ell(\rho) \geq s, \quad \ell = 1, \ldots, L \tag{12.7}$$

$$\lambda_e^I(\rho) \geq s, \quad e = 1, \ldots, M \tag{12.8}$$

$$c(\rho) = c_0 \tag{12.9}$$

$$V(\rho) = V_0 \tag{12.10}$$

The values obtained for the simultaneous optimization of both macroscopic Λ_ℓ and microscopic λ_e^I buckling load factors are shown in Fig. 12.11b, and the optimized designs are displayed in Fig. 12.11a. As a reference, the values of a topology optimization (TO) with a lower bound on the density $\rho = 0.001$ are given.

To achieve a stiff (small compliance) design like C_1, thick solid structures are needed. Little material is left for the lattice part, which results in low local volume and a small minimal microscopic buckling load factor. Hence, only (12.8) and (12.9) are active, and (12.7) remains inactive.

For less restrictive (larger) compliance bounds, less solid material is necessary, and material is redistributed to the lattice region (C_2). Thus, the microscopic buckling load factor can be improved. When it reaches the value of the macroscopic one,

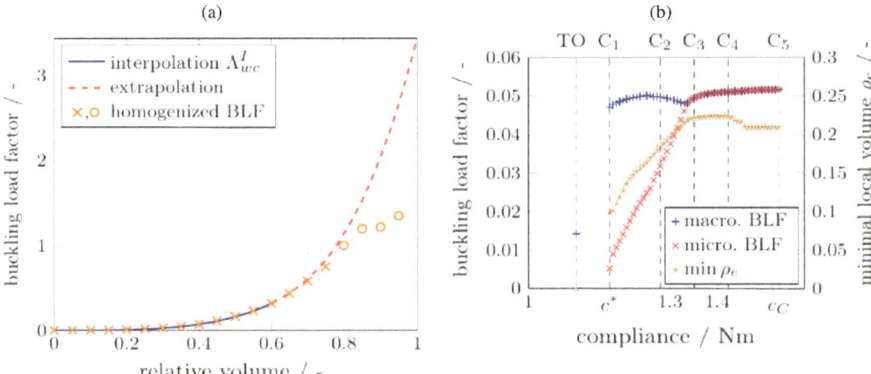

Fig. 12.11 **a** Homogenized worst-case buckling load factors and the interpolating curve. For high relative volumes, numerical issues arise (marked by an o). Thus, only values up to 0.6 are interpolated, and quadratic extrapolation is applied above. **b** Optimized function values of concurrent load factor maximization of pure macro- and pure microscopic buckling with compliance and volume constraint. Up to a compliance of 1.33, only the microscopic buckling load factor is active. For higher compliance values, the macroscopic buckling load factor dominates. Selected designs are shown in Fig. 12.12a

(12.7) becomes active (C_3). Therefore, raising only the microscopic buckling load factor further yields no improvement in the objective, as the macroscopic load factor will define the value of the slack variable. For this reason, both micro- and macroscopic buckling load factors are raised simultaneously: the first by increasing the lattice density especially in the upper region of the design domain and the second by stiffening the exoskeleton (C_4, C_5).

To validate the optimized results, we dehomogenize design C_4, i.e., reinterpret the design field by a lattice structure. Figure 12.12b shows different buckling modes: the one associated with a macroscopic buckling mode, i.e., deflection of the structure as a whole (low-frequency), and the other associated with a microscopic mode, i.e., deflection of the lattice (high-frequency). The microscopic buckling load factor stems from modes at the boundary. These modes cannot be approximated well by homogenization theory, which ignores boundary effects. We therefore also search for the smallest load factor associated with an interior mode, i.e., a mode that does not exhibit deflection at the structure's boundary.

The macroscopic load factor is determined to be 0.0507 and differs by less than 1% from the predicted macroscopic load factor of 0.0511 in the optimized homogenized design. A gap of 10% exists between the interior microscopic load factor of the dehomogenized design of 0.0563 and the predicted one of 0.0511. This is expected, as the predicted microscopic load factor is based on our worst-case model, which assumes the worst stress type and orientation (see (12.4)).

It should be noted that the microscopic buckling load factor predicted by the worst-case model is smaller than the actual microscopic buckling load factor, and hence our worst-case model acts as a safeguard against pure microscopic buckling in

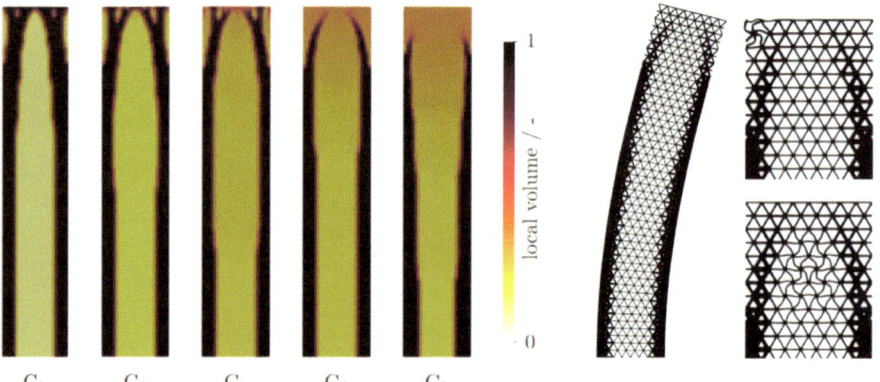

Fig. 12.12 Left: Optimized designs for the marked data in Fig. 12.11b. To increase the microscopic buckling resistance, the optimizer reduces lattice porosity. Center: First macroscopic buckling mode for design C4 dehomogenized with eight lattice cells. Right: Magnification of the first microscopic mode and first interior mode for dehomogenization as in the center

this example. Naturally, such an observation is generally only valid up to remaining discretization and interpolation errors, as well as the error introduced by homogenization.

Using the methods from Chap. 15, the method was extended in several ways [14]. First, instead of asymptotic homogenization based on the finite element method, we use a numerical homogenization scheme based on beam theory to upscale the lattice's properties, which can also capture nonlinear effects (see Chap. 15). Second, we extended our model and method to a three-dimensional setting. This required, among other things, to capture the buckling responses of the base cell with respect to applied stresses from a six-dimensional space.

In this instance, for the numerical examples, we limit the design variable, i.e., the lattice volume ρ, to the interval $[0.018, 0.20]$. The lower bound is motivated by manufacturability, while the upper bound is given such that the application of a beam model remains physically reasonable. Moreover, we globalize the local buckling constraints, i.e., we replace the local constraints (12.8) with:

$$\frac{1}{M} \sum_{e=1}^{M} \left(\max(v - \lambda_e^I(\rho_e, \sigma_e(\rho)), 0) \right)^2 < \varepsilon \qquad (12.11)$$

whereby this function is the normalized sum of the squared violations of Eq. 12.8, motivated by the method of least squares in regression analysis. Due to the maximum function, only load factors that are below the given threshold v contribute to the function value. The derivative of this function is computed using only a single adjoint equation, in which the right-hand side is a sum of the adjoint right hand sides of

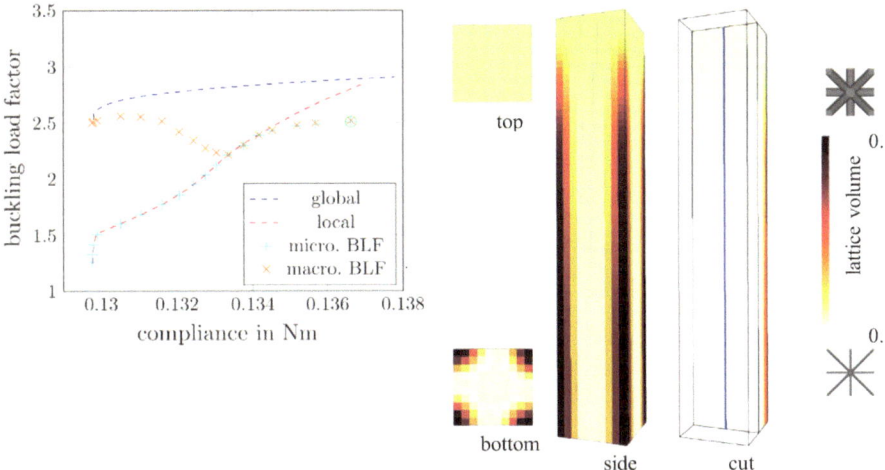

Fig. 12.13 a Optimized function values for compliance minimization under a volume constraint (Eq. 12.10) and constraints on both the macroscopic (Eq. 12.7) and microscopic (Eq. 12.11) buckling load factors. The design of the data marked by a green circle is shown in (**b**). **b** Macroscopic stability is maintained by a thick lattice at the vertical edges, while microscopic stability is achieved by a graded lattice structure. The design at the central vertical line (blue) is examined in Fig. 12.14

the local constraints [6]. Thus, only one adjoint problem must be solved, which drastically reduces the computational effort.

We perform the optimization with a column-shaped design domain with a square cross section and a side-to-height ratio of 1 : 6. At the top, we apply a distributed force of 1 N. Thus, the macroscopic load factor directly represents the critical load. We fix all degrees of freedom for nodes in a central region of the bottom face, while all other nodes at the bottom are only fixed in the vertical direction. These boundary conditions are chosen to prevent stress peaks in the corners of the bottom face.

The function values obtained are shown in Fig. 12.13a. From left to right, first the microscopic load factor increases. This comes with a decrease in the macroscopic buckling load factor, which is already inactive. The two curves meet at a compliance of approximately 0.1333, i.e., the microscopic limit load is equal to the macroscopic critical load. Requiring higher buckling loads comes at the cost of mechanical compliance. The largest achievable critical load is 2.51 N, and for larger values the problem becomes infeasible.

The design for this data is shown in Fig. 12.13b. There is some thick lattice at the vertical edges, which increases the macroscopic stability while the lattice in the middle of the structure is smoothly graded from bottom to top (see Fig. 12.14). As the stress increases from bottom to top, the lattice becomes thicker to achieve an almost homogeneous buckling load factor. The minimal lattice rod diameter is 0.6 mm, while at the top it is 0.76 mm, which is an increase of more than 50% with respect to the relative lattice volume.

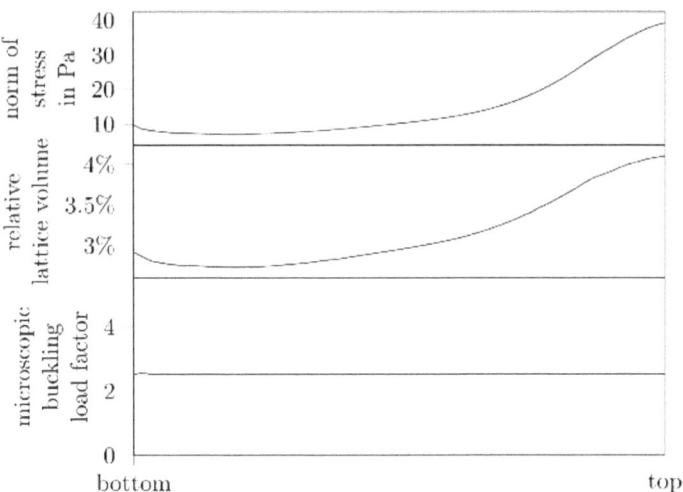

Fig. 12.14 Norm of stress, design, and microscopic buckling load factor at the central vertical line of the design shown in Fig. 12.13b. The gradation of the stress leads to a graded lattice design to achieve a nearly homogeneous microscopic buckling load factor

The gradation of the lattice structure clearly shows the advantage of a homogenization-based model for the load factor, as the approach of choosing a larger lower bound for the relative lattice volume to meet stability requirements everywhere can lead to a waste of material in regions of low stress.

The workflows described in this article are only examples for designing multifunctional optimized components utilizing the capabilities offered by additive manufacturing. One possible extension in the case of optimized lattice structures is to work with more than one parametrized cell type. A generalization of the two-scale design method to such a setting based on the sequential global programming concept was established under the umbrella of the CRC814 and first published in [33]. Another extension would consist in combining the techniques proposed in this chapter to come up with a cellular design made up of base cells using multiple materials (e.g., a very soft and a stiff phase). A consequent realization of the offline computation of properties of building blocks and their interpolation in macroscopic optimization frameworks would require techniques as described in [31]. Finally, a more flexible optimization of the candidate structures for the base cell itself could be of interest when moving from buckling stability to other physical quantities. An example of how this could be achieved in a robust manner is given in [34, 35], where a feature mapping approach is utilized to design meta-materials [36].

References

1. Achtziger, W., Bendsøe, M.: Optimal topology design of discrete structures resisting degradation effects. Struct. Optim. **17**, 74–8 (1999)
2. Bendsøe, M., Kikuchi, N.: Generating optimal topologies in structural design using a homogenization method. Comput. Methods Appl. Mech. Eng. **71**, 197–224 (1988)
3. Bendsøe, M., Sigmund, O.: Topology Optimization: Theory, Method and Applications, 2nd edn. Springer, Berlin (2003)
4. Birkhoff, G., Schultz, M., Varga, R.: Piecewise hermite interpolation in one and two variables with applications to partial differential equations. Numer. Math. **11**, 232–56 (1968)
5. Bluhm, G., Sigmund, O., Wang, F., Poulios, K.: Nonlinear compressive stability of hyperelastic 2D lattices at finite volume fractions. J. Mech. Phys. Solids **137**(103851) (2020)
6. Clausen, A.: Topology Optimization for Additive Manufacturing. Ph.D. Thesis, Technical University of Denmark (DTU) (2016)
7. Collette, Y., Siarry, P.: Multiobjective Optimization: Principles and Case Studies. Springer Science & Business Media (2004)
8. Greifenstein, J., Stingl, M.: Simultaneous material and topology optimization based on topological derivatives. In: IFIP Conference on System Modeling and Optimization, pp. 118–27. Springer, Berlin (2013)
9. Greifenstein, J., Stingl, M.: Simultaneous optimization of build orientation and topology in layered manufacturing. In: Proceedings of 5th International Conference on Additive Technologies, pp. 194–201 (2014)
10. Greifenstein, J., Stingl, M.: Robust optimal component design under consideration of local material defects. In: Proceedings of 6th International Conference on Additive Technologies, pp. 246–50 (2016)
11. Greifenstein, J., Stingl, M.: Simultaneous parametric material and topology optimization with constrained material grading. Struct. Multidiscip. Optim. **54**, 985–98 (2016)
12. Greifenstein, J., Stingl, M.: Topology optimization with worst-case handling of material uncertainties. Struct. Multidiscip. Optim. **61**, 1377–97 (2020)
13. Hübner, D., Gotterbarm, M., Kergaßner, A., et al.: Topology optimization in additive manufacturing considering the grain structure of inconel 718 using numerical homogenization. In: Proceedings of 7th International Conference on Additive Technologies, pp. 102–11 (2018)
14. Hübner, D., Herrnböck, D., Wein, F., Mergheim, J., Steinmann, P., Stingl, M.: Lattice structure optimization in additive manufacturing using numerical homogenization based on beam models. Arch. Appl. Mech. (2023)
15. Hübner, D., Rohan, E., Lukeš, V., Stingl, M.: Optimization of the porous material described by the biot model. Int. J. Solids Struct. **156**, 216–33 (2019)
16. Hübner, D., Stingl, M.: On a combined geometry and multimaterial optimization approach for the design of frame structures in the context of additive manufacturing. In: Proceedings of 7th International Conference on Additive Technologies (Maribor), pp. 112–9 (2018)
17. Hübner, D., Wein, F., Stingl, M.: Two-scale optimization of graded lattice structures respecting buckling on micro- and macroscale. Struct. Multidiscip. Optim. **66**, 163 (2023)
18. Kufner, T., Leugering, G., Semmler, J., Stingl, M., Strohmeyer, C.: Simulation and structural optimization of 3d timoshenko beam networks based on fully analytic network solutions. ESAIM: Math. Model. Numer. Anal. **52**, 2409–2431 (2018)
19. Leugering, G., Nazarov, S., Schury, F., Stingl, M.: The Eshelby theorem and application to the optimization of an elastic patch. SIAM J. Appl. Math. **72**, 512–34 (2012)
20. Liu, J., Gaynor, A., Chen, S., et al.: Current and future trends in topology optimization for additive manufacturing. Struct. Multidiscip. Optim. 1–27 (2018)
21. Mergheim, J., Breuning, C., Burkhardt, C., et al.: Additive manufacturing of cellular structures: multiscale simulation and optimization. J. Manuf. Process. 275–290 (2023)
22. Miettinen, K.: Nonlinear multiobjective optimization, vol. 12. Springer Science & Business Media (1999)

23. Neves, M.: Symbolic computation to derive a linear-elastic buckling theory for solids with periodic microstructure. Int. J. Comput. Methods Eng. Sci. Mech. **20**, 523–39 (2019)
24. Rodrigues, H., Guedes, J., Bendsøe, M.: Hierarchical optimization of material and structure. Struct. Multidiscip. Optim. **24**, 1–10 (2002)
25. Schury, F., Greifenstein, J., Leugering, G., Stingl, M.: On the efficient solution of a patch problem with multiple elliptic inclusions. Optim. Eng. **16**, 225–46 (2014)
26. Schury, F., Stingl, M., Wein, F.: Efficient two-scale optimization of manufacturable graded structures. SIAM J. Sci. Comput. **34** (2012)
27. Semmler, J., Pflug, L., Stingl, M.: Material optimization in transverse electromagnetic scattering applications. SIAM J. Sci. Comput. **40**, B85–B109 (2018)
28. Sivapuram, R., Dunning, P., Kim, H.: Simultaneous material and structural optimization by multiscale topology optimization. Struct. Multidiscip. Optim. **54**, 1267–81 (2016)
29. Stingl, M., Kočvara, M., Leugering, G.: A sequential convex semidefinite programming algorithm with an application to multiple-load free material optimization. SIAM J. Optim. **20**, 130–155 (2009)
30. Svanberg, K.: A class of globally convergent optimization methods based on conservative convex separable approximations. SIAM J. Optim. **12**, 555–573 (2002)
31. Valentin, J., Hübner, D., Stingl, M., Pflüger, D.: Gradient-based two-scale topology optimization with b-splines on sparse grids. SIAM J. Sci. Comput. **42**, B1092–B1114 (2020)
32. Verein zur Förderung der Software openCFS. Opencfs. https://opencfs.org/ (n.d)
33. Vu, B., Lukeš, V., Stingl, M., Rohan, E.: A sequential global programming approach for two-scale optimization of homogenized multiphysics problems with application to biot porous media (2023). arXiv:2301.11852 [cs.CE]
34. Wein, F., Dunning, P., Norato, J.: A review on feature-mapping methods for structural optimization. Struct. Multidiscip. Optim. **62**, 1597–638 (2020)
35. Wein, F., Stingl, M.: A combined parametric shape optimization and ersatz material approach. Struct. Multidiscip. Optim. **57**, 1297–315 (2018)
36. Wormser, M., Wein, F., Stingl, M., Körner, C.: Design and additive manufacturing of 3d phononic band gap structures based on gradient based optimization. Materials **10**, 1125 (2017)
37. Xia, L., Breitkopf, P.: Multiscale structural topology optimization with an approximate constitutive model for local material microstructure. Comput. Methods Appl. Mech. Eng. **286**, 147–167 (2015)

Michael Stingl has been a full professor in Applied Mathematics at the Friedrich-Alexander-Universität Erlangen-Nürnberg since 2014. His main research interest is in algorithmic optimization with a particular focus on processes and applications arising in the context of material and topology optimization. Up to now he (co-)authored more than 80 research publications. His work has been supported by several DFG and EU grants.

Chapter 13
DEM Simulation of the Powder Application in Powder Bed Fusion

Vasileios Angelidakis, Michael Blank, Eric J. R. Parteli, Sudeshna Roy, Daniel Schiochet Nasato, Hongyi Xiao, and Thorsten Pöschel

13.1 Introduction

The packing behavior of powders is significantly influenced by various types of inter-particle attractive forces, including adhesion and non-bonded van der Waals forces [4, 7, 8, 19, 41, 43]. Alongside particle size and shape distributions, the inter-particle interactions, particularly frictional and adhesive forces, play a crucial role in determining the flow behavior and consequently the packing density of the powder layer. The impact of various types of attractive forces on the packing density of powders with different materials and particle size distributions remains largely unexplored and requires further investigation. Accurately comprehending these effects through experiments while considering specific particle size distributions and material properties poses significant challenges. To address these challenges, we employ Discrete Element Method (DEM) simulations to characterize the packing behavior of fine powders. We can demonstrate quantitative agreement with experimental results by incorporating the appropriate particle size distribution and using an adequate model of attractive particle interactions. Furthermore, our findings indicate that both adhesion, which is modeled using the Johnson-Kendall-Roberts (JKR) model [14], and van der Waals interactions are crucial factors that must be taken into account in DEM simulations.

Characterization of the packing structure is essential to achieve simulation-driven optimization of the layer quality. While local packing density is commonly employed to characterize granular packings, relying solely on this parameter is inadequate for identifying structural defects in disordered packings [33]. This is due to the fact that numerous packing arrangements can exhibit the same local density. Conversely, local structural anisotropy is a fundamental characteristic of non-crystalline packings and

V. Angelidakis · M. Blank · E. J. R. Parteli · S. Roy · D. S. Nasato · H. Xiao · T. Pöschel (✉)
Friedrich-Alexander-Universität Erlangen-Nürnberg, Institute of Multiscale Simulation of Particulate Systems, Cauerstraße 3, 91058 Erlangen, Germany
e-mail: thorsten.poeschel@fau.de

D. Drummer and M. Schmidt (eds.), *Progress in Powder Based Additive Manufacturing*, Springer Tracts in Additive Manufacturing,
https://doi.org/10.1007/978-3-031-78350-0_13

plays a significant role in crucial mechanical properties within disordered packings, such as jamming [34], plasticity [33], and shear band formation [10, 11, 42]. We use local structural anisotropy as a measure to characterize the structures of the deposited particles. This approach is threshold-free and provides a meaningful distribution that accurately reflects the variations in the packing structure within a deposited powder layer. It holds particular relevance in discerning the structural differences between homogeneous packing observed in non-cohesive powders and the heterogeneous packing tendencies prevalent in highly cohesive powders.

The heat transfer within the powder bed is significantly impacted by the quality of the final powder layer, including aspects such as the packing density and surface profile [44]. It is worth noting that many DEM simulations commonly disregard the influence of temperature on inter-particle interaction forces. Nevertheless, it is widely recognized that temperature can have a substantial impact on the mechanical behavior of particle interactions. The temperature dependence of viscosity is one of the most important variables for polymers, which is given by an Arrhenius type equation [6, 13]. As the temperature approaches higher values, nearing the melting point, the particles within a stressed state experience a reduction in stiffness. Consequently, this leads to increased deformation and larger overlaps between particles due to the compressive forces acting upon them. Luding et al. [20] introduced a new discrete model for the sintering of particulate materials with a temperature-dependent elastic modulus.

The role of particle shape coupled with its thermal properties is hardly ever explored in the literature and is thus a subject with great potential. We propose an extension of the DEM that incorporates heat transfer capabilities by combining a multisphere algorithm with a thermal discrete particle model, thereby enabling the simulation of non-spherical particles. To showcase the applicability of this model, we conduct simulations of a powder spreading process utilizing irregularly shaped Polyamide 12 (PA12) powder particles. This demonstration illustrates the effectiveness of the proposed model in capturing the behavior of non-spherical particles in real-world scenarios.

Improving the quality of the layer during powder spreading can be achieved by various means. In addition to exploring the thermal, cohesive, and overall mechanical properties of the powder material itself, an optimization of the steps of the powder application process can result in improved layer characteristics. To this end, the efficiency of a vibrating recoating mechanism has been explored with respect to the layer density and surface roughness.

13.2 Particle-Based Tool for Powder Spreading

The first particle-based numerical tool was developed by Parteli [28] based on an existing DEM solver [15] to model the dynamics of geometrically complex particles subjected to dynamic boundary conditions. Snapshots of the simulation setup with complex shaped particles and dynamic boundary conditions are shown in Fig. 13.1.

Fig. 13.1 Snapshots of a simulation of particles with complex geometric shapes. Particles constructed with the multisphere method are inserted into the system with dynamic boundary conditions that mimic the device used in additive manufacturing [28]

This setup mimics the device used for powder spreading in additive manufacturing. In this numerical setup, the moving boundaries (walls) are modeled by triangular meshes, which can be imported into the solver. The device consists of a rake for powder application (which moves from left to right in Fig. 13.1) and a building tank (central area) placed on top of a vertically adjustable platform. The simulation starts by releasing approximately 6500 particles from a short distance above the platform, as shown in Fig. 13.1a. After falling under gravity, the settled particles are transported on the platform as the rake moves from left to right, as shown in Fig. 13.1b–d. This small setup helps us to address several questions relevant to the real powder spreading process, such as the impact of particle shape and size distribution on the flowability of the powder within the device. This, in turn, allows us to understand the surface profile and packing density of the powder bed.

In 2016, Parteli and Pöschel [29] established a real scale setup for the first time to simulate powder spreading using a roller as the spreading device, as shown in Fig. 13.2a. The numerical tool based on the DEM accounts for a realistic description of inter-particle forces, particle size distributions, and complex geometric shapes of the powder particles, which play a major role in the static and dynamic characteristics of granular systems. With this numerical setup, the authors predicted the powder packing and surface roughness profile dependent on the recoating velocity and the powder particle size distribution.

13.3 Geometrically Complex Particles

The first step to achieve reliable numerical simulations of the powder spreading application in additive manufacturing is the accurate representation of the complex geometric shapes of the particles. While preliminary studies were done by Parteli [27, 28], this representation was first accomplished by Parteli and Pöschel [29] in the context of powder spreading for powder bed fusion using the multisphere method, which consists of building clumps of spherical particles to model the complex shape

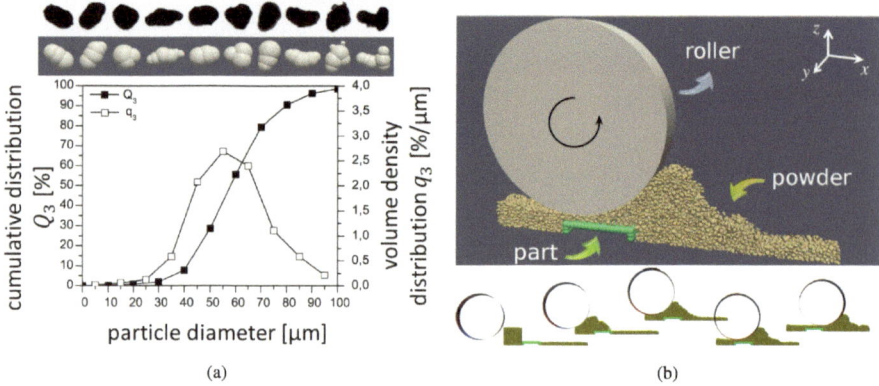

Fig. 13.2 **a** (Top) Light microscope images of commercially available PA12 powder particles (first row) and corresponding particle models using the multisphere method (second row) for implementation in the DEM. (Bottom) Cumulative distribution and volume density distribution as a function of the particle diameter. **b** Snapshot of the simulation indicating the main elements of the powder application process [29]

of the target. Within the DEM, there are various simulation techniques available for modeling non-spherical particles. One of the oldest and most versatile techniques is the multi-sphere approach, which involves rigidly connecting a group of spheres to create irregular particles [1, 3, 16]. In the multisphere approach, each individual subsphere serves the purpose of contact detection, enabling an accurate interaction between particles. However, when it comes to calculating inertial characteristics and integrating particle motion, the entire collection of spheres is considered as a single particle. This simplification reduces the computational complexity and facilitates efficient simulations. The fidelity of the multisphere particle's morphology, or how closely it approximates the shape of the target particle, is determined by the number and size of the spheres used in its construction. A higher number of spheres or smaller sphere sizes yield greater morphological fidelity, providing a better representation of the target shape. However, this also increases the computational cost associated with interactions between neighboring particles, as more contact points need to be considered. Therefore, the choice of the number and size of spheres in a multisphere particle involves a trade-off between morphological accuracy and computational efficiency, and researchers must carefully consider these factors based on the specific requirements of their simulation to strike an appropriate balance. Overall, the multisphere approach offers a flexible and widely used method for modeling non-spherical particles within DEM simulations, providing a compromise between accuracy and computational efficiency.

Figure 13.2a shows some examples in which target shapes were obtained from SEM images of commercially available PA12 powder as described in [29]. The translational and rotational motion of the multisphere particles is driven by the forces acting on the constituent spheres, which include gravity, inter-particle forces, and the forces due to particle and wall interactions. The small size particles $(1-100\,\mu m)$ involved in the process require attractive interaction forces to be modeled and incorporated into the DEM. We developed a particle-based model that takes contact forces and inter-particle attractive interactions into account. The contact forces are modeled using the Cundall and Strack model [5] considering elastic and dissipative forces in the normal collision direction and in the tangential direction. This model was extended by Parteli et al. [31] to incorporate attractive particle interaction forces. This improved DEM model takes bonded adhesion contacts and non-bonded van der Waals forces into account. The simulation predictions of the solid fraction of powders covering a wide range of size distributions were quantitatively in agreement with the experimental results [31].

Figure 13.2b shows the simulation snapshot indicating the main components of the powder spreading setup implemented for numerical simulation using the DEM model. The roller moves in the positive x direction with a translational velocity because it rotates in a counterclockwise direction with a constant rotational velocity.

13.3.1 Influence of Particle Size

The packing behavior of the deposited powder layer is strongly dependent on the transport dynamics of the powder over the substrate. In ideal powder spreading conditions, the goal is to enhance the production speed while maintaining flat, densely packed powder layers. To investigate this behavior, we analyzed the roughness of the powder bed as a function of the roller velocity V_R as shown in Fig. 13.3. We observe a clear increase in the bed roughness with the increase in the roller's translational velocity as represented by the blue squares in Fig. 13.3.

Fig. 13.3 Roughness of the deposited powder layer as a function of the translational velocity of the roller V_R [29, 30]

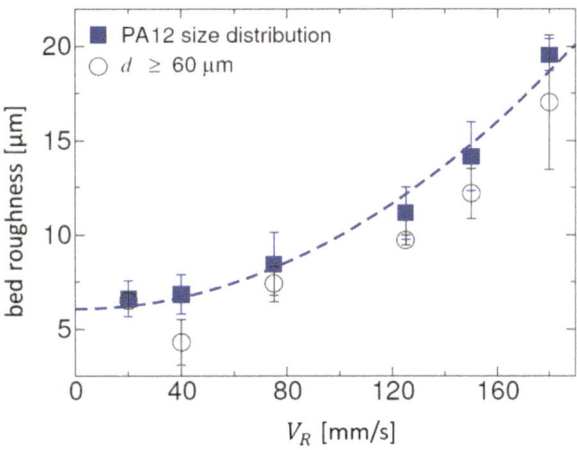

To further analyze how the particle size distribution affects the roughness of the bed, we modified the size distribution shown in Fig. 13.2b by removing all particles of diameter smaller than 60 μm. The surface roughness resulting from particles of size $d \geq 60$ μm is shown as a function of the roller's translational velocity, represented by the circles in Fig. 13.3. Overall, we achieve a smaller roughness with the larger particle size distribution, and the roughness increases in the same fashion with the roller's velocity as it does for the original size distribution.

Note that the attractive inter-particle forces (adhesion and non-bonded van der Waals forces) lead to the formation of large agglomerates in the powders associated with the blue squares in Fig. 13.3. For very small particles < 30 μm, the gravitational forces are dominated by the cohesive forces, and thus the particles are transported as irregular clumps and deposited on the surface. This results in a higher roughness than the original size distribution of the powder. Thus, a way to decrease the surface roughness of the deposited powder layers in additive manufacturing is to modify the particle size distribution of the powder by filtering out the smaller particles from the system.

13.3.2 Influence of Particle Shape

The impact of the real particle shape on the powder layer quality has been explored in detail by Nasato et al. [25] using multisphere reconstruction of 3D images of the powder particles obtained using the CT-REX tomograph as shown in Fig. 13.4. Samples from commercial Polyamide 11 (PA11), modified PA11, PA12, and reconstructed images from PA12, named S1, S2, S3, and S4, respectively, are considered for this study.

The surface roughness is measured for all four samples for different recoating velocities of the powder spreading tool. It is interesting to note from Fig. 13.5a that for every recoating velocity, the surface roughness is the highest for either the spheres or the sample S2 (rounded PA11). Samples S1, S3, and S4 have similar results concerning the roughness within the standard deviation, while it should be noted that samples S3 and S4 have a smaller roughness than sample S1, particularly for the higher recoating velocity of 250 mm/s. Samples S3 and S4 have the lowest value of aspect ratio and elongation ratio that was measured. Thus, it is speculated that the smaller surface roughness of the S3 and S4 samples is related to the alignment of the particles during the powder spreading process, leading to a smoother surface. An analysis of the packing density results for the different powder samples is shown in Fig. 13.5b. For a recoating velocity of up to 200 mm/s, samples S1 to S4 achieve higher packing densities than the sphere sample. The sample using spheres only reveals a higher packing density for the high recoating velocity of 250 mm/s. This is explained by the better flowability of the spheres at high velocities compared to the non-spherical particles. Thus, a competitive mechanism influences the quality of the powder layer in the recoating process. Spherical particles generally have better flowability and are thus preferred for additive manufacturing applications. However,

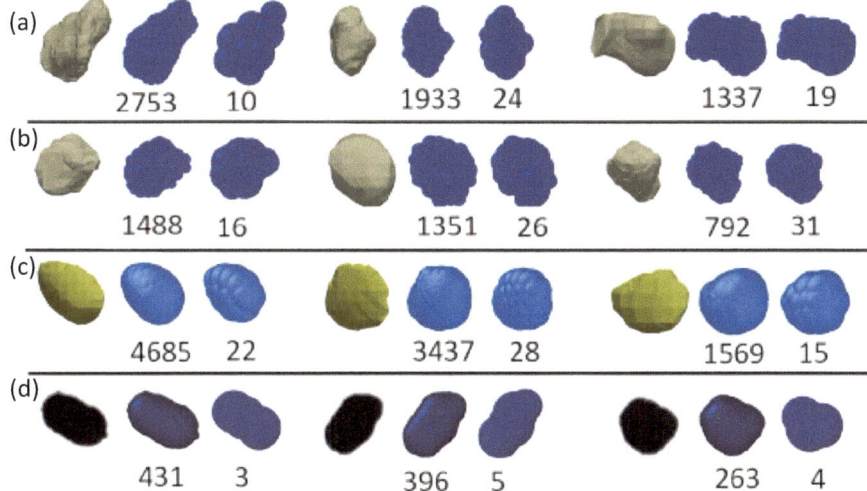

Fig. 13.4 Multisphere reconstruction of particles used in additive manufacturing. From left to right: original template (SEM image), all spheres before optimization, and optimized multisphere representation. **a** S1 sample (PA11), **b** S2 sample (PA11 powder rounded through precipitation), **c** S3 sample (PA12), and **d** S4 sample (PA12 reconstructed from SEM images) [24, 25]

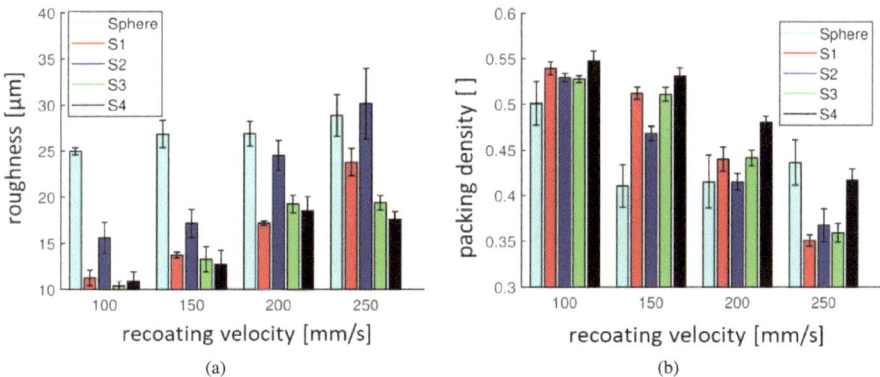

Fig. 13.5 a Roughness and **b** packing density of the deposited powder layer as a function of the translational velocity of the recoating blade V_R [25]

this only applies for high recoating velocities and for lower recoating velocities, the elongated shape of the particles plays a dominant role in maintaining higher packing densities.

13.4   Attractive Interactions Between Particles

It is crucial to not only consider repulsive contact forces but also attractive forces arising from van der Waals interactions when accurately modeling the dynamics of powder particles. Van der Waals forces are a type of intermolecular force that results from fluctuations in electron distributions, creating temporary dipoles and inducing attractive forces between particles. In this section, our focus is on investigating the effects of quantum mechanically based van der Waals forces and tensile forces arising from surface adhesion on the macroscopic behavior of powder particles on the packing structure of powder. While van der Waals forces are generally associated with intermolecular interactions at the atomic or molecular level, their influence can extend to macroscopic particles of small size ranges [31].

13.4.1   Significance of Attractive Forces

In cooperation with experimental validation, it must always be verified that the models used for DEM simulations lead to results that are consistent with the physical reality [40]. Therefore, the numerical tool, which is extended to include a more detailed description of particle geometry and the electrostatic inter-particle interactions, was first tested for simulations of a granular system in a simple geometry by Parteli et al. [31]. They verified the bulk packing density of fine glass powders in the diameter range $(4-52)\,\mu m$ by comparing the results with experiments and DEM simulations. The numerical setup is demonstrated in Fig. 13.6a, where silica glass beads are deposited in a rectangular box. We adopt periodic boundary conditions in the x and y directions and a frictional wall at $z = 0$. The height of the box is considered large enough to produce packings of depth larger than $30\langle d\rangle$. To validate our models, we obtained qualitative agreement between the experimental and numerical results, taking into account attractive forces for particle interaction, which include adhesion force and non-bonded van der Waals forces.

We performed various sets of simulations assuming different particle interaction force models, namely (i) viscoelastic interaction (no cohesion in Fig. 13.6b), (ii) adhesive and viscoelastic interaction (adhesion in Fig. 13.6b), and (iii) the full model, including viscoelastic, adhesive, and van der Waals interaction (adhesion and non-bonded in Fig. 13.6b). The results are shown in Fig. 13.6b. The packing density obtained from the experimental results was compared to the given sets of simulations. For the pure viscoelastic interaction model, the packing density obtained is almost independent of the mean particle size, $\langle d\rangle$, with a slight tendency of increasing packing density for smaller particles. This may be explained by the geometric effects of smaller particles filling the pore space. The simulation results with the viscoelastic interaction models agree with the experimental results for large particle sizes, $\langle d\rangle \gtrsim 40\,\mu m$ but disagree for smaller particle sizes. Simulations incorporating JKR-type adhesive forces reveal a decay in ϕ with decreasing $\langle d\rangle$. A good agreement is obtained

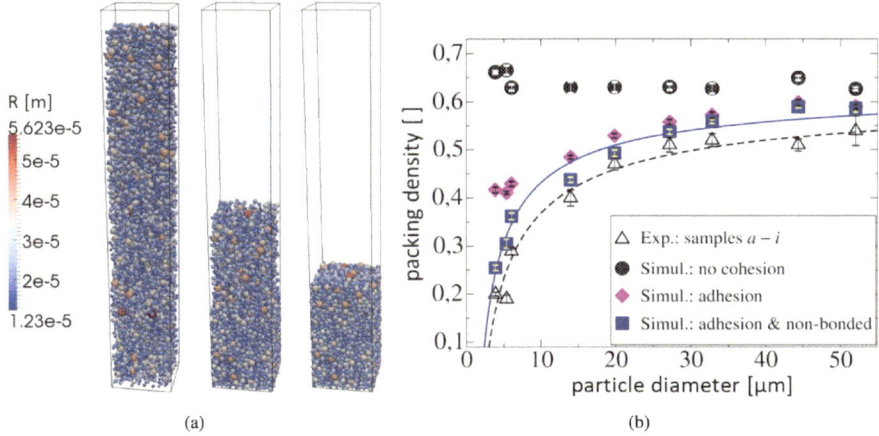

Fig. 13.6 a Numerical simulation of the powder packing of size distribution corresponding to sample i with radius varying from $10-55\,\mu$m and R representing the radius of the particles and **b** packing density as a function of the average particle diameter [31]

with the experiment for $\langle d \rangle \gtrsim 20\,\mu$m, although the data diverge for smaller particles. Simulations with the full model, including viscoelastic, JKR-adhesive, and van der Waals interactions allowed us to reproduce the experimentally measured packing fraction for the entire particle size interval $4\,\mu$m $\lesssim \langle d \rangle \lesssim 52\,\mu$m. Thus, from this we conclude that, for a predictive simulation of fine powder behavior, both adhesive and van der Waals forces are essential and should be considered in DEM simulations as neglecting any of these contributions in simulations of fine powders may lead to unreliable results.

13.4.2   Effect of Inter-Particle Cohesion on Packing Structure

Inter-particle cohesion plays a key role in the structure and density of the deposited powder layer, and thus, in the overall quality of the end product [12, 35]. Recently, we explored the structural anisotropy, packing density, and surface profile roughness of the deposited powder layer in spreading processes with varying cohesion using DEM simulations [38]. The local density and surface roughness of the produced layers were also quantified to identify the influence of cohesion on the formation of surface texture. Remarkably, based on the statistics of the local anisotropy and the surface roughness, we observe a synchronized transformation of the structural anisotropy and the surface roughness profile with increasing cohesion. In particular, the structure changes from homogeneous to heterogeneous, the layer surface roughness increases, and its peaks change shape, around $Bo \approx 10$, where Bo is a measure of inter-particle cohesion strength in a dimensionless form. In the following subsections, we highlight our main observations based on the aforementioned characteristics of powder layer packings.

Figure 13.7a illustrates an exemplary deposited layer of highly cohesive particles, characterized by the Bond number ($Bo \approx 30$). On examination, we observe a heterogeneous structure comprising both densely packed and loosely packed regions. This variation in packing density highlights the complexity and non-uniformity of particle arrangements within the deposited layer. The presence of regions with varying packing densities within the deposited layer of highly cohesive particles renders the overall packing density inadequate as a sole characterization metric. In order to address the challenge of characterizing the structural inhomogeneity in the deposited layer, we propose a novel method that utilizes a threshold-free measure based on the local structural anisotropy. The method of characterizing the structural inhomogeneity using a threshold-free measure based on the local structural anisotropy, measured by Q_k, and calculated from Voronoi tessellation, is explained in Roy et al. [38]. As stated by Rieser et al. [34], in regions of dense and homogeneous packing, the values of Q_k tend to exhibit random fluctuations, resulting in a Gaussian-shaped distribution. However, in regions characterized by inhomogeneity and high anisotropy, the distribution of Q_k deviates from the Gaussian shape. In such cases, both highly positive and highly negative values of Q_k appear at the tails of the distribution.

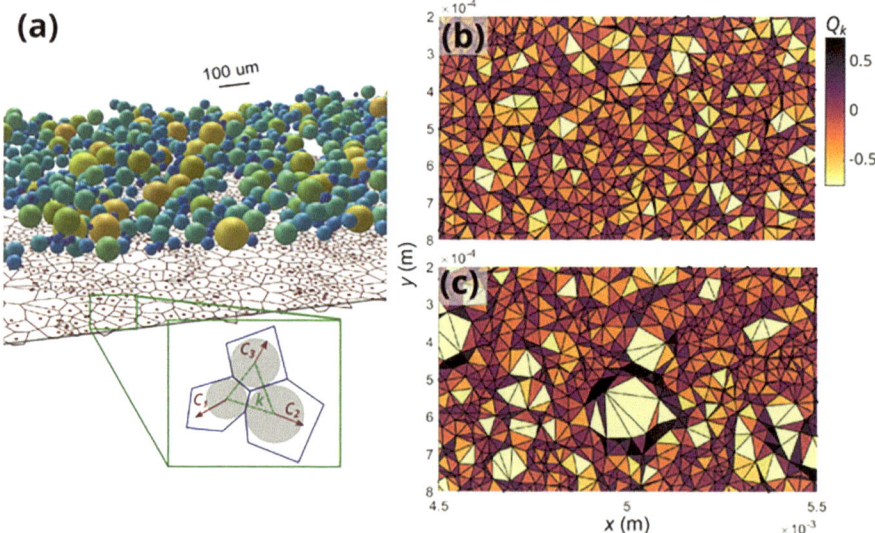

Fig. 13.7 Characterizing the structural anisotropy in the deposited layer. **a** Granular packing of the powder layer for $Bo \approx 30$. The points represent the projections of the center points of the particles in the plane. The corresponding 2D Voronoi tessellation is also shown. Also shown in **a** is a schematic representation of the particle packing with the superimposed Voronoi tessellation (blue) and Delaunay triangles (green). The vectors C_p (red) point from the center of the particles to the centroids of the Voronoi cells. The calculated Q_k for **b** $Bo = 0$ and **c** $Bo \approx 30$, where the Delaunay triangles are colored by the corresponding values Q_k [36]

Figure 13.8a shows the distribution of Q_k for different Bond numbers (Bo), providing visual evidence of this behavior. The non-Gaussian shape observed in the distribution indicates the presence of inhomogeneous regions with pronounced anisotropy, where the particle arrangements deviate significantly from random fluctuations. This analysis enables the characterization and differentiation of regions with varying degrees of structural complexity, providing valuable insights into the packing behavior as influenced by the Bond number. Referring to Fig. 13.8b, we observe interesting transition behavior of the standard deviation and skewness of the Q_k distribution for different Bond numbers (Bo). Initially, up to $Bo \approx 10$, the standard deviation remains relatively constant, indicating a consistent level of structural anisotropy. Similarly, the skewness also exhibits minimal variation within this range. However, beyond $Bo \approx 10$, both the standard deviation and skewness start to exhibit a monotonic increase and decrease, respectively. This suggests a significant shift in structural anisotropy for cohesive materials, with $Bo \approx 10$ serving as a transition point. For higher cohesion values ($Bo > 10$), particles have the ability to maintain larger voids during the spreading process, resulting in a more pronounced anisotropic structure. The observed fluctuations in the trends for higher Bo values, even within repeated simulations, can be attributed to the inherent high anisotropy of the structures. These fluctuations highlight the complexity and variability of particle arrangements in cohesive systems with enhanced anisotropy. By analyzing the standard deviation and skewness of the Q_k distribution, we gain valuable insights into the transition in structural anisotropy for cohesive materials and the impact of different Bond numbers on the packing behavior.

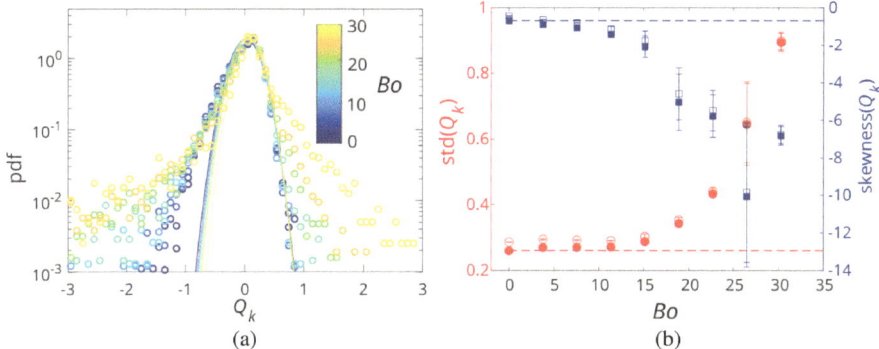

(a)   (b)

Fig. 13.8 **a** Probability density of normalized divergence of center-to-centroid vectors for the quasi-2D packing of powder deposited for different Bo. $Q_k > 0$ regions are more densely packed than their surroundings; hence, we call these regions overpacked. $Q_k < 0$ regions are more loosely packed than their surroundings and are therefore labeled underpacked. **b** Standard deviations (red circles) and skewness (blue squares) versus Bo. We also compared the standard deviations and skewness of the distributions with the divergence calculated from the condition without 2D projections (hollow points) and the condition with 2D projections (solid points), as explained in [36]. Error bars reflect standard errors in three simulation repetitions. The dashed lines indicate the standard deviation and the skewness of the distribution Q_k for $Bo = 0$ [36]

13.5 Thermal and Mechanical Properties

This section aims to expand our understanding of the various aspects involved in powder application and gain a more comprehensive insight into the process dynamics by combining the thermal and mechanical effects of the process and material parameters on the powder application process. To incorporate heat transfer capabilities into our DEM model for simulating non-spherical particles, we have extended the existing framework by merging a multisphere algorithm with a thermal discrete particle model [26]. The heat transfer for multi-sphere particles includes conduction, convection, and radiation. Conduction occurs when a clump overlaps with another clump or a boundary, while convection and radiation occur for the exposed portion of a clump's surface with the background. The total thermal transfer rate is expressed as the sum of the heat flux from conduction, convection, and radiation. To demonstrate the effectiveness of our extended DEM model with heat transfer capabilities, we conducted simulations of a powder spreading process using PA12 powder particles of irregular shapes.

13.5.1 Multi-sphere Particle Generation

In this study, multi-sphere particles were generated to approximate the shape of PA12 powder particles. To achieve this, an image-informed particle generation procedure was implemented, utilizing scanning electron microscopy (SEM) images of PA12 particles as a Ref. [23]. The procedure involved extruding the SEM images into three-dimensional (3D) representations, which served as the target geometry for generating realistic multi-sphere particles. The generation of multi-sphere particles was performed using a recently developed method based on the Euclidean transform of 3D images. The open-source software CLUMP [2] was employed, which offers various techniques for the multisphere generation. For adequately approximating the target particle shapes, a total of 10 sub-spheres were used per particle. The choice of the sub-sphere count was determined to strike a balance between accurately representing the target particle morphology and maintaining computational efficiency. By employing the image-informed particle generation procedure and the multi-sphere approach with 10 sub-spheres per particle, the study aimed to create realistic representations of PA12 powder particles for subsequent simulations and analysis. The different irregular particle shapes that are included in our studies are from light microscope images of commercially available PA12 as illustrated in Fig. 13.2a.

13.5.2   Simulation of Powder Spreading of Multisphere Particles with Thermal Properties

A small part of the powder bed (width 0.75 mm) is simulated using periodic boundary conditions in the y-direction. The contact model used for this simulation is the Hertz-Mindlin model combined with viscous damping [22]. In this numerical setup, the substrate is assumed to be smooth. We insert multispheres of 10 different irregular shapes generated from images of commercially available PA12 particles as shown in Fig. 13.9. The multispheres are inserted in front of the spreader tool until the total bulk particle volume equals $0.7\,\mathrm{mm}^3$, which is sufficient material to create a powder layer of 5 mm length, 0.75 mm width, and 100 μm height. We allow the particles to settle and relax under the effect of gravity before we start the spreading process. These simulations begin with particle temperatures set at $T_{\mathrm{part}} = 393\,\mathrm{K}$, and the powder is spread over a pre-heated plate held at $T_{\mathrm{plate}} = 430\,\mathrm{K}$, as previously documented in [17, 18, 32]. After the particles have settled down and the system is relaxed, the spreading process starts by moving the tool at a constant speed v mm/s. Particles reaching the end of the powder bed (at $x = 5$ mm) are not considered in the analysis. Finally, the simulation is stopped when the system is static, i.e., the kinetic energy of the system is sufficiently low. Thus, the maximum simulation time is chosen as $t_{\mathrm{max}} = 1$ s, taking into account the time for initial settling and relaxation of particles, the spreading time, and the time for the system to reach a static state. A detailed description of the setup is provided in [37].

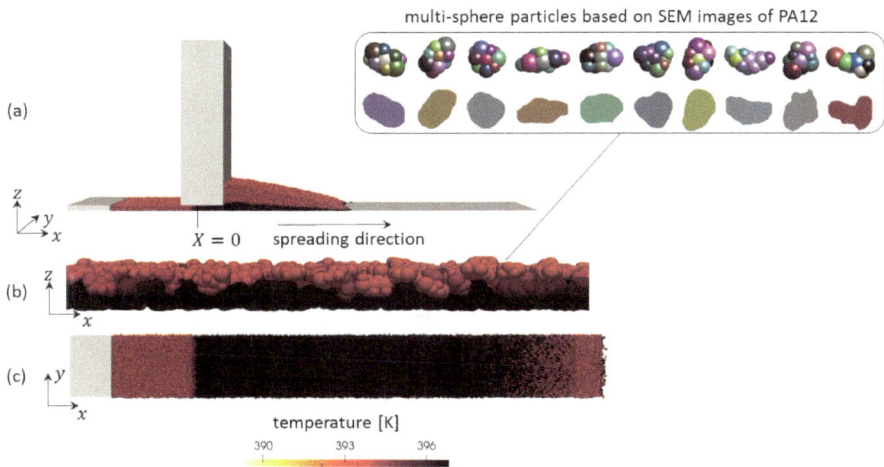

Fig. 13.9   a Powder spreading setup with a blade speed $v = 10$ mm/s, showing **b** front view after 0.08 s and **c** top view after 1.0 s of the spread powder layer [37]

13.5.3 Packing Density Analysis with Thermal Influence

The elastic modulus is dependent on the temperature as shown in [20]

$$E^*(T_{ij}) = \frac{E_0^*}{\alpha}\left[1 + (\alpha - 1)\tanh\left(\frac{T_m - T_{ij}}{\Delta T}\right)\right] \tag{13.1}$$

where $T_m = 451\,$K is the melting temperature, $\Delta T = 20\,$K is the temperature interval over which phase change occurs, and α is the parameter defining the dependency of the elastic modulus on the temperature. The $E^*(T_{ij})$ model proposed by Luding et al. [20] conforms to $\alpha = 2$, which needs further experimental validation. Note that according to Eq. 13.1, an extreme softening of the material of up to 80 % occurs at a large value of α and at an extreme temperature of the particles, close to their melting point. However, in our powder spreading setup, such a large temperature is not achieved by particles without sintering.

We characterized the packing density of the powder layer subjected to varying blade velocity and the dependence of the particle stiffness on temperature. In Fig. 13.10, we observe that the packing density is significantly influenced by the blade velocity. The packing density corresponding to a lower blade speed of 50 mm/s in Fig. 13.10a is lower than that in Fig. 13.10b for a blade speed of 250 mm/s. Further, the packing density is more sensitive to the values of α at a lower blade speed of $v = 50\,$mm/s, whereby it should be noted that the packing density is more sensitive to a smaller value of α, i.e., there is a significant difference in packing fraction between $\alpha = 2$ and 20 as opposed to between the values for $\alpha > 20$. A lower blade speed allows more time for inter-particle heat exchange, and thus the value α plays a significant role in softening the material. This leads to a change in the packing density of the powder layer.

In Fig. 13.11a and b, we compare the packing density for different blade speeds. The real modulus of PA12 powder is on the order of 10^8 Pa. However, for the sake

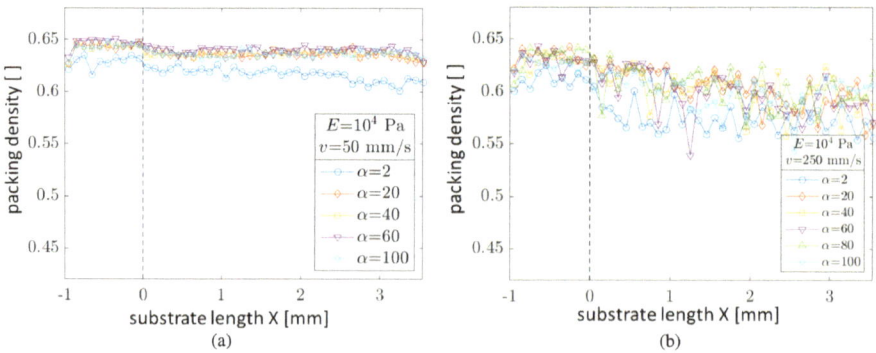

Fig. 13.10 Packing density as a function of substrate length X for different coefficient α for **a** $v = 50\,$mm/s and **b** $v = 250\,$mm/s for initial elastic modulus $E_0^* = 10^4$ Pa

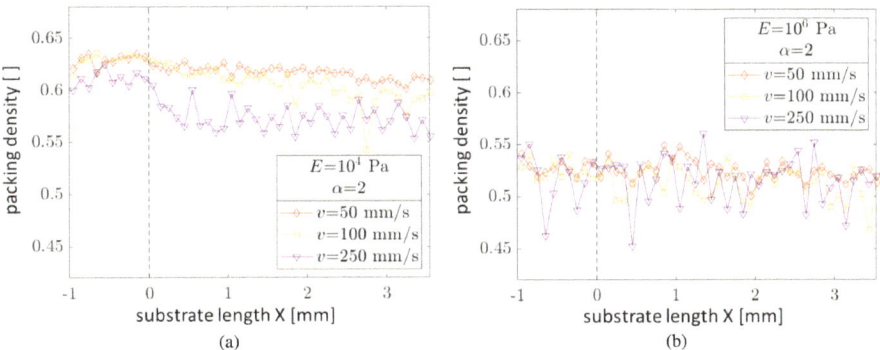

Fig. 13.11 Packing fraction for various blade speeds v for a different initial elastic modulus **a** $E_0^* = 10^4$ Pa and **b** $E_0^* = 10^6$ Pa for $\alpha = 2$

of computational efficiency, we use a lower elastic modulus in our simulations. To further test the effect of the elastic modulus on the packing density of the powder layer, we simulate the powder spreading process using two different values of the initial elastic modulus, E_0^*: 10^4 Pa and 10^6 Pa, respectively, while keeping $\alpha = 2$ as constant. We observe that the packing density is less sensitive to blade velocity for materials with a higher elastic modulus. This is because stiff particles allow less interparticle compression, and thus the packing density is less sensitive to the blade speed. For soft particles, the packing density is influenced by the flow dynamics of the material, which depends on the blade speed.

13.6 Vibrating Recoating Mechanism

Improving the layer quality can be achieved via modification of the powder application process [9, 21]. Nasato et al. [23, 39] explored the applicability of vibrating recoating mechanisms as a technique that can be tailored according to material characteristics, such as particle size, shape, and cohesion, to produce layers of improved characteristics. They performed DEM simulations using cohesive, multisphere particles of realistic shapes of PA12 particles to explore powder spreading systems using two spreading tools, namely a blade and a roller. Figure 13.12 shows the simulation system.

Vibration applied on the PA12 material via the recoating tool was used as a means of compacting the powder particles during their application to form the layered bed. The influence of the vibration characteristics on the layer density and surface roughness was investigated. The vibration frequency was varied to values of 100, 250, 500, and 1000 Hz, while the amplitude was varied to 2, 5, 10, and 20 μm, for recoating velocities of 100, 150, 200, and 250 mm/s.

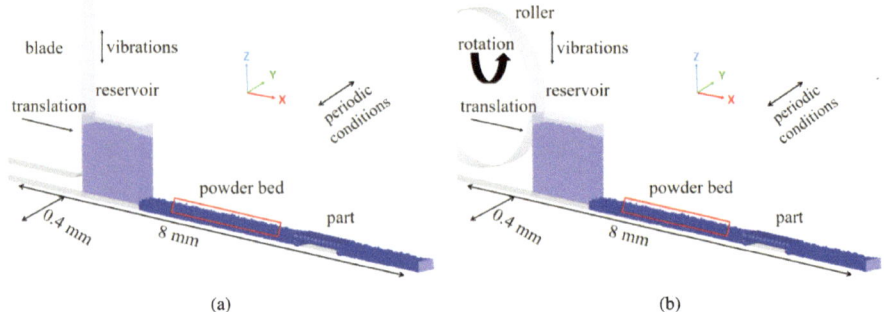

(a)   (b)

Fig. 13.12 Powder spreading simulation setup using **a** a blade and **b** a roller recoating tool. Red boxes indicate the powder bed region where porosity is evaluated after the layer is recoated. Porosity is additionally evaluated on the top of the part [23]

Figure 13.13 demonstrates that for a recoating velocity of 100 mm/s, considering a large frequency (i.e., 1000 Hz) combined with intermediate amplitude (10 μm) leads to a major improvement in the powder bed density, reducing the porosity by 13.2%. However, for larger amplitudes, no significant change in the porosity is observed which indicates an optimum between 10 and 20 μm for the amplitude. In

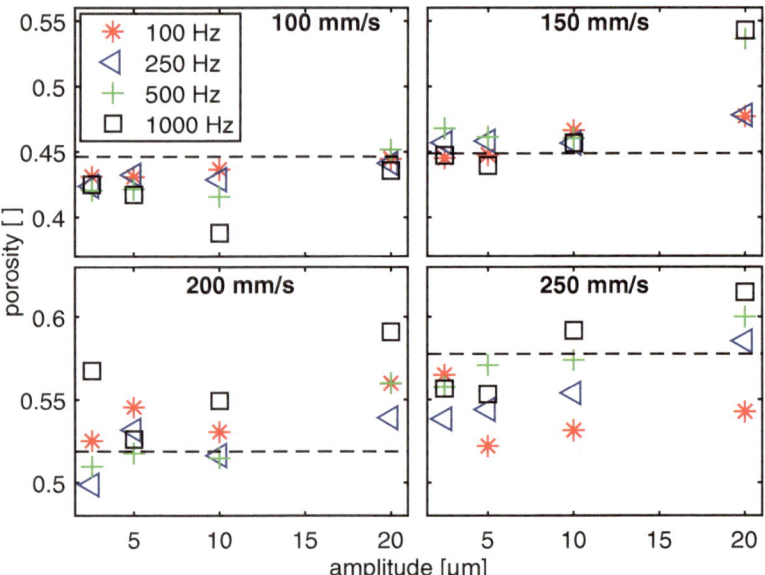

Fig. 13.13 Porosity of granular layer obtained for different amplitudes (x-axis) and frequencies. The powder spreading tool is a blade, operating with a translational velocity from 100 mm/s (top left) to 250 mm/s (bottom right). The dashed line indicates the reference case where no vibration was applied. The porosity was measured on the powder recoated on the top of the granular bed [23]

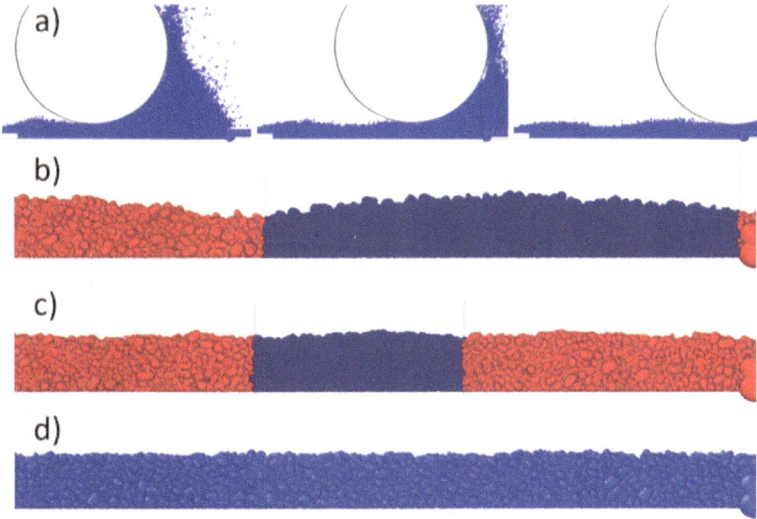

Fig. 13.14 Formation of bumps on the surface of the powder bed as a result of the vibrating mechanism. **a** Snapshots taken at different stages of the recoating process show the formation of the bumps for **b** recoating velocity 250 mm/s, frequency 100 Hz and amplitude 20 µm **c** recoating velocity 100 mm/s, frequency 100 Hz and amplitude 20 µm. The distance between bumps D_b is 2.5 and 1.0 mm for cases (**b**) and (**c**), respectively. The particles marked in dark blue indicate a region between the periodic bump, and correspond to D_b. **d** No bump could be identified for recoating velocity 100 mm/s, frequency 250 Hz and amplitude 20 µm [23]

all cases, increased vibration energy seems to disturb the powder layer, and leads to a loosening of the granular bed, as seen e.g., for 1000 Hz and 20 µm. This effect can also be observed for a recoating velocity of 150 mm/s, where the porosity of the granular bed increases by 20.9% for 1000 Hz and 20 µm.

The influence of vibrations on the recoating mechanism using a roller was investigated by Nasato et al. [23]. They found that vibrations increased the porosity of the recoated layer in all cases for recoating velocities above 150 mm/s, which they attributed to being an artifact of the roller geometry, largely influencing the powder bed. The roller introduces an artificial bumpiness on the layer surface, as shown in Fig. 13.14, which is, however, minimized for small recoating velocities, such as 100 mm/s.

13.7 Conclusions

We developed a numerical tool specifically designed for particle-based DEM simulations in the context of powder application within additive manufacturing devices. Our tool takes the intricate geometric shapes of the particles involved into account. To assess the capabilities of our model, we utilized it to investigate the transport of powder particles using a roller as coating system. Through our simulations, we were able to discern that the process velocity has a significant influence on the packing characteristics of the applied powder. The findings of our study highlight the importance of process speed in determining how the powder particles pack together during the coating process. This understanding can be valuable in optimizing the additive manufacturing process, allowing for improved control over the final properties and quality of the manufactured objects. By incorporating complex particle geometric shapes into our simulations, our numerical tool provides a more realistic representation of the powder application process. This enables us to gain insights into the underlying mechanisms governing particle transport and packing, aiding in the development of more efficient and effective additive manufacturing techniques.

Subsequently, we expand upon our existing DEM model by introducing inter-particle attractive forces. We emphasized the importance of considering two essential contributions to the particle interaction model: adhesion and van der Waals forces. Adhesion is incorporated into our model using the Johnson-Kendall-Roberts (JKR) theory, while van der Waals interactions are also taken into account. We assert that neglecting either of these contributions leads to significant deviations in the simulation results, particularly for fine powders. Such deviations render the DEM method unreliable in accurately capturing the packing behavior.

Furthermore, our investigation considered the influence of powder cohesion on the structural anisotropy of the deposited layers. Structural anisotropy refers to the variation in the spatial distribution of particles within the layers and provides insights into the overall organization and packing behavior of the powder particles during the spreading process. Understanding the effect of powder cohesion on structural anisotropy is important for optimizing processes where particle arrangement plays a significant role, such as in additive manufacturing, powder compaction, or coating applications.

Finally, the focus of the research was to further explore the powder application process through process simulations. In particular, the thermal and mechanical effects of the process and material parameters on the powder application process are the primary research objectives. The integration of the multisphere algorithm with the thermal discrete particle model into the DEM framework during this research phase represents a significant advancement in simulating non-spherical particles with heat transfer capabilities. This extension opens up new possibilities for studying a wide range of systems, such as granular materials, powders, and particulate flows, where particle shape and thermal effects are of critical importance.

The addition of vibrations during powder application was investigated via DEM simulations with non-spherical, cohesive particles as a new application system. The effect of vibration and process characteristics was explored. Small frequency and amplitude, combined with small recoating velocity, lead to denser granular beds of reduced porosity. Large frequency and amplitude, however, lead to a vibro-fluidized state of the particles and granular beds with loose configuration characterized by an increased porosity. For practical applications, the choice of frequency and amplitude should be considered in combination with a specific translational velocity of the recoating process.

References

1. Abou-Chakra, H., Baxter, J., Tüzün, U.: Three-dimensional particle shape descriptors for computer simulation of non-spherical particulate assemblies. Adv. Powder Technol. **15**(1), 63–77 (2004)
2. Angelidakis, V., Nadimi, S., Otsubo, M., Utili, S.: CLUMP: a code library to generate universal multi-sphere particles. SoftwareX **15**, 100735 (2021)
3. Cabiscol, R., Finke, J.H., Kwade, A.: Calibration and interpretation of dem parameters for simulations of cylindrical tablets with multi-sphere approach. Powder Technol. **327**, 232–245 (2018)
4. Castellanos, A.: The relationship between attractive interparticle forces and bulk behaviour in dry and uncharged fine powders. Adv. Phys. **54**(4), 263–376 (2005)
5. Cundall, P.A., Strack, O.D.: A discrete numerical model for granular assemblies. Geotechnique **29**(1), 47–65 (1979)
6. Glasstone, S., Laidler, K.J., Eyring, H.: The theory of rate processes; the kinetics of chemical reactions, viscosity, diffusion and electrochemical phenomena. Technical report, McGraw-Hill Book Company (1941)
7. Götzinger, M., Peukert, W.: Dispersive forces of particle-surface interactions: direct AFM measurements and modelling. Powder Technol. **130**(1–3), 102–109 (2003)
8. Götzinger, M., Peukert, W.: Particle adhesion force distributions on rough surfaces. Langmuir **20**(13), 5298–5303 (2004)
9. Haeri, S., Wang, Y., Ghita, O., Sun, J.: Discrete element simulation and experimental study of powder spreading process in additive manufacturing. Powder Technol. **306**, 45–54 (2017)
10. Harrington, M., Durian, D.J.: Anisotropic particles strengthen granular pillars under compression. Phys. Rev. E **97**(1), 012904 (2018)
11. Harrington, M., Xiao, H., Durian, D.J.: Stagnant zone formation in a 2D bed of circular and elongated grains under penetration. Granular Matter **22**, 1–9 (2020)
12. He, Y., Hassanpour, A., Bayly, A.E.: Linking particle properties to layer characteristics: discrete element modelling of cohesive fine powder spreading in additive manufacturing. Addit. Manuf. **36**, 101685 (2020)
13. Jagota, A., Scherer, G.W.: Viscosities and sintering rates of a two-dimensional granular composite. J. Am. Ceram. Soc. **76**(12), 3123–3135 (1993)
14. Johnson, K.L., Kendall, K., Roberts, A.: Surface energy and the contact of elastic solids. Proc. R. Soc. London. A. Math. Phys. Sci. **324**(1558), 301–313 (1971)
15. Kloss, C., Goniva, C., Hager, A., Amberger, S., Pirker, S.: Models, algorithms and validation for opensource dem and CFD-dem. Prog. Comput. Fluid Dyn., Int. J. **12**(2–3), 140–152 (2012)
16. Kodam, M., Bharadwaj, R., Curtis, J., Hancock, B., Wassgren, C.: Force model considerations for glued-sphere discrete element method simulations. Chem. Eng. Sci. **64**(15), 3466–3475 (2009)
17. Laumer, T., Stichel, T., Schmidt, M.: Influence of temperature gradients on the part properties for the simultaneous laser beam melting of polymers. In: Proceedings of Laser in Manfacturing Conference 2015, June 22–June 25, 2015 Munich, Germany (2015)
18. Li, C., Snarr, S.E., Denlinger, E.R., Irwin, J.E., Gouge, M.F., Michaleris, P., Beaman, J.J.: Experimental parameter identification for part-scale thermal modeling of selective laser sintering of pa12. Addit. Manuf. **48**, 102362 (2021)

19. Li, Q., Rudolph, V., Peukert, W.: London-van der Waals adhesiveness of rough particles. Powder Technol. **161**(3), 248–255 (2006)
20. Luding, S., Manetsberger, K., Müllers, J.: A discrete model for long time sintering. J. Mech. Phys. Solids **53**(2), 455–491 (2005)
21. Marchais, K., Girardot, J., Metton, C., Iordanoff, I.: A 3D dem simulation to study the influence of material and process parameters on spreading of metallic powder in additive manufacturing. Comput. Part. Mech. **8**(4), 943–953 (2021)
22. Müller, P., Pöschel, T.: Collision of viscoelastic spheres: compact expressions for the coefficient of normal restitution. Phys. Rev. E **84**(2), 021302 (2011)
23. Nasato, D.S., Briesen, H., Pöschel, T.: Influence of vibrating recoating mechanism for the deposition of powders in additive manufacturing: discrete element simulations of polyamide 12. Addit. Manuf. **48**, 102248 (2021)
24. Nasato, D.S., Heinl, M., Hausotte, T., Pöschel, T.: Numerical and experimental study of the powder bed characteristics in the recoated bed of the additive manufacturing process. In: Wriggers, P., Bischoff, M., Oñate, E., Owen, D.R.J., Zohdi, T. (eds.) PARTICLES 2017. Proceedings of the V International Conference on Particle-based Methods. 26–28 September 2017, Hannover, Germany, pp. 429–439. Barcelona (2017). CIMNE
25. Nasato, D.S., Pöschel, T.: Influence of particle shape in additive manufacturing: discrete element simulations of polyamide 11 and polyamide 12. Addit. Manuf. **36**, 101421 (2020)
26. Ostanin, I., Angelidakis, V., Plath, T., Pourandi, S., Thornton, A., Weinhart, T.: Rigid clumps in the mercurydpm particle dynamics code. Comput. Phys. Commun. (in press)
27. Parteli, E.: Using LIGGGHTS for performing DEM simulations of particles of complex shapes with the multisphere method. In: Proceedings, DEM6-6th International Conference on Discrete Element Methods and Related Techniques, Golden USA (2013)
28. Parteli, E.J.: DEM simulation of particles of complex shapes using the multisphere method: application for additive manufacturing. In: AIP Conference Proceedings, vol. 1542, pp. 185–188. American Institute of Physics (2013)
29. Parteli, E.J.R., Poeschel, T.: Particle-based simulation of powder application in additive manufacturing. Powder Technol. **288**, 96–102 (2016)
30. Parteli, E.J.R., Pöschel, T.: Particle-based simulations of powder coating in additive manufacturing suggest increase in powder bed roughness with coating speed. In: European Physical Journal. Web of Conferences, vol. 140, p. 15013 (2017)
31. Parteli, E.J.R., Schmidt, J., Blümel, C., Wirth, K.-E., Peukert, W., Pöschel, T.: Attractive particle interaction forces and packing density of fine glass powders. Nat.-Sci. Rep. **4**, 6227 (2014)
32. Peyre, P., Rouchausse, Y., Defauchy, D., Régnier, G.: Experimental and numerical analysis of the selective laser sintering (SLS) of PA12 and PEKK semi-crystalline polymers. J. Mater. Process. Technol. **225**, 326–336 (2015)
33. Richard, D., Ozawa, M., Patinet, S., Stanifer, E., Shang, B., Ridout, S., Xu, B., Zhang, G., Morse, P., Barrat, J.-L., et al.: Predicting plasticity in disordered solids from structural indicators. Phys. Rev. Mater. **4**(11), 113609 (2020)
34. Rieser, J.M., Goodrich, C.P., Liu, A.J., Durian, D.J.: Divergence of Voronoi cell anisotropy vector: a threshold-free characterization of local structure in amorphous materials. Phys. Rev. Lett. **116**(8), 088001 (2016)
35. Roy, S., Shaheen, M.Y., Pöschel, T.: Effect of cohesion on structure of powder layers in additive manufacturing. Granular Matter **25**(4), 68 (2023)
36. Roy, S., Xiao, H., Angelidakis, V., Pöschel, T.: Structural fluctuations in thin cohesive particle layers in powder-based additive manufacturing. Addit. Manuf. (submitted) (2023)
37. Roy, S., Xiao, H., Angelidakis, V., Pöschel, T.: Thermal modelling of temperature distribution in metal additive manufacturing (2023). (in preparation)
38. Roy, S., Xiao, H., Shaheen, M.Y., Pöschel, T.: Local structural anisotropy in particle simulations of powder spreading in additive manufacturing. In: Casablanca International Conference on Additive Manufacturing, pp. 139–149. Springer, Berlin (2022)
39. Schiochet Nasato, D., Pöschel, T., Parteli, E.J.R.: Effect of vibrations applied to the transport roller in the quality of the powder bed during additive manufacturing. In: Drstvenšek, I.,

Drummer, D., Schmidt, M. (eds.) Proceedings of the 6th International Conference on Additive Technologies iCAT 2016, Nürnberg Nov. 29–30, pp. 260–265. Interesansa, Ljubljana (2016)

40. Schmidt, J., Parteli, E.J.R., Uhlmann, N., Wörlein, N., Wirth, K.-E., Peukert, W.: Packings of micron-sized spherical particles—insights from bulk density determination, X-ray microtomography and discrete element simulations. Adv. Powder Technol. **31**, 2293–2304 (2020)
41. Severson, B., Keer, L.M., Ottino, J.M., Snurr, R.Q.: Mechanical damping using adhesive micro or nano powders. Powder Technol. **191**(1–2), 143–148 (2009)
42. Xiao, H., Ivancic, R.J., Durian, D.J.: Strain localization and failure of disordered particle rafts with tunable ductility during tensile deformation. Soft Matter **16**(35), 8226–8236 (2020)
43. Yu, A.-B., Bridgwater, J., Burbidge, A.: On the modelling of the packing of fine particles. Powder Technol. **92**(3), 185–194 (1997)
44. Zhao, Y., Aoyagi, K., Cui, Y., Yamanaka, K., Chiba, A.: Multiscale heat transfer affected by powder characteristics during electron beam powder-bed fusion. Powder Technol. **421**, 118438 (2023)

Thorsten Pöschel has headed the Institute for Multiscale Simulation of Particle Systems. Prior to his appointment at the Friedrich-Alexander-Universität Erlangen-Nürnberg, he was a Professor of Theoretical Physics at the University of Bayreuth working in the field of Statistical Mechanics and Kinetic Theory of Granular Gases. In laser-based powder bed fusion, the focus of the simulation is on the powder flow behavior as well as the melt pool dynamics.

Chapter 14
Macroscopic Modeling, Simulation, and Optimization

Christian Burkhardt, Dominic Soldner, Paul Steinmann, and Julia Mergheim

14.1 Introduction

During the repeated melting and cooling of (initially) powder material in powder bed based additive manufacturing (AM), high temperatures and large temperature gradients arise. These lead to thermal expansion and contraction of the material, for both metals and polymers and to shrinkage during crystallization in the case of semi-crystalline polymers. The heterogeneous temperature field, temperature-dependent mechanical behavior, and heterogeneous crystallization result in warpage and residual stresses during powder bed fusion (PBF). These effects are observed in the PBF of metals, using a laser beam (PBF-LB/M) or an electron beam (PBF-EB/M) [5, 31], and in the laser based PBF of polymers (PBF-LB/P) [1].

The residual stresses and the warpage influence the final mechanical and geometrical part properties. The aim is to replace costly experimental trial-and-error approaches with numerical simulations to minimize the described effects. The characteristics of powder bed based additive manufacturing (AM) make process modeling and simulation using the Finite Element method (FEM) challenging. The complex material behavior including phase-, temperature-, and rate dependencies must be considered. Furthermore, various physical phenomena occur on very different lengths and time scales.

This chapter provides insights concerning the modeling and simulation of beam-based additive processes. A macroscopic simulation framework is introduced, which is based on the FEM. It includes various specifications to capture the process characteristics, model the complex material behavior of metals and polymers during the process, and make the complex process simulations feasible. In recent years,

C. Burkhardt · D. Soldner · P. Steinmann · J. Mergheim (✉)
Friedrich-Alexander-Universität Erlangen-Nürnberg Institute of Applied Mechanics,
Egerlandstraße 5, 91058, Erlangen, Germany
e-mail: julia.mergheim@fau.de

© The Author(s) 2025
D. Drummer and M. Schmidt (eds.), *Progress in Powder Based Additive Manufacturing*, Springer Tracts in Additive Manufacturing,
https://doi.org/10.1007/978-3-031-78350-0_14

extensive research was conducted on the simulation of powder bed based AM for metals and polymers as shown in the reviews [17, 30] for metals or [7, 11, 29] for polymers. Thermo-mechanical approaches were used to simulate the exposure of a few layers or small components using adaptive spatial [4] and temporal [46] discretization techniques. In these high-fidelity simulations, locally resolved evolutions of thermal and mechanical quantities can be accurately captured [8, 40]. Due to the extensive numerical effort, high-fidelity process simulation requires high-performance computers, and, even then, the feasible part sizes are limited. In recent years, the research focus has thus been directed towards the analysis of suitable physical and mathematical model reduction techniques. It was shown that mathematical reduction techniques, as presented in [44, 47], are only partially applicable to the simulation of AM processes. Hence, the research focus shifted to physical model reduction techniques. One example of this is layer agglomeration [20], in which the numerical layer thickness is artificially increased compared to the experimental one. The scaled layer thickness requires an adjustment of the heat source modeling such as scaling the beam dimensions or by reduced models like so-called flash heating [6].

While the aforementioned challenges in the modeling and simulation of powder bed based AM mostly also apply to polymers, there are some fundamental differences. For instance, the solidification mechanism for semi-crystalline polymers differs from metals. In contrast to a rapid solidification after beam exposure as is the case for metals, (non)-isothermal crystallization occurs with a significant time offset after beam exposure and is accompanied by volumetric shrinkage. To predict part warpage and residual stresses, the evolution of the degree of crystallization has to be considered in combination with a thermo-mechanical model. The former is often computed using the Nakamura model, which was extended in [45, 48] to account for remelting during PBF-LB/P while the latter is often modeled by adapting approaches from polymer curing [1].

In this chapter, the main features of our macroscopic thermo-mechanical simulation framework are described. The underlying continuum model is introduced in Sect. 14.2. Section 14.3 presents the simulation approach for the PBF-LB/M and PBF-EB/M process, including layer agglomeration and flash heating. Two process simulation examples are provided to illustrate the influence of the part geometry on the prevailing temperature and resulting stresses, as well as on the resulting microstructure of Ti-6Al-4V. The polymer-specific modeling approaches are presented in Sect. 14.4, including an adapted crystallization model that accounts for the remelting of crystallized material. With this model, the dependence of the degree of crystallization on the component geometry is investigated. Further, since the resulting degree of crystallization influences the mechanical behavior, the corresponding mechanical model is briefly described. The insights gained are summarized, a conclusion is drawn, and further research topics are identified in Sect. 14.5.

14.2 Continuum Modeling

Following a continuum theory approach, all considered quantities of interest, such as the temperature ϑ or displacements $\mathbf{u}$ are considered as continuous fields for all appearing phases, i.e., powder, and molten and solid material. On this macroscopic scale, the powder phase is represented in a homogenized fashion. To compute the temperature field over a domain $\mathcal{B}$, the balance of energy, equipped with boundary conditions is solved in an adapted form [23],

$$\dot{h}(\vartheta)\,\rho = \operatorname{div}(\lambda(\vartheta)\,\nabla\vartheta) + Q + \mathcal{D} \quad \text{in } \mathcal{B} , \qquad (14.1)$$

where h denotes the specific enthalpy, and the mass density and thermal conductivity are represented by ρ and λ, respectively. The heat source Q accounts for the energy provided by the laser- or electron beam, and the heat source due to dissipation or crystallization is represented by $\mathcal{D}$.

To compute the displacement field and the residual stresses, the balance of linear momentum

$$\operatorname{div} \boldsymbol{\sigma} = \mathbf{0} \quad \text{in } \mathcal{B} , \qquad (14.2)$$

is considered, where $\boldsymbol{\sigma}$ corresponds to the Cauchy stress. Constitutive equations will be specified later for metals and polymers to determine the stresses $\boldsymbol{\sigma}$ in terms of the strains, strain rate, temperature, degree of crystallization, and material properties.

These balance equations are discretized in space by means of FEM and in time using a two-stage S-DIRK scheme [15]. The numerical tool that was developed to solve the aforementioned equations relies on the h-adaptive FEM framework provided by the open-source C++ library deal.II [3]. Further, the material phases of powder, solid, and melt are incorporated by assigning every element to its so-called material ID, which can change unidirectionally during the simulation from powder to melt, and bidirectionally from melt to solid. Both the adaptivity and the treatment of the material phases are illustrated in Fig. 14.1.

The part geometry can be simulated in two ways: (i) by powder elements that are being molten, not considering any information on the part geometry during meshing (which can potentially lead to a high mesh density along the part boundary), or (ii), by including the part geometry during meshing and introducing elements that possess powder properties but are not allowed to change their status to melt or solid. These elements are denoted as the surrounding powder in Fig. 14.1. This approach allows for resolving the part geometry while reducing the number of elements.

One key characteristic of powder bed based AM is the differences in the temporal and spatial length scales, ranging from µs to h for the former, and µm to cm for the latter. The adaptive space and time discretizations are the first steps to reducing the numerical cost associated with the solution of the balance equations. To further reduce the computational effort, various reduction techniques can be employed. In the sequel, two exemplary approaches are briefly discussed, the first of which concerns the energy input, while the second applies a multi-rate time integration approach.

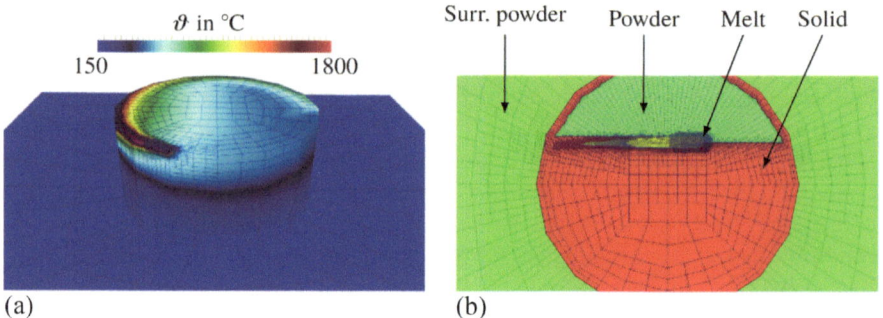

Fig. 14.1 Illustration of the process simulation to account for different phases and mesh adaptivity as proposed in [46]. The pictures show the temperature and phase distribution during the manufacturing of a pin on a base plate (in **a** only solid and melt material is shown). Every element has its so-called material ID assigned (see different colors in **b**), i.e., powder, melt, or solid

14.2.1 Energy Input

The energy input of the laser or electron beam is represented by a volumetric heat source based on a Gaussian distribution [18] and further preintegrated in time [21]. An energetically consistent heat input independent of the spatial discretization is thus ensured [45]. The factor $\tau = \frac{v_b \Delta t}{r_b}$ is introduced that relates the traveled distance of the beam with a velocity v_b within a time step Δt to the beam radius r_b. For larger values of τ, the scanning process is no longer accurately resolved although the larger time step sizes significantly reduce the computational cost. The influence of τ (and therefore of the time step size) on the resulting temperatures differs for polymers and metals due to their different heat conductivities. This is illustrated by the single line example presented in Fig. 14.2 for PA12 via PBF-LB/P and Ti-6Al-4V via PBF-EB/M with varying time step sizes. In this visual overview, the temperature profiles along an evaluation line parallel to the scanning line below the powder bed surface immediately after the end of the exposure are shown. Materials with a higher thermal conductivity such as metals lead to larger deviations in the temperatures than materials with a lower thermal conductivity such as polymers. Therefore, a significantly larger time step size can be chosen for the process simulation of polymers compared to the simulation of metals without introducing large inaccuracies.

14.2.2 Multi-time-Stepping

Multi-rate integration methods were first introduced for AM simulations in [46]. Their effectiveness in reducing the computation time of process simulations has been confirmed in subsequent works [19].

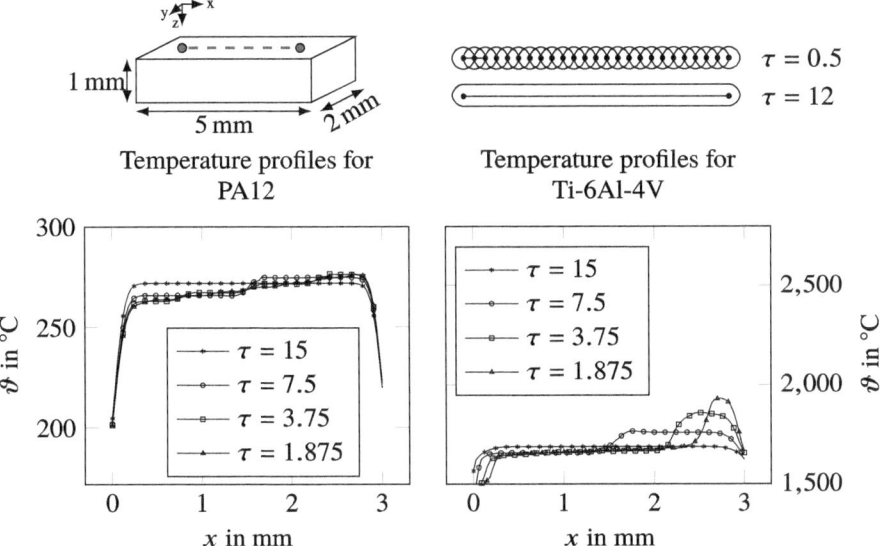

Fig. 14.2 Effects of the time-integrated heat source for a single line scan path for PA12 and Ti-6Al-4V (see [46]). The relative line length τ is illustrated for two different values where $\tau = 0.5$ coincides with a standard temporal discretization of the scan path, while $\tau = 12$ computes the energy input of 24 time instances following the standard approach within one time step. This approach was proposed by [21] and has been extended to a generalized scheme in [45]

The main idea is to apply domain decomposition in combination with distinct time discretizations for the respective subdomains. Suppose a continuum body $\mathcal{B}$ is decomposed into subdomains $\mathcal{B}_a$ and $\mathcal{B}_b$ along an interface Γ. For instance, subdomain $\mathcal{B}_a$ could represent the region that is subject to energy input, potentially requiring a finer time discretization than the remaining domain $\mathcal{B}_b$. The subdomains are then linked via constraint terms along Γ, for which various approaches exist. In [46], the scheme proposed in [34] was adopted, rendering a mixed implicit-explicit scheme. The subdomains are integrated using an implicit time integration scheme and the interface term along Γ, representing a boundary condition for the subdomains, is treated explicitly, using weakly imposed constraints for the temperature and heat flux. This renders a completely decoupled solution procedure.

However, to utilize the implemented Runge-Kutta scheme, this decomposition is only introduced during the energy input and the respective subdomains are merged together in the time between successive layers. For the case of PBF-LB/P of PA12, this procedure is illustrated in Fig. 14.3. Due to the explicit treatment of the interface term along Γ, a material parameter and mesh size-dependent critical time step size exists and estimates of it were presented in [46] for different material parameters and mesh sizes. In [24], an implicit approach for the interface terms which is not limited by a critical time step size was investigated and is currently being further developed.

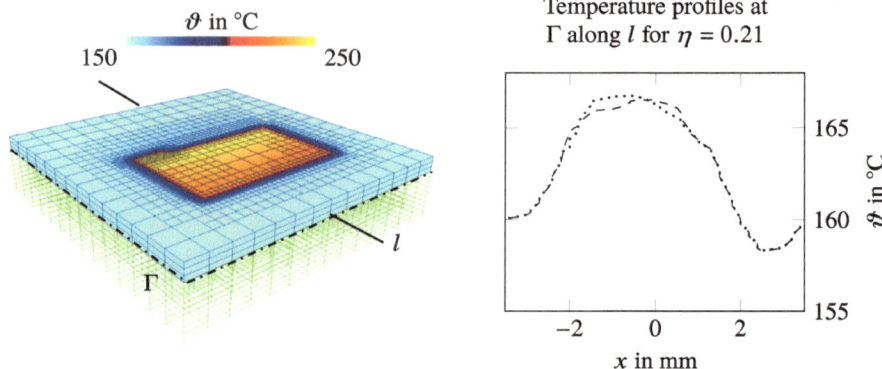

Fig. 14.3 Illustration of domain decomposition for the balance of energy as presented in [46]. The two subdomains (solid and wireframe) are connected at the interface Γ. While the solid domain is solved with $\tau = 5$, a ten times larger time step is used for the wireframe domain. At the end of the energy input, the two subdomains are merged and the temperature profile along the line l is given on the right. The differences due to the explicit treatment of the interface term appear acceptable and an averaging procedure is used upon domain merging

14.3 Simulations of PBF-LB/M and PBF-EB/M Processes

PBF of metals can be further subdivided into laser beam based (PBF-LB/M) and electron beam based (PBF-EB/M) based on the respective heat source. In both cases, a three-dimensional part is created by layer-wise melting of the desired cross-section on top of a building plate. The PBF-LB/M process is usually conducted without preheating and operates in a shielding gas atmosphere to reduce oxidation. The local heat input of the laser beam results in high-temperature gradients in the beam vicinity. In contrast, the electron beam operates in a vacuum chamber where highly energetic electrons are accelerated by an electric field. This leads to an increased penetration depth with a maximum intensity below the powder bed surface, resulting in a greater layer thickness and higher scan velocities compared to the PBF-LB/M process. A heat input model following [22] is used for the simulation of the PBF-EB/M process which takes these effects into account.

14.3.1 Single Scan Lines

First, the thermal simulation model was validated for the PBF-EB/M process via single scan lines. For this purpose, experimentally determined and simulated melt pool lifetimes for different line energies were compared, as shown in Fig. 14.4. The simulated time spans match their experimental counterparts.

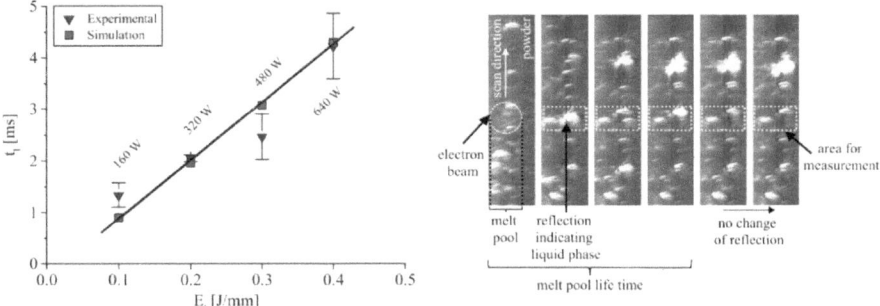

Fig. 14.4 Comparison of melt pool lifetimes experimentally measured and numerically predicted for different line energies E_l for PBF-EB/M as presented in [42]. Reproduced according to the terms of the CC BY 4.0 license [42]

In [42], simulated melt pool dimensions were also successfully compared with experimental measurements for lower line energies in the conduction regime, although the melt pool dynamics and evaporation were not considered.

14.3.2 Current Findings

For the simulation of metals, the thermo-mechanical equations are solved in a staggered fashion based on the findings in [43]. A thermo-elasto-visco-plastic constitutive law at finite strains making use of the logarithmic strain space is applied [9]. The viscous part in the plastic regime captures the rate-dependent material behavior for elevated temperatures while stress relaxation based on the resetting approach [12] is considered for temperatures close to the melting point. The simulation framework introduced for PBF-LB/M is validated using thermal quantities such as the melt pool dimensions and maximum temperatures in the conduction regime and mechanical quantities such as residual stresses as presented in [10].

14.3.2.1 Efficient Simulation Setting

Due to the extensive numerical effort required for AM simulations, a numerical study of the influences of the temporal and spatial discretizations is conducted to obtain an efficient but sufficiently accurate simulation setting. Therefore, the scanning of a single line of 1 mm length over 6 layers is simulated with varying spatial and temporal discretizations. The resulting thermal and mechanical quantities are compared at evaluation point A which is located in the center of the scan line on top of the second layer.

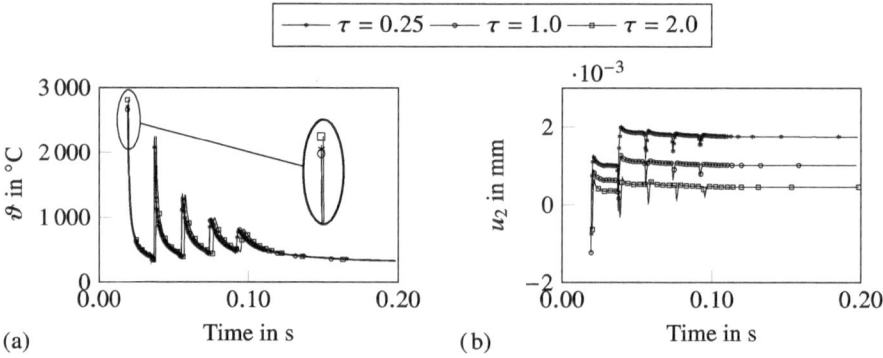

(a)   Time in s   (b)   Time in s

Fig. 14.5 Sensitivity study of thermal and mechanical quantities for the single line example of Ti-6Al-4V over 6 layers for different temporal discretizations. The evolution of **a** temperature and **b** displacement u_2 (in the negative scanning direction) over time for values of $\tau = 0.25$, $\tau = 1.0$, and $\tau = 2.0$ is displayed. While the temperature just shows minor deviations for the investigated temporal discretizations, the displacements u_2 decrease with increasing τ. Adjusted and reproduced according to the terms of the CC BY 4.0 license [10]

For various time step sizes, hence various values of τ for a constant beam velocity, the resulting temperatures and displacements in e_2-direction (negative scanning direction) at point A are shown in Fig. 14.5. While the temperatures differ just in their maximum values for the investigated time step sizes the displacements show greater discrepancies. The high thermal conductivity of metals requires a fine temporal resolution. Especially the length of the time interval with temperatures near the melting point has proven to be crucial for the final deformation [10].

The influence of the spatial discretization is illustrated in Fig. 14.6 for three refinement levels of the finite element mesh. The maximum refinement level R is a measure of the smallest element edge length of the h-adaptive finite elements. Both the temperatures and the displacements u_2 closely align with the two finer spatial discretizations $R = 5$ and $R = 4$. For $R = 3$, the deviations increase as shown in Fig. 14.5b.

The distributions of the von Mises stress in the e_2-e_3-plane, as depicted in Fig. 14.7, allow a similar conclusion. While the different spatial discretizations result in similar stress distributions, the results for the temporal discretizations show significant deviations. However, compared to the displacements, the stresses seem to be less sensitive to increased time step sizes. Further sensitivity studies are presented in [10].

14.3.2.2   Simulation of PBF-EB/M Process of Lattice Structure

Once the efficient simulation setting is defined, it can be used to simulate a unit cell of a cellular material. Cellular materials are characterized by an outstanding stiffness-to-weight ratio, and AM processes with their high geometrical freedom

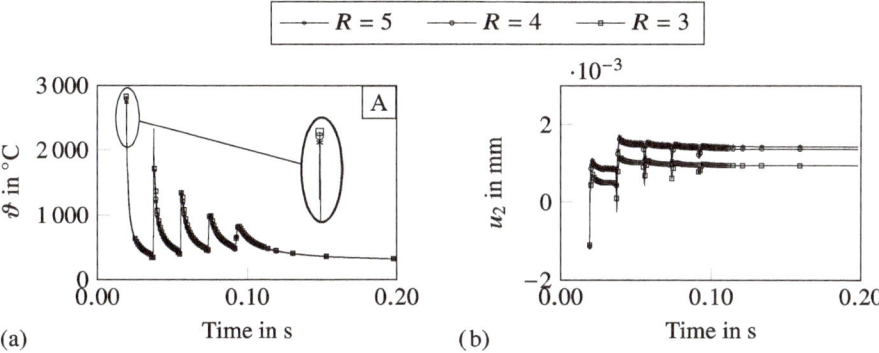

Fig. 14.6 Sensitivity study of thermal and mechanical quantities for the single line example of Ti-6Al-4V over 6 layers for different temporal discretizations. The evolution of **a** temperature and **b** displacement u_2 over time for the values of the maximum refinement level $R = 5$, $R = 4$, and $R = 3$. Minor deviations can be observed in the temperatures of all three discretizations. The displacements u_2 for the coarsest discretization level $R = 3$ are lower compared to the results of the more refined cases. Adjusted and reproduced according to the terms of the CC BY 4.0 license [10]

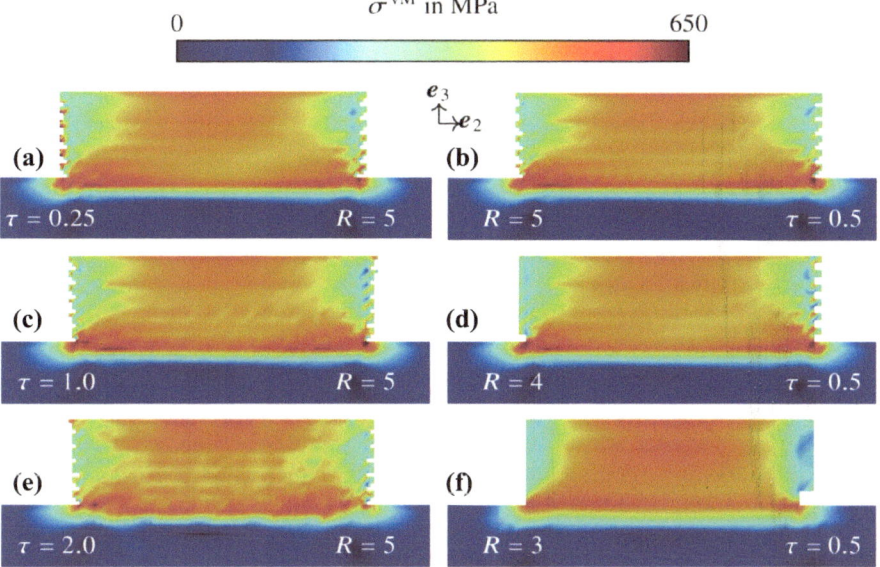

Fig. 14.7 Distributions of von Mises stress in the e_2-e_3-plane for different temporal **a, c, e** and spatial **b, d, f** discretizations after final cooling. The plots on the left side (**a, c, e**) represent the results with an increasing time step size at an identical minimum element edge length of $R = 5$. The results on the right side **b, d, f** are determined with an identical time step size $\tau = 0.5$ with increasing minimum element edge length. The scanning direction is from right to left. Adjusted and reproduced according to the terms of the CC BY 4.0 license [10]

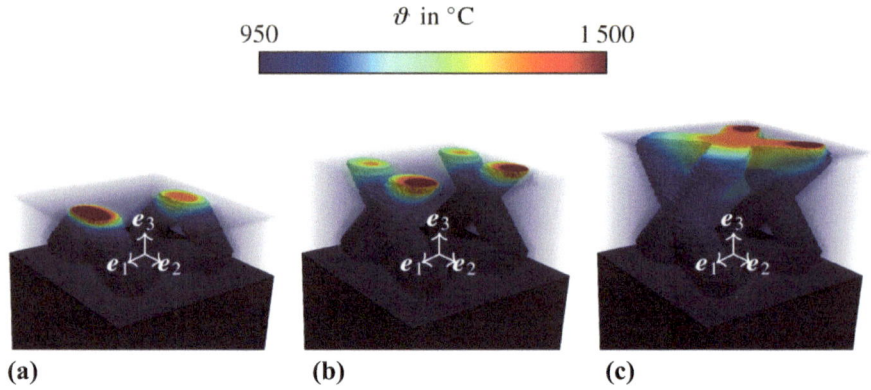

Fig. 14.8 Temperature distribution during the PBF-EB/M process for the exposure of layers **a** 44, **b** 75, and **c** 94. The electron beam follows a standard cross snake hatching strategy along the e_1-direction and is rotated by 90° after each layer. The powder material is indicated in a transparent manner. Reproduced according to the terms of the CC BY 4.0 license [10]

are ideal for their manufacturing. The geometry of the part built via PBF-EB/M is depicted in Fig. 14.8. It consists of 100 layers of 50 μm height and has an edge length of 4 mm [32].

The electron beam with a power of 200 W, a radius of 200 μm, and a velocity of $0.5 \, \text{m s}^{-1}$ follows a cross snake hatching pattern. The scanning direction is rotated by 90° after each layer as depicted in Fig. 14.8. The exposure of the last layer is followed by a cooling period of 1 h to reach room temperature. The electron beam is represented by the heat source model following [22]. The elastic behavior of the material CMSX-4 is assumed to be cubic symmetric, and the plastic behavior follows isotropic von Mises plasticity.

In this example, two typical effects that occur in PBF-EB/M and PBF-LB/M processes are illustrated: the temperature gradient mechanism (TGM) [31, 49] and overheating. The former is, among other effects, responsible for the development of residual stresses during AM processes. The local energy input results in large temperature gradients within the part, leading to a heterogeneous thermal strain distribution. During the subsequent cooling period, the thermal expansion turns into contraction resulting in high tensile stresses. This effect repeats itself for further layers with varying degrees of impact depending on the stiffness of the surrounding material. This can be observed in the final stress distribution σ_{22} close to the building plate after the cooling phase, as shown in Fig. 14.9b. For the evaluation of the residual stresses, point C, at the X-cross-section in the e_2-e_3-plane, and point D, which is located on the strut centerline in layer 68, are introduced. Comparing the evolutions of the resulting stress σ_{22} at points C and D in Fig. 14.10b shows that especially the cooling period causes the final stress magnitude. The stresses at both points prior to the cooling phase are comparatively small due to the high preheating temperature. The difference in the final stress magnitudes for points C and D can be explained

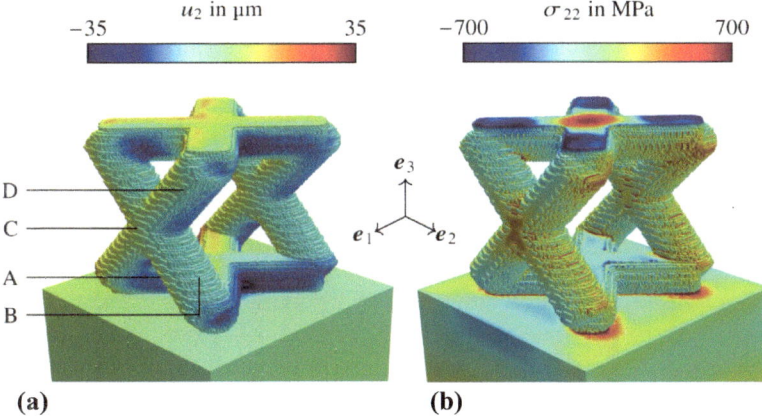

Fig. 14.9 **a** Distribution of the displacement u_2 immediately before the final cooling phase with the evaluation points A, B, C, and D. Points A and B are located on top of layer 25 within the melted area at the interface to the powder. Both points have the same distance to the strut centerline. Points C and D are located on the strut centerline at the indicated positions on layers 48 and 68, respectively. **b** Distribution of stress σ_{22} after the final cooling phase. Adjusted and reproduced according to the terms of the CC BY 4.0 license [10]

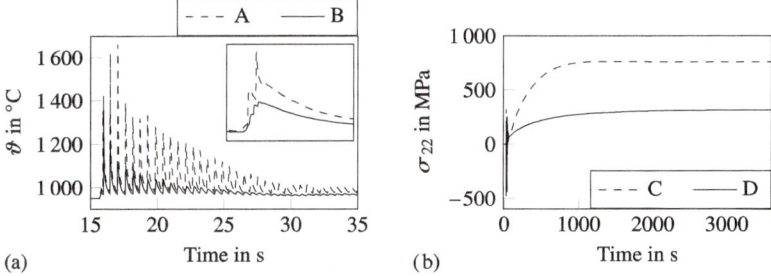

Fig. 14.10 Evolutions of **a** temperatures at points A and B, and **b** of the stress σ_{22} at points C and D over time. The location of the evaluation points A, B, C, and D is indicated in Fig. 14.9. The exposure of different layers can be distinguished by the temperature peaks. The stress curve shows that the cooling phase, in particular, is decisive for the resulting stress amplitude

by their surroundings. Close to point C, there is a lot of stiff solid material at the intersection of the struts, whereas point D is surrounded by softer powder material and shows significantly smaller stresses.

The overheating effect which describes (for tilted structures) higher temperatures in certain regions compared to others can be illustrated by the temperature evolutions at points A and B. Both points are located on top of the 25th layer with an identical distance to the strut centerline. Point A is on the inside of the inclination, while B is on the outside. The temperatures at point A are considerably higher compared to those at point B due to the powder material underneath that displays a lower thermal conductivity compared to solid material. Furthermore, the heat source in the following layers is closer to the location of point A compared to point B, which increases the temperature difference of both points. Further evaluations are provided in [32].

14.3.2.3   Conic Shape with Narrow Cylinders

In this example, the dependence of the microstructure on the part geometry is analyzed. For this purpose, a structure with varying cross-sections, consisting of conical shapes with narrow cylinders in between (see Fig. 14.11a) is investigated. For the thermal simulation, a reduced model is introduced, which allows simulations at part scale. Subsequently, the temperatures are used to predict the microstructure of Ti-6Al-4V and the results are compared to experimental measurements of the structure fabricated by PBF-LB/M. In order to limit the numerical effort, a layer agglomeration technique is employed [20] and only the lower third of the structure is simulated [33]. The layer height in the simulation is artificially scaled by a factor of $n_1 = 2$ compared to the experiment. As the numerical effort to resolve the scanning lines individually via the Goldak heat source is extensive, a flash heating approach following [6] is used in which the total energy is applied in a homogeneous fashion over the cross-section of the part in the current powder layer in a reduced time period Δt^s. Taking layer agglomeration into account, the heat source follows as

$$Q = \frac{U n_1}{\Delta t^s V_1},\tag{14.3}$$

with the V_1 denoting the volume of the lumped meta-layer. The flashing time is chosen based on the process parameters and the geometry of the part and a suitable value for the current example was found to be $\Delta t^s = 0.1$ ms.

In the first step, the simulated temperatures are compared to experimental measurements via the interlayer temperature as depicted in Fig. 14.11b. The interlayer temperature is the maximum surface temperature right before the next powder layer is applied. The measurements were conducted by the Institute of Photonic Technologies (LPT) at FAU using a infrared camera VarioCAM HD head 600 (Infratex, Dresden, Germany) with a calibrated temperature range between -40 and $2000\,^\circ$C [33]. As the emission coefficient is temperature- and surface-dependent, the measured values are interpreted relatively and not as absolute values. The simulated trend of interlayer temperatures closely aligns with the experimental measurements. For the experiments and simulations, the interlayer temperature increases and decreases in a similar fashion to that of the cross-section area of the part. Due to the increasing cross-section in the building direction, the heat is distributed over a larger area than that of the underlying solid. The lower thermal conductivity of the powder compared to solid material results in increasing interlayer temperatures while the opposite effect can be observed for narrowing cross-sections.

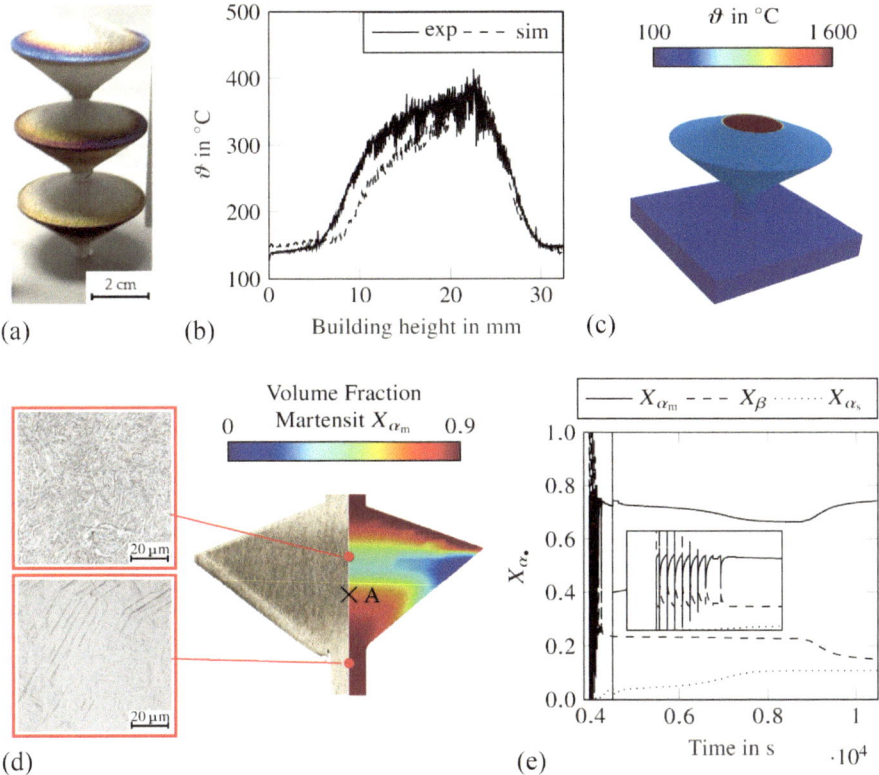

Fig. 14.11 Reduced thermal simulation. **a** Finished part consisting of three conic shapes with narrow cylinders in between. **b** Comparison of measured and simulated interlayer temperature for the first 650 layers, representing a third of the structure. **c** Temperature distribution during the exposure in the simulation. **d** Section of the built structure on the left and simulated distribution of the martensitic α_m-phase fraction on the right along the build direction with the evaluation point A. **e** Evolution of the β-, martensitic α_m- and stable α_s-phase fractions over time at the evaluation point A

These geometric effects affect not only the interlayer temperature but also other part properties such as the microstructure. Therefore, based on the simulated temperature field, a phenomenological microstructure model for Ti-6Al-4V [36] is employed. The solid phase is subdivided into volume-averaged phase fractions X of the β-phase, the martensitic α_m-phase, and the stable α_s-phase. The phase transformation rates $\dot{X}$ are based on the current phase fractions X, the temperatures ϑ, and the temperature rates $\dot{\vartheta}$. At high cooling rates, e.g., during the transformation of the β-phase into the martensitic α_m-phase, the instantaneous phase transformations are described mathematically by Karush-Kuhn-Tucker (KKT) conditions. At lower cooling rates, the diffusion-based transformations that take place are modeled by modified logistic differential equations. The solid phase transitions for Ti-6Al-4V occur in the temperature range from 20 to 1000 °C. For this temperature interval, the deviations of the flash heating approach method compared to the Goldak model are marginal.

The resulting microstructure from the experiment in the e_1-e_3-plane of the lower third of the specimen is depicted in Fig. 14.11d on the left and the simulated one on the right. Furthermore, the evaluation point A is introduced. In the simulation results, a martensit-dominated microstructure is shown in the narrow cylinders. A similar microstructure, an acicular martensitic structure, can also be observed in the experimental optical micrographs samples. In the area of larger cross-sections, there is a martensitic α_m microstructure together with combined β and stable α_s.

The evolutions of the phase fractions at point A over time are shown in Fig. 14.11e. While the temperature changes at point A are large during the exposure period, the microstructure is dominated by the instantaneous transformation between the martensitic α_m-phase and the β-phase. During the remaining process, the diffusion-based transformations to the stable α_s-phase progress. The final phase fractions at point A after cooling to room temperature are $X_{\alpha_m} = 0.74$, $X_{\alpha_s} = 0.16$, and $X_\beta = 0.1$, respectively. More details about the experimental study and the numerical results are provided in [33].

14.4 Simulation of the PBF-LB/P Process

In powder bed fusion of polymers using a laser beam (PBF-LB/P), the building chamber is kept at a temperature within the so-called process window and the top surface is heated up close to the melting temperature. In contrast to metal AM, where material solidifies rapidly after melting, the molten material in PBF-LB/P slowly crystallizes during the manufacturing process. In addition, crystallization can be present at a constant temperature (isothermal) or during cooling at a specific rate (non-isothermal), which also presents a difference to the modeling of metal AM.

In the initial process model, it was assumed that crystallization begins at the end of the built job and that molten and powder materials reside next to each other until the cooling stage of the process. However, it was shown in e.g., [14, 45] that crystallization occurs before the start of the cooling stage, and may even be induced only after a few layers during the build job.

Further, the sensitivity to small temperature variations is more severe for PBF-LB/P compared to metal AM. To this end, a proportional–integral–derivative (PID)-controlled radiation boundary condition model was proposed in [45] to replicate the heating mechanism present in PBF-LB/P machinery. It was shown that the use of a constant temperature value for the radiation boundary condition does not guarantee reaching the prescribed built temperature on the top surface in the building chamber.

Finally, as presented in Fig. 14.2, due to the lower thermal conductivity of PA12, the time window for the integrated energy input (denoted by τ) can be larger than that of metals, thereby allowing for a significant reduction in the numerical effort while keeping the number of errors at a moderate level.

It should be noted that all experimental work presented in this chapter and in the related publications [41, 45, 46, 48] was carried out by the Institute of Polymer Technology (LKT) at the Friedrich-Alexander-Universität Erlangen-Nürnberg (FAU). Within the framework of a further cooperation with the Bayerisches Laserzentrum GmbH (blz) and the Institute of Photonic Technologies (LPT) at the FAU, we investigated the possibility of multi-material PBF-LB/P [26].

14.4.1 Adapted Crystallization Model

To model the crystallization behavior of semi-crystalline polymers, the Nakamura model is often employed, as shown in [27, 35, 37, 50]. In its rate formulation, the relative degree of crystallization α can be computed by

$$\dot{\alpha} = K(\vartheta)\, G(\alpha) \geq 0, \qquad (14.4)$$

where K is linked to the crystallization rate and G is termed the Nakamura function [27, 37]. It is apparent that this model can only account for a monotonic increase in the degree of crystallization. During PBF-LB/P, the temperature is reduced due to the deposition of fresh powder material, which is kept at a lower temperature, and reheating due to the heating system and the energy provided by the laser beam. During the deposition of the colder polymer powder, crystallization may be induced, followed by remelting. The model from Eq. 14.4 clearly cannot account for this.

Hence, in [45, 48] a modified model has been proposed, that allows researchers to account for the described temperature variations. In Fig. 14.12, the original model as in Eq. 14.4 and the modified model are compared for a prescribed temperature

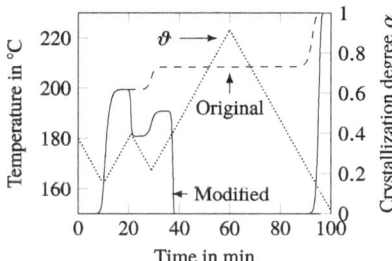 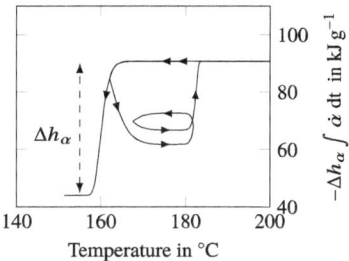

Fig. 14.12 Comparison of the Nakamura model with the modified approach proposed in [45, 48]. On the left, the prescribed temperature profile is indicated by the dotted line and the resulting evolution of the relative degree of crystallization by the dashed and solid lines for the Nakamura and modified approaches, respectively. The plot on the right illustrates the corresponding accumulated heat absorption and release per unit mass, indicating energetic consistency of the modified approach. The term Δh_α denotes the total specific energy that is released during crystallization. For the cooling rates considered, this value is assumed to be constant

evolution. The temperature evolution is composed of repeated cooling and heating, while also exceeding the melting temperature. In this case, partially crystallized material would undergo re-melting. This cannot be captured by Eq. 14.4, where the degree of crystallization would rest at a certain plateau until the temperature reduces again.

In contrast the modified approach allows for partial and complete remelting of crystallized material. On the right in Fig. 14.12, the energetic consistency of the modified approach is visible, where the accumulated heat release and absorption is plotted.

14.4.2 Validation

In this section, some results from the validation of the temperature and degree of crystallization simulations are presented. In [45], experimentally obtained temperatures at the surface and bulk were used for validation and the results are presented in Fig. 14.13. For the validation, rectangular prisms with different sizes of the cross-sectional area, i.e., $5 \times 5\,mm^2$, $10 \times 10\,mm^2$, and $20 \times 20\,mm^2$ were built using a meander scan strategy. The surface temperatures were measured using an infrared thermographic system. The maximum temperatures at the top surface for different combinations of laser beam power and scan velocity were compared to numerically

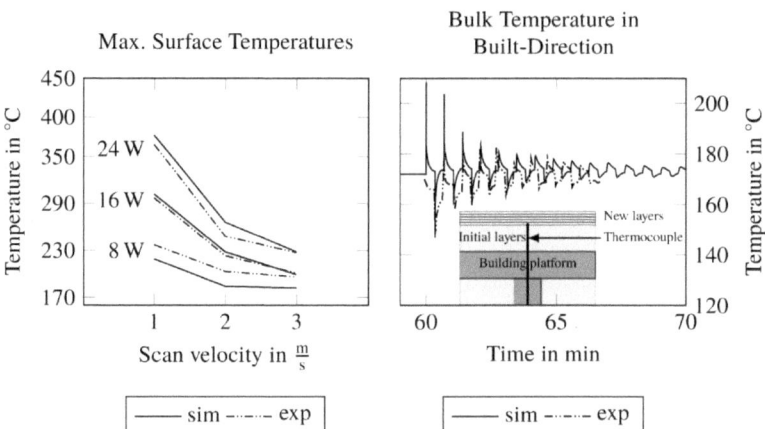

Fig. 14.13 Validation of the numerical tool using measurements of surface and bulk temperature in the build direction during the manufacture of a rectangular prism as discussed in [45]. On the left, the maximum temperatures were measured experimentally by a thermo-camera. On the right, the temperatures in the built direction were measured using a thermocouple with a fixed relative position at the height of the initial part layer

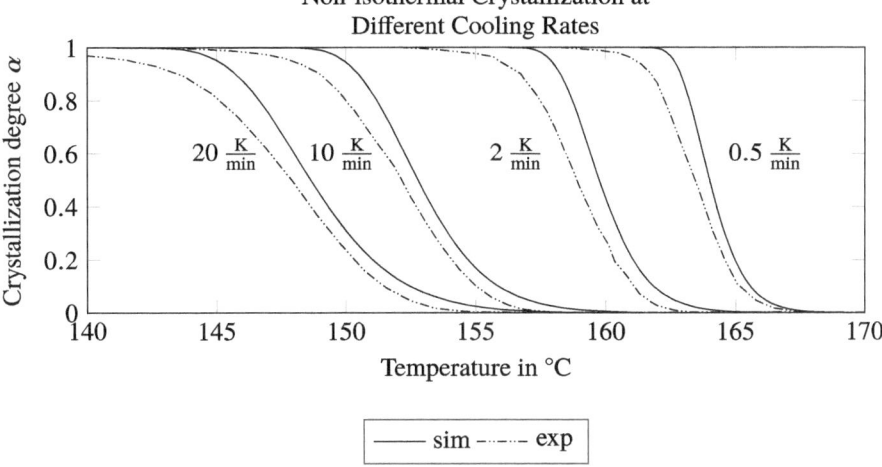

Fig. 14.14 Evolution of the relative degree of crystallization for non-isothermal cooling at different rates as presented in [45], using experimental data and parameters for the Nakamura model as identified in [50]

predicted values. It should be noted that the energy input for the numerical experiments was modeled using a relative line length of up to $\tau = 70$ for a beam velocity of $3\,\mathrm{m\,s^{-1}}$. For the case of $10 \times 10\,\mathrm{mm^2}$, the results are presented in Fig. 14.13 on the left and show a good agreement.

For the measurements in the bulk direction, a thermocouple was positioned at the height of the initial powder layer and kept constant. A beam power of $16\,\mathrm{W}$ and a beam velocity of $2\,\mathrm{m\,s^{-1}}$ was used to fabricate parts with dimensions of $20 \times 20 \times 1\,\mathrm{mm^3}$. The results, depicted in Fig. 14.13 on the right, also show good agreement between the measurements and numerical predictions.

The parameter identification of the Nakamura crystallization model has been carried out in [50], which has been subsequently used and extended in [45] for the process simulation. The evolution of the degree of crystallization is plotted in Fig. 14.14 for one-element tests with a prescribed temperature evolution.

14.4.3 Current Findings

Further investigations were performed with the presented simulation tool and published in [48]. For instance, a geometric dependency of the evolution of temperature and, therefore, crystallization was shown, as illustrated in Fig. 14.15. When comparing the temperature and degree of crystallization at the Points A and B, it

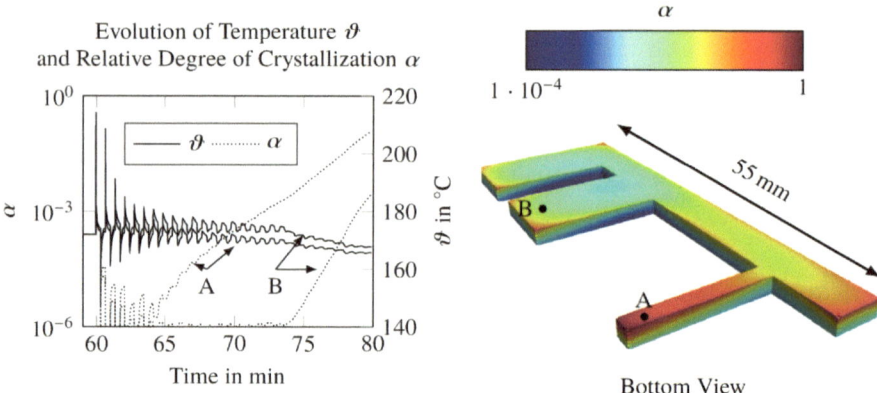

Fig. 14.15 Influence of part geometry on temperature and crystallization evolution [48]. Thin part features cool down faster, leading to an increased crystallization rate. The dotted lines refer to the relative degree of crystallization α and solid lines to the temperature ϑ, evaluated at the points A and B respectively (located in the built direction at the height of the initial layer), as indicated in the contour plot on the right showing the bottom view of the part

becomes apparent that thinner features cool down faster, leading to an accelerated crystallization evolution.

Since crystallization is accompanied by a volumetric shrinkage, heterogeneous crystallization may lead to distortion and residual stresses. To study these effects, a visco-elastic-visco-plastic mechanical model at finite deformations is currently being developed. The modeling approach is based on the formulation of mechanical stress power on a thermal intermediate configuration as proposed in [28] and equipped with free energies as proposed in [16]. The model includes non-linear isotropic and kinematic hardening, as well as a general non-linear visco-elastic contribution [38, 39] and a non-linear relaxation time function, inspired by [2]. The capabilities of the model are depicted in Fig. 14.16. As an example, the experimental results for a step-relaxation experiment at a temperature of 120 °C and the corresponding model calibration are shown on the left. To include the effect of crystallization while retaining a thermodynamically consistent formulation, both the relaxation time and the plastic contribution depend on the degree of crystallization, while the elastic contributions do not (see e.g., [25]).

For a cyclic shear test of one element, the effect of a sudden increase in the mechanical properties at the onset of crystallization (see e.g., [13], which was confirmed by experiments performed by the Institute of Polymer Technology (LKT) at the FAU) is presented in Fig. 14.16 on the right. Molten material is kept at a constant temperature of 168 °C which leads to isothermal crystallization. Initially, the material is in a fully molten state and unable to build up stresses. At the onset of crystallization (here $\alpha \approx 1$ %) the overall stiffness increases significantly due to an increase in viscosity and yield stress. The final calibration of the model is ongoing work and will be part of a future publication.

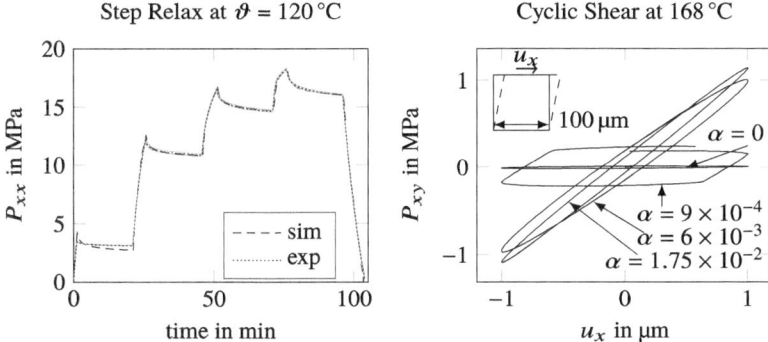

Fig. 14.16 On the left, the capabilities of the visco-elastic-visco-plastic model at finite strains are illustrated for a step-relaxation test at 120 °C, which is one of the measurements conducted as part of a collaboration with the Institute of Polymer Technology (LKT) at the FAU that was used for the model calibration. The behavior during crystallization is accounted for by a dependency of the viscous and plastic contribution on the degree of crystallization. On the right, a model problem for a cyclic shear test illustrates the effect of crystallization on the model response with a sharp increase of overall stiffness at the onset of crystallization [13]. The calibration of the latter is ongoing work

14.5   Summary and Outlook

In this chapter, the recent progress in macroscopic process simulations of powder bed based additive manufacturing of metals and polymers was presented. The simulation approach utilizes the Finite-Element method for solving the equations of the thermo-mechanical problem and is equipped with adaptive discretization in space and time, as well as multi-rate integrators and physical model reduction techniques, such as layer agglomeration and time-integrated energy input. This allows for drastically reducing the high computational cost of high-fidelity models. The numerical tool accounts for several process characteristics such as the energy input by the beam, phase change from powder to melt and, subsequently, solid material, a PID-controlled radiation boundary condition that mimics the heating system, and the deposition of fresh powder by an element birth method. The different phases, i.e., powder, melt, and solid, are modeled in a continuum mechanical fashion as homogenized material parameters where every element is assigned to one of the phases.

The differences between PBF-LB/M, PBF-EB/M, and PBF-LB/P from a simulative point of view were discussed. For metals, an elasto-visco-plastic material model using a logarithmic strain measure with temperature-dependent material parameters is used in a one-way coupling scheme. Based on the simulation of a lattice unit cell of CMSX-4, the known effects of powder bed based processes such as overheating and TGM could be shown. Furthermore, by using an integrated heat input model combined with a modified beam exposure time, i.e., flash heating, the microstructure of Ti-6Al-4V could be predicted at part scale and was compared with experimental measurements showing qualitatively good agreement. Geometric dependencies of the microstructure could be identified, which are attributed to different temperature histories.

For the process simulation of semi-crystalline polymers, a visco-elastic-visco-plastic material model at finite deformation was chosen due to the considerable volumetric shrinkage during crystallization. Further, a monolithic solution scheme was employed due to the sensitivity of the process, even to small temperature variations. For the modeling of crystallization, the Nakamura model was extended for application in PBF-LB/P to account for remelting. The thermal model was validated by comparing the computed maximum surface temperatures and bulk temperatures for cubical cross-sections with different edge lengths to experimental measurements. Further, the crystallization model was validated at the point level through comparisons with data from the literature. The model was then used to investigate the dependence of the temperature evolution, and thus of the degree of crystallization, on geometrical feature sizes. It was shown that PA12 displays a geometry dependency that is similar to the findings of the investigations of the microstructure of Ti-6Al-4V. Furthermore, the first results of the mechanical material parameter identification for PA12 produced by PBF-LB/P were shown on a step relaxation test.

Further work includes the parameter identification for the mechanical model of PA12 during crystallization. For metals, the next steps primarily involve investigating further reduction techniques. In particular, the adaptation of the flash heating method to represent local effects by scan path adjustments represents an active field of research.

References

1. Amado Becker, A.: Characterization and prediction of SLS processability of polymer powders with respect to powder flow and part warpage. ETH Zürich (2016)
2. Amin, A., Lion, A., Sekita, S., Okui, Y.: Nonlinear dependence of viscosity in modeling the rate-dependent response of natural and high damping rubbers in compression and shear: experimental identification and numerical verification. Int. J. Plasticity **22**(9), 1610–1657 (2006)
3. Arndt, D., Bangerth, W., Feder, M., Fehling, M., Gassmöller, R., Heister, T., Heltai, L., Kronbichler, M., Maier, M., Munch, P., Pelteret, J.-P., Sticko, S., Turcksin, B., Wells, D.: The deal.II library, version 9.4. J. Numer. Math. **30**(3), 231–246 (2022)
4. Baiges, J., Chiumenti, M., Moreira, C.A., Cervera, M., Codina, R.: An adaptive finite element strategy for the numerical simulation of additive manufacturing processes. Additive Manuf. **37**, 1–22 (2020)
5. Bartlett, J.L., Li, X.: An overview of residual stresses in metal powder bed fusion. Additive Manuf. **27**, 131–149 (2019)
6. Bayat, M., Klingaa, C.G., Mohanty, S., De Baere, D., Thorborg, J., Tiedje, N.S., Hattel, J.H.: Part-scale thermo-mechanical modelling of distortions in laser powder bed fusion—analysis of the sequential flash heating method with experimental validation. Additive Manuf. **36**, 101508 (2020)

7. Brighenti, R., Cosma, M.P., Marsavina, L., Spagnoli, A., Terzano, M.: Laser-based additively manufactured polymers: a review on processes and mechanical models. J. Mater. Sci. **56**(2), 961–998 (2021)
8. Bruna-Rosso, C., Mergheim, J., Previtali, B.: Finite element modeling of residual stress and geometrical error formations in selective laser melting of metals. Proc. Inst. Mech. Engineers, Part C: J. Mech. Eng. Sci. **235**(11), 2022–2038 (2020)
9. Burkhardt, C., Soldner, D., Mergheim, J.: A comparison of material models for the simulation of selective beam melting processes. Procedia CIRP **94**, 52–57 (2020)
10. Burkhardt, C., Steinmann, P., Mergheim, J.: Thermo-mechanical simulations of powder bed fusion processes: accuracy and efficiency. Adv. Modeling Simul. Eng. Sci. **9**(1), 9 (2022)
11. Das, A., Chatham, C.A., Fallon, J.J., Zawaski, C.E., Gilmer, E.L., Williams, C.B., Bortner, M.J.: Current understanding and challenges in high temperature additive manufacturing of engineering thermoplastic polymers. Addit. Manuf. **34**, 101218 (2020). https://doi.org/10.1016/j.addma.2020.101218
12. Denlinger, E.R., Irwin, J., Michaleris, P.: Thermomechanical modeling of additive manufacturing large parts. J. Manuf. Sci. Eng. **136**(6), 061007 (2014). https://doi.org/10.1115/1.4028669
13. Descher, S., Wünsch, O.: Simulation framework for crystallization in melt flows of semi-crystalline polymers based on phenomenological models. Archive Appl. Mech. **4** (2022)
14. Drummer, D., Greiner, S., Zhao, M., Wudy, K.: A novel approach for understanding laser sintering of polymers. Additive Manuf. **27**, 379–388 (2019)
15. Ellsiepen, P., Hartmann, S.: Remarks on the interpretation of current non-linear finite element analyses as differential–algebraic equations. Int. J. Numer. Methods Eng. **51**(6), 679–707 (2001). https://doi.org/10.1002/nme.179
16. Felder, S., Vu, N.A., Reese, S., Simon, J.W.: Modeling the effect of temperature and degree of crystallinity on the mechanical response of Polyamide 6. Mech. Mater. **148**, 103476 (2020). https://doi.org/10.1016/j.mechmat.2020.103476
17. Galati, M., Iuliano, L.: A literature review of powder-based electron beam melting focusing on numerical simulations. Additive Manuf. **19**, 1–20 (2018)
18. Goldak, J., Chakravarti, A., Bibby, M.: A new finite element model for welding heat sources. Metallurgical Trans. B **15**(2), 299–305 (1984)
19. Hodge, N.E.: Towards improved speed and accuracy of laser powder bed fusion simulations via representation of multiple time scales. Addit. Manuf. **37**, 101600 (2021). https://doi.org/10.1016/j.addma.2020.101600
20. Hodge, N.E., Ferencz, R.M., Solberg, J.M.: Implementation of a thermomechanical model for the simulation of selective laser melting. Comput. Mech. **54**(1), 33–51 (2014). https://doi.org/10.1007/s00466-014-1024-2
21. Irwin, J., Michaleris, P.: A line heat input model for additive manufacturing. J. Manuf. Sci. Eng. **138**(11), 111004 (2016). https://doi.org/10.1115/1.4033662
22. Klassen, A., Bauereiß, A., Körner, C.: Modelling of electron beam absorption in complex geometries. J. Phys. D: Appl. Phys. **47**(6), 065307 (2014)
23. Krabbenhoft, K., Damkilde, L., Nazem, M.: An implicit mixed enthalpy–temperature method for phase-change problems. Heat Mass Transf. **43**, 233–241 (2007). https://doi.org/10.1007/s00231-006-0090-1
24. Krieg, N.: Multi-time-stepping using a nitsche-based interface formulation applied to the heat equation. Master's thesis, Friedrich-Alexander-Universität Erlangen Nürnberg (2022)
25. Landgraf, R.: Modellierung und Simulation der Aushärtung polymerer Werkstoffe. Dissertation, TU Chemnitz (2015). https://nbn-resolving.org/urn:nbn:de:bsz:ch1-qucosa-187720
26. Laumer, T.: Realization of multi-material polymer parts by simultaneous laser beam melting. J. Laser Micro/Nanoeng. **10**(2), 140–147 (2015)
27. Levy, A.: Robust numerical resolution of Nakamura crystallization kinetics. Int. J. Theor. Appl. Math. **3**(4), 143 (2017)
28. Lion, A.: Constitutive modelling in finite thermoviscoplasticity: a physical approach based on nonlinear rheological models. Int. J. Plasticity **16**(5), 469–494 (2000)

29. Lupone, F., Padovano, E., Casamento, F., Badini, C.: Process phenomena and material properties in selective laser sintering of polymers: a review. Materials **15**(1), 183 (2021)
30. Markl, M., Körner, C.: Multiscale modeling of powder bed–based additive manufacturing. Annu. Rev. Mater. Res. **46**(1), 93–123 (2016)
31. Mercelis, P., Kruth, J.P.: Residual stresses in selective laser sintering and selective laser melting. Rapid Prototyping J. **12**(5), 254–265 (2006)
32. Mergheim, J., Breuning, C., Burkhardt, C., Hübner, D., Köpf, J., Herrnböck, L., Yang, Z., Körner, C., Markl, M., Steinmann, P., Stingl, M.: Additive manufacturing of cellular structures: multiscale simulation and optimization. J. Manuf. Processes **95**, 275–290 (2023)
33. Nahr, F., Rasch, M., Burkhardt, C., Renner, C., Baumgaertner, B., Hausotte, T., Koerner, C., Steinmann, P., Mergheim, J., Schmidt, M., Markl, M.: Geometrical influence on material properties of ti6al4v parts in powder bed fusion. J. Manuf. Mater. Process. submitted (2023)
34. Nakshatrala, P.B., Nakshatrala, K.B., Tortorelli, D.A.: A time-staggered partitioned coupling algorithm for transient heat conduction. Int. J. Numer. Methods Eng. **78**(12), 1387–1406 (2009). https://doi.org/10.1002/nme.2524
35. Neugebauer, F., Ploshikhin, V., Ambrosy, J., Witt, G.: Isothermal and non-isothermal crystallization kinetics of polyamide 12 used in laser sintering. J. Thermal Anal. Calorimetry **124**(2), 925–933 (2016)
36. Nitzler, J., Meier, C., Müller, K.W., Wall, W.A., Hodge, N.E.: A novel physics-based and data-supported microstructure model for part-scale simulation of laser powder bed fusion of ti-6al-4v. Adv. Modeling Simul. Eng. Sci. **8**(1), 1–39 (2021)
37. Patel, R.M., Spruiell, J.E.: Crystallization kinetics during polymer processing—analysis of available approaches for process modeling. Polym. Eng. Sci. **31**(10), 730–738 (1991)
38. Reese, S., Govindjee, S.: Theoretical and numerical aspects in the thermo-viscoelastic material behaviour of rubber-like polymers. Mech. Time-Dependent Mater. **1**, 357–396 (1997)
39. Reese, S., Govindjee, S.: A theory of finite viscoelasticity and numerical aspects. Int. J. Solids Struct. **35**, 3455–3482 (1998)
40. Riedlbauer, D.: Thermal and Thermomechanical Modeling and Simulation of Selective Beam Melting Processes: Thermische und Thermomechanische Modellierung und Simulation Von Selektiven Strahlschmelzprozessen. Schriftenreihe Technische Mechanik, Lehrstuhl für Technische Mechanik (2019)
41. Riedlbauer, D., Drexler, M., Drummer, D., Steinmann, P., Mergheim, J.: Modelling, simulation and experimental validation of heat transfer in selective laser melting of the polymeric material PA12. Comput. Mater. Sci. **93**, 239–248 (2014). https://doi.org/10.1016/j.commatsci.2014.06.046
42. Riedlbauer, D., Scharowsky, T., Singer, R.F., Steinmann, P., Körner, C., Mergheim, J.: Macroscopic simulation and experimental measurement of melt pool characteristics in selective electron beam melting of Ti-6Al-4V. The Int. J. Adv. Manuf. Technol. **88**(5–8), 1309–1317 (2017)
43. Riedlbauer, D., Steinmann, P., Mergheim, J.: Thermomechanical finite element simulations of selective electron beam melting processes: performance considerations. Comput. Mech. **54**(1), 109–122 (2014)
44. Soldner, D., Brands, B., Zabihyan, R., Steinmann, P., Mergheim, J.: A numerical study of different projection-based model reduction techniques applied to computational homogenisation. Comput. Mech. 1–13 (2017)
45. Soldner, D., Greiner, S., Burkhardt, C., Drummer, D., Steinmann, P., Mergheim, J.: Numerical and experimental investigation of the isothermal assumption in selective laser sintering of pa12. Additive Manuf. **37**, 101676 (2021)
46. Soldner, D., Mergheim, J.: Thermal modelling of selective beam melting processes using heterogeneous time step sizes. Comput. Math. Appl. **78**(7), 2183–2196 (2019)
47. Soldner, D., Steinmann, P., Mergheim, J.: Reducing computational cost of the simulation of selective laser sintering of PA12 (2016)
48. Soldner, D., Steinmann, P., Mergheim, J.: Modeling crystallization kinetics for selective laser sintering of polyamide 12. GAMM-Mitteilungen **44**, 9 (2021)

49. Withers, P., Bhadeshia, H.: Residual stress. part 2—nature and origins. Mater. Sci. Technol. **17**(4), 366–375 (2001)
50. Zhao, M., Wudy, K., Drummer, D.: Crystallization kinetics of polyamide 12 during selective laser sintering. Polymers **10**(2), 168 (2018)

Paul Steinmann was appointed professor at the University of Kaiserslautern in 1997. In 2007, he took over the Institute of Applied Mechanics (LTM) at Friedrich-Alexander-Universität Erlangen-Nürnberg. He is currently the Director of the UK Glasgow Computational Engineering Center (GCEC). His research interests range from ma-terial modeling to multiscale methods, multi-physics problems, non-standard con-tinua, configurational failure mechanics, biomechanics, and general advances in finite element and discretization methods.

Julia Mergheim associated with the Institute of Applied Mechanics (LTM) at Friedrich-Alexander-Universität Erlangen-Nürnberg, is a researcher whose expertise encompasses Nonlinear Finite Element Method (FEM), Multiscale Material Modeling, and Fracture and Damage Mechanics.

Chapter 15
Mesoscopic Modeling and Simulation of Properties of Additively Manufactured Metallic Parts

Zerong Yang, Ludwig Herrnböck, Matthias Markl, Julia Mergheim, Paul Steinmann, and Carolin Körner

15.1 Introduction

The conventional process development for powder bed fusion (PBF) processes targets the manufacturing of defect-free parts, e.g., with low porosity, no layer binding faults and cracks, or high dimensional accuracy. These topics are addressed for laser powder bed fusion (PBF-LB/M) in Chap. 8 and electron beam powder bed fusion (PBF-EB) in Chap. 7 for different alloys. However, although a part is defect-free, the microstructure and consequently the mechanical properties vary locally in complex geometries manufactured with standard process strategies. The microstructure is mainly influenced by the local solidification conditions during manufacturing. Epitaxial crystal growth occurs in combination with a natural grain selection along the whole part due to different crystal growth rates depending on the local transient temperature field and the crystal orientation. Depending on the PBF process and the manufacturing conditions, nucleation plays an important role during microstructure formation regarding the grain size and orientation and there are two different strategies to circumvent or benefit from these effects. In the first strategy, the processing conditions are modified in such a way that the solidification conditions are almost equal regardless of the part geometry (see Chap. 7). The second strategy is to tailor the local solidification conditions to achieve a beneficial local microstructure, e.g., to exploit a material anisotropy to ensure the orientation with the highest stiffness in the part-loading direction. Both strategies are costly and time-consuming regarding pure experimental development.

This chapter describes a numerical simulation chain to predict the mechanical properties of metallic parts ranging from thermal via microstructural to

Z. Yang · L. Herrnböck · M. Markl · J. Mergheim · P. Steinmann · C. Körner (✉)
Friedrich-Alexander-Universität Erlangen-Nürnberg, Chair of Materials Science and Engineering for Metals, Martensstraße 5, 91058 Erlangen, Germany
e-mail: carolin.koerner@fau.de

© The Author(s) 2025
D. Drummer and M. Schmidt (eds.), *Progress in Powder Based Additive Manufacturing*, Springer Tracts in Additive Manufacturing,
https://doi.org/10.1007/978-3-031-78350-0_15

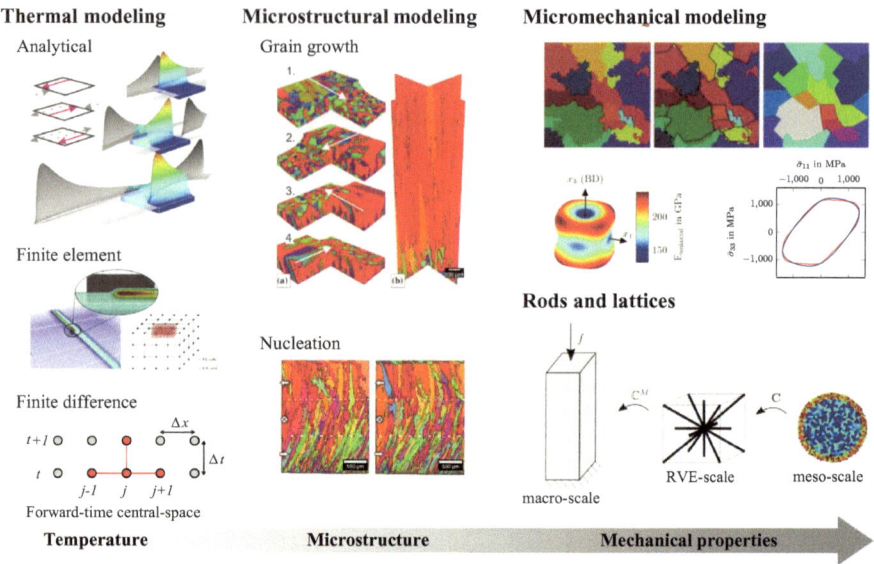

Fig. 15.1 Graphical overview of the applied numerical models described in Sect. 15.2. The subfigures are reproduced under the terms of the CC BY 4.0 license: Thermal modeling—analytical [27], Thermal modeling—finite element [28], Microstructural modeling—grain growth & Micromechanical modeling [23], Microstructural modeling—nucleation [26], Thermal modeling—finite difference recreated in analogy to [5]. The rods and lattices are reproduced with permission [32]

micromechanical modeling and simulation covering all necessary effects from the solidification conditions via crystal growth to mechanical responses (see Fig. 15.1).

Section 15.2 recaptures the applied numerical methods and models. First, the thermal, microstructural, and micromechanical models are presented.

Subsequently, the theory of geometrically exact rods is briefly presented with the aim of simulating PBF-manufactured cellular structures in a numerically efficient way.

In Sect. 15.3, the presented simulation chain is applied to different scenarios ranging from bulk material to cellular structures, where single rods up to large crash absorbers are considered.

The chapter concludes with an outlook on further research topics.

15.2   Numerical Methods

This section describes the numerical methods applied. The manufacturing process is represented by thermal models (Sect. 15.2.1) coupled to a microstructural model (Sect. 15.2.2) used to obtain the grain structure and texture of PBF-manufactured metal parts. Subsequently, a micromechanical model (Sect. 15.2.3) uses the microstructural simulation results as inputs for studying the resulting mechanical behavior. Finally, homogenization methods for the evaluation of the effective mechanical response of rods and lattices are introduced (Sect. 15.2.4).

15.2.1   Thermal Models

It is well known that the microstructure of PBF-manufactured parts is highly dependent on thermal conditions. The temperature is modeled by approximating the powder bed as a dense material and only considering heat conduction while neglecting other effects such as phase transitions, melt pool dynamics, evaporation, or the random powder bed. The governing model equation, considering heat conduction in an isotropic medium with an external heat source, is written as [4]

$$c_p \frac{\partial T}{\partial t} = \frac{\partial H(T)}{\partial t} = -\frac{\lambda}{\rho} \left(\frac{\partial^2 T}{\partial x^2} + \frac{\partial^2 T}{\partial y^2} + \frac{\partial^2 T}{\partial z^2} \right) + \frac{Q}{\rho}, \qquad (15.1)$$

where c_p, λ and ρ are the specific heat capacity at constant pressure, thermal conductivity, and mass density of the material, respectively. These three parameters are assumed as constant. T is temperature, t is time, H is enthalpy, Q is the energy source term, and x, y, z are Cartesian coordinates.

Depending on the application of the thermal model on the PBF process, different approaches were employed to solve Eq. (15.1), namely an analytical solution [27], a finite element (FE) [28], and a finite difference (FD) [46] method. Each of these approaches has advantages and disadvantages which are highlighted in the following paragraphs.

15.2.1.1   Analytical Solution

The analytical solution is based on Rosenthal's solution for the temperature distribution caused by a moving point heat source with constant speed across the surface [42]. This analytical solution applies some important assumptions and simplifications, e.g., constant material properties, a constant rate of heat input from the energy source, and a heat transfer purely by conduction.

Furthermore, the stationary coordinate system is transformed by applying a moving coordinate system with the origin at the center of the heat source [27]. In the context of PBF, the point heat source refers to a laser or an electron beam.

The melt pool size obtained by the analytical solution has been validated for different materials (e.g., IN718, CMSX-4, and AA5456) and different beam sources (e.g., laser and electron beams) [15, 27, 36]. The analytical approach has advantages in terms of its computational efficiency for simple scanning strategies and simple geometries but will be very complex for—or even not applicable to—complex scanning patterns.

15.2.1.2 Finite Element Method

The FE method described in Chap. 14 is adopted to solve the nonlinear Eq. (15.1) [28]. The temporal domain is discretized by a two-stage S-DIRK Runge-Kutta method [6] and the spatial domain is discretized by the FEs using the open source library deal.II [2]. Due to a very localized heat-affected zone in PBF processes, an adaptive mesh refinement and coarsening mechanism is employed to reduce the numerical cost [28]. For a more detailed description of the FE solver, readers are referred to Chap. 4 and [28, 45]. The advantages of the FE method include that it can be used on different scales from single lines to multiple layers by different heat inputs per line or layer and that the the FE mesh can be reused in subsequent micromechanical simulations. However, the FE method is rather expensive compared to the other two approaches.

15.2.1.3 Finite Difference Method

A finite difference (FD) method is further employed to solve Eq. (15.1) [46]. The FD solver adopts a forward-time and central-space approach, which is one of the conventional numerical techniques for solving differential equations [12, 22]. The enthalpy method is used to deal with the nonlinear behavior of specific heat during melting and solidification, i.e., the enthalpy is treated as an independent variable other than temperature [20]. The enthalpy evolves with time and temperature and is converted from the $H - T$ relation.

The $H - T$ relation can be determined experimentally or numerically by, for example, using CALPHAD methods [1]. The benefit of using the FD method is a reduced computational effort for complex geometries and complex scanning strategies achieving a sufficient degree of accuracy [46].

15.2.2 Microstructural Model

The microstructure, mainly the grain structure, is modeled based on the 3D cellular automaton (CA) algorithm proposed by [8]. A wealth of information can be extracted from the 3D CA simulation results, such as concerning the texture, grain morphology, grain boundary alignment, etc. These results are suitable and sufficient as input for micromechanical simulations, as described in Sect. 15.2.3. Meanwhile, the 3D CA method shows strong competitive advantages over other popular methods such as phase field and Kinetic Monte Carlo in terms of the balance of simulation output and computational cost [14, 29].

In the CA model, individual grains are outlined by the superposition of octahedra, ignoring details such as dendrite shapes. One growing dendrite is modeled by an octahedron with the threefold rotation axes representing the ⟨100⟩ growth directions. Its orientation in three-dimensional space is represented by the rotation of the octahedron around three Euler angles. The octahedron is coupled to a distinct simulation cell and grows with velocity v_{oct}. It is assumed that the growth velocity v_{oct} only depends on the local thermal undercooling $\Delta T = T_{liq} - T$ with T_{liq} the liquidus temperature. There are different ways to estimate $v_{oct}(\Delta T)$, e.g., based on phase field simulations [26] or the theoretical model for directional solidification [30], commonly known as the KGT model. Since the KGT model is computationally expensive, the computational cost is reduced by fitting the results with polynomial [9], quadratic [10], or power functions [13]. The application examples in this chapter use $v(\Delta T) = 1 \cdot 10^{-4} \cdot \Delta T^2$ [27, 28, 46]. Once the facet of the growing octahedron reaches the center of a neighboring cell, that cell is captured by the dendrite where a new octahedron is initiated. The newly-emerging octahedron inherits the orientation of the parent one, and its position is determined according to predefined geometrical rules [8, 10]. All octahedra grow independently at a rate that depends on the local undercooling. Readers interested in the CA model presented here can refer to the original work by Gandin et al. [7, 8, 10, 38] and our previous publications [27–29, 31, 46, 47] for a more detailed description and discussion.

The CA model contains a nucleation model [26] to represent the formation of new nuclei during melt pool solidification. In the nucleation model, a shadow grid for nucleation in addition to the normal CA grid is required due to the competitive growth of grains. The shadow grid shares the same temperature field and rule of growth velocity as the normal CA grid. Nuclei with random orientation are initiated in the shadow grid in front of the liquid-solid interface on the side of the undercooled melt. Once the size of a nucleus exceeds a predefined critical radius, the corresponding nucleation cell is applied to the normal CA grid and subsequently deleted from the shadow grid, i.e., this nucleus survived and is now a new grain. All nuclei in the shaded grid that do not exceed the critical radius before the corresponding cell is captured by an octahedron in the normal CA grid do not survive and are removed from the shaded grid. Although the nucleation model described above is rather simple, it has been demonstrated to be very helpful in elucidating the effect of nucleation on the resulting microstructure of PBF-manufactured parts [26].

15.2.3 Micromechanical Model

The grain structure and texture of additively manufactured materials strongly influence their mechanical behavior, both in the elastic and the inelastic regions. The resulting elastic properties may range from isotropy to transversal and orthotropic symmetry. In the same way, the grain structure and texture may influence the yield criterion. The grain structure is obtained from a CA, modeling the grain growth and material orientation during the manufacturing process as introduced in Sect. 15.2.2. The CA serves as input to the definition of the mechanical behavior of the material. The mechanical behavior on the mesoscale, capturing the texture and grain growth, is modeled by means of geometrically linear gradient-enhanced crystal plasticity. The strain gradient extension is based on the non-integrability of plastic strains resulting in geometrically necessary dislocation densities. Computational homogenization is applied to determine the resulting elastic and plastic properties of PBF-manufactured metals [23].

The mechanical behavior of each grain is modeled with the gradient-enhanced crystal plasticity model introduced in [23–25]. There, the linear strain field ε is decomposed additively into an elastic and a plastic contribution. The plastic strain is composed of α crystallographic slips γ^α, where each slip system α is fully described by a slip plane and a slip direction. As an example, we name the face-centered cubic Inconel 718, which has 12 independent slip systems. Introducing the dependency of the material behavior on the gradients of the plastic slips is motivated by geometrically necessary dislocations, which are the non-integrability of plastic distortions [35]. Following the principle of virtual power, the balance of linear momentum and a microforce balance are derived. Both equations are coupled by the so-called Schmid stresses and are solved for displacements and plastic slips. The energetically conjugated pairs are elastic strains ε^e and stresses σ, as well as plastic slips γ^α and corresponding gradient-extended Schmid stresses π^α.

To overcome the problem of searching for active slip systems, the slips are regularized in time, applying a creep law. However, by proper choice of parameters, the solution converges towards the time-independent case. Since the plastic slips are considered as a degree of freedom due to the gradient enhancement, boundary conditions have to be prescribed to them on each grain boundary. Two standard boundary conditions are either of the Neumann type, namely micro-free, or micro-hard, which is a Dirichlet-type boundary condition. The gradient enhancement allows for defining a more sophisticated grain boundary model. Therefore, we define the free energy between two grains and derive a constitutive equation for the micro-tractions between two grains, depending on their plastic slips [23].

To identify the homogenized material properties of the additively manufactured material, concepts from multi-scale computational homogenization are utilized. Therefore, a representative volume element (RVE) is constructed and solved under appropriate boundary conditions. For the present case, these are periodic boundary conditions for the fluctuations of the primary unknowns and anti-periodic boundary conditions for the tractions and micro-stresses. The RVE must be constructed in such a way that it represents an arbitrary volume of the bulk material in a sufficiently precise manner. By applying perturbation techniques, a fourth-order macroscopic elasticity tensor is evaluated [33]. Further, uni- and biaxial loads up to a predefined amount of plastification in the fully resolved bulk leads to a generic yield surface in the six-dimensional stress space describing the plastic behavior of the bulk material. A yield function may be fitted to the results. A similar procedure is utilized in Sect. 15.3.3 to obtain a yield surface for geometrically exact beams.

15.2.4 Rods and Lattices

Periodic lattices represent a widespread application, enabled by the versatility of additive manufacturing processes. To estimate the effective mechanical response of periodic lattice structures, multi-scale computational homogenization is a well-established procedure. This approach derives the effective macroscopic mechanical response by homogenizing the mechanical behavior of a representative volume element (RVE) of the lattice. In particular, here, the RVE of the lattice is modeled utilizing the theory of geometrically exact rods [43, 44]. The developed framework enables the consideration of the texture and material orientation in each grain within the rods. The grain structure strongly influences the elastic and inelastic properties of the individual rods and ultimately the homogenized elastic behavior of the cellular material [32]. Further, we are able to consider both the core and the shell. It is later shown, that the shell displays different properties to the core.

In the context of this research project, each strut in the lattice is modeled by geometrically exact beams. The theory of geometrically exact beams does not only allow for arbitrary deformation but also considers the rotation of each cross-section. Thus, two different strain measures are introduced: the translational strains $\mathbf{v}_0$ and the rotation strains $\mathbf{k}_0$. Further, we simply denote them as strains and curvature. Their energetic conjugates are forces $\mathbf{n}_0$ and moments $\mathbf{m}_0$, collectively denoted as stress resultants. Within the nonlinear beam theory, the elastic and elastoplastic constitutive model, relating the stress resultants to the strains and curvatures, does not only take the beam's underlying material behavior into account but also the geometry of each cross-section. It is therefore of crucial importance to correctly describe the constitutive relation considering both aspects. A framework to evaluate the struts' constitutional behavior was presented in [3] and extended in [17]. Further, [17, 18] provides the possibility to derive the elastoplastic properties described by a yield criterion and a hardening behavior. There, a generic yield surface and hardening matrix are fitted to results from uni- and biaxial loading of the rod's cross-section.

To capture the homogenized properties of an additively manufactured lattice, a multi-scale computational homogenization framework over three different scales is introduced: firstly, the macro-scale representing a component made of the cellular material and modeled by classical continuum mechanics, secondly, the RVE-scale representing the cellular material whose struts are modeled by geometrically exact beams and, finally, the meso-scale representing the grain structure within one strut, again modeled in terms of classical continuum mechanics. Figure 15.2 illustrates the three different scales considered in this contribution. With perturbation techniques one may derive a macroscopic elasticity tensor $\mathbb{C}^M$ [33]. The elasticity tensor of each strut $\mathbf{C}$ is evaluated following [3] on the meso-scale.

With the presented multi-scale approach, large structural parts may be solved in a reasonable time. However, in [19] the authors have shown that the homogenization approach is of no use when structural instabilities such as buckling appear on the RVE-scale. In such cases, the lattice must be solved in its full resolution to facilitate capturing large deformations within the RVE-scale and elastoplasticity [17, 18].

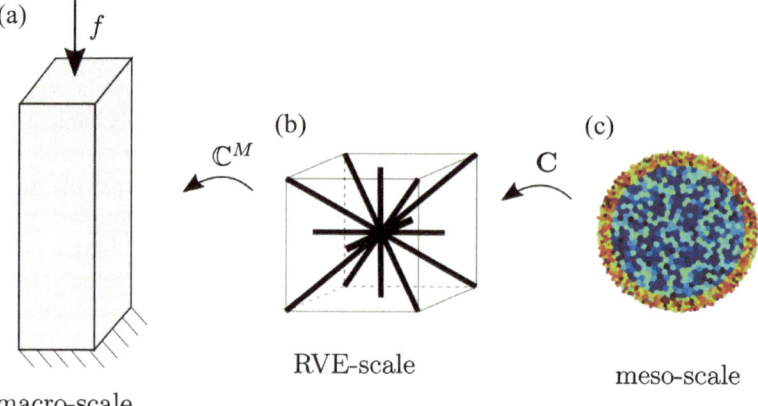

Fig. 15.2 Visualization of the three different scales. The cross-section of the strut with resolved grain structure (**c**) builds the meso-scale and returns the rod's elasticity matrix **C** to the RVE-scale (**b**). The RVE-scale is homogenized to obtain the elasticity matrix $\mathbb{C}^M$ for the macro-scale (**a**). Reproduced with permission [32]

15.3 Applications

This section presents application examples of the numerical models described in Sect. 15.2. First, the application on bulk (Sect. 15.3.1) and cellular materials (Sect. 15.3.2) is demonstrated, respectively. In each section, the microstructural simulation results were validated by comparison with experiments. Subsequently, methods from multiscale homogenization were used to derive the elastic and plastic behavior of additively manufactured metallic parts. These examples demonstrate the significant impact of the microstructure on the elastic homogenization properties. Finally, the theory of geometrically exact rods is used to solve elastoplastic lattice structures in full resolution (Sect. 15.3.3). After considering the isotropic plasticity, the crystalline nature of a strut is taken into account and a framework is presented to capture the crystal plasticity within beams.

15.3.1 Application on Bulk Materials

This subsection presents the application of the simulation chain on parts with simple and large geometries, i.e., on bulk materials. Microstructural simulation results should be experimentally validated before being fed to the micromechanical simulation. Therefore, the experimental validation of the microstructural simulation is presented first.

Following this, the use of the microstructural simulation for micromechanical simulation is demonstrated.

Table 15.1 Summary of experimental parameters used to validate the microstructural model for bulk materials [27]

	Value	Unit
Scanning strategy	Cross-snake hatching, 90° rotation after each layer	–
Beam power	800	W
Beam velocity	5	m/s
Line offset	100	μm
Preheat temperature	950	°C
Layer thickness	50	μm

To validate the simulation-predicted microstructure, cuboid IN718 samples of $15(W) \times 15(L) \times 20(H)$ mm^3 were created using an Arcam A2 PBF-EB machine [27]. The experimental parameters are summarized in Table 15.1.

The setup for the microstructural simulation mimicked the experimental conditions, except that the build height was 10 mm instead of 20 mm. The temperature field obtained by the analytical model was used for the microstructural simulation. It is worth mentioning that the nucleation model was not developed before the publication of [27]. Therefore, the microstructural simulation was performed without considering nucleation. Nevertheless, we will see that, as follows, the texture and grain structure predicted by the CA model without nucleation agree well with the PBF-EB experimental results. For details on the simulation setup, the reader is referred to [27].

The validation is primarily performed by comparing the simulation results with the electron backscatter diffraction (EBSD) measurements on the longitudinal and horizontal sections. Figure 15.3a–b show a comparison of the simulation-predicted grain structure with the experimental one in the longitudinal sections.

For both results, the base is dominated by a wide variety of randomly oriented grains. Within the first 500 μm of the build, the misoriented grains are overtaken, and columnar grains start to emerge. In the upper part of the build, almost only the {100} oriented columnar grains survive. One may notice that there are many small new grains in the EBSD micrograph, and these are thought to arise from powder impurities or discrepancies in powder deposition. These grains hinder the growth of already well-oriented grains, resulting in a very fine columnar microstructure. Although this effect has not been considered in the microstructural model so far, this comparison confirms the simulation results for the height of the grain selection zone, as well as for the columnar growth of the remaining grains. Figure 15.3c–d provide a comparison of the horizontal sections at the height of 10 mm located in the columnar grain zone. An excellent agreement is also achieved from this comparison in terms of the grain morphology and orientation of predominantly grown grains.

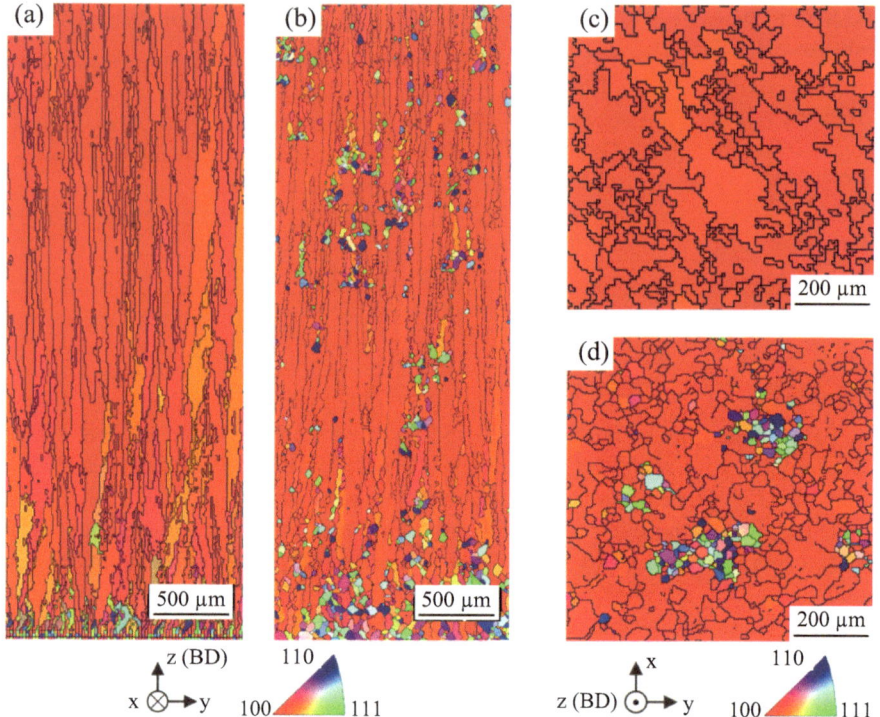

Fig. 15.3 Comparison of the simulation-predicted microstructure (**a, c**) with EBSD measurements (**b, d**) using a longitudinal section in the middle of the part (**a, b**) and a cross section at a height of 10 mm (**c, d**). Reproduced under the terms of the CC BY-NC-ND 4.0 license [27]

The above comparison provides a solid basis for the use of microstructural simulation results in micromechanical simulations. In order to evaluate the mechanical properties of the bulk materials, the microstructural simulation results are geometrically simplified [23]. Figure 15.4 displays the steps from the texture resulting from the microstructural simulation to the final FE mesh used in the multi-scale homogenization framework.

A slightly different material orientation is assigned to each grain, and the bulk stiffness tensor is computed by making use of perturbation techniques.

One may eventually visualize the spatial variation of the averaged stiffness tensor of multiple RVEs with slightly different material orientations in each grain by Young's modulus surface, as shown in Fig. 15.5 [41].

The macroscopic material behavior is clearly transversal isotropic but shows a slight secondary anisotropy in the building plane. This anisotropy results from the applied scan strategy [23]. Further, it is evident that the highest values of Young's modulus are found in the space diagonals. Ultimately, the RVE allows for the computation of a bulk yield surface described by the macro yield function. In Fig. 15.5, a section of the five-dimensional yield surface in the six-dimensional stress space is

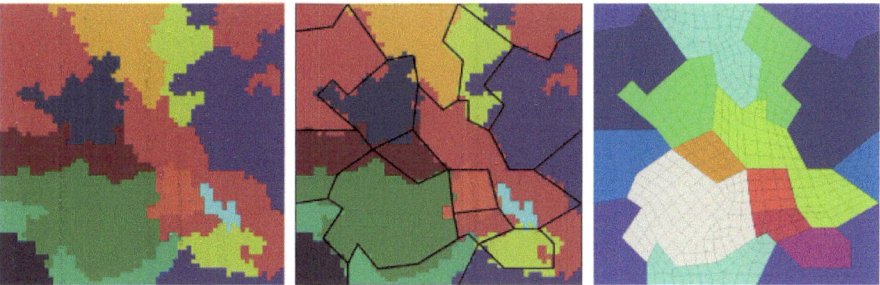

Fig. 15.4 Design of the RVE. On the left, the RVE resulting from the CA is displayed. In the center, an approximation and, finally, on the right the discretized simplification for the evaluation of the macroscopic properties is visualized. The RVE must have periodic boundaries. Reproduced according to the conditions of the CC BY 4.0 license [23]

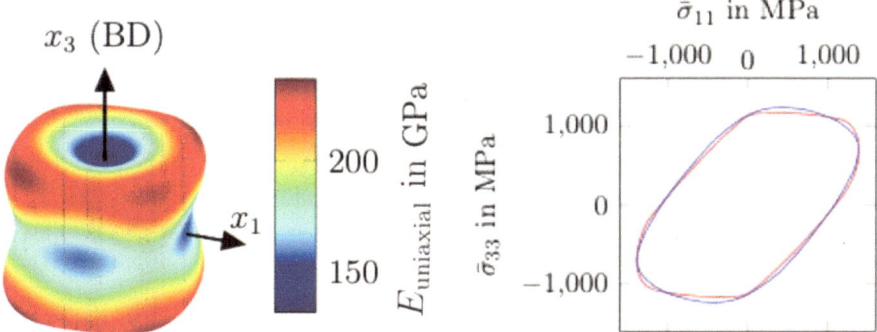

Fig. 15.5 Left: Young's modulus surface depicting the spatial variation of Young's modulus for the RVE shown in Fig. 15.4. Right: averaged yield surface of the RVE (red) and macroscopic fit (blue). Reproduced according to the conditions of the CC BY 4.0 license [23]

visualized. An averaged yield surface is displayed in red and a macroscopic fit to the yield surfaces is evaluated in blue. Further insights are provided in [23–25].

15.3.2   Application on Cellular Materials

Similar to the previous subsection, this subsection begins with the experimental validation of the microstructural simulation of thin cylindrical rods representing the smallest base unit of cellular structures. Given that the microstructural simulations were validated for thin structures, the numerical tool was further used to study the effect of the PBF-EB scanning strategy on the grain structure and texture of thin tilted lattice struts.

Table 15.2 Summary of the experimental parameters used to validate the microstructural model for cellular materials [28]

	Value	Unit
Scanning strategy	cross-snake hatching, 90° rotation after each layer	–
Beam power	300	W
Beam velocity	0.5	m/s
Line offset	100	μm
Preheat temperature	900	°C
Layer thickness	50	μm

In the end, the mechanical properties of PBF-manufactured cellular materials are evaluated based on the multi-scale homogenization framework introduced in Sect. 15.2.4.

The samples used for validation experiments on cellular materials were much smaller than those used in validation experiments on bulk materials and thus thin cylindrical rod samples, $\varnothing 2.9 \times 8\,mm^3$, of CMSX-4 were prepared [28]. The experimental parameters are summarized in Table 15.2. It should be pointed out that the ultra-slow beam velocity of 0.5 m/s was used to avoid cracks [28, 37]. Here, the temperature field was provided by the FE thermal model, and the nucleation model was not applied. For details on the experimental and simulation setup, the reader is referred to [28].

Similarly, the comparison was first made on the longitudinal sections, as shown in Fig. 15.6. It can be seen that, in the 3D view, the grains grow circularly towards the center of the cylinder. This phenomenon is confirmed by the longitudinal section of the experimental sample, which displays a cylindrical shell section on the microstructure. Along the entire build height, new grains are constantly emerging from the powder rim that grow inwardly and upwardly towards the center of rotation. The cylinder core is dominated by columnar {100} grains displaying a distinct fiber texture.

In Fig. 15.7, the micrographs and grain sizes based on the horizontal sections at the upper end of the cylinder are compared. Figure 15.7a–b shows that there are three distinct regions in the numerically predicted and experimentally obtained micrographs. At the edge of the cylinder, fine needle-like grains grow inward, namely, in the shell region. The grains from this region bend upward and pierce the cross-section, and they are visible as evenly distributed grains in the adjacent region, namely, the transition region. Well-oriented {100} grains are aligned in the build direction, dominating the center of the cylinder, namely, the core region. The grain size was measured in various regions of the simulated and experimental micrographs using different methods, namely interception and planimetry. As shown in Fig. 15.7c–d, the grain size measurements also display a good match between the simulation and experiment, with only a slight overestimation by the simulation. This slight mismatch was considered to be due to the neglect of nucleation in the simulation [28].

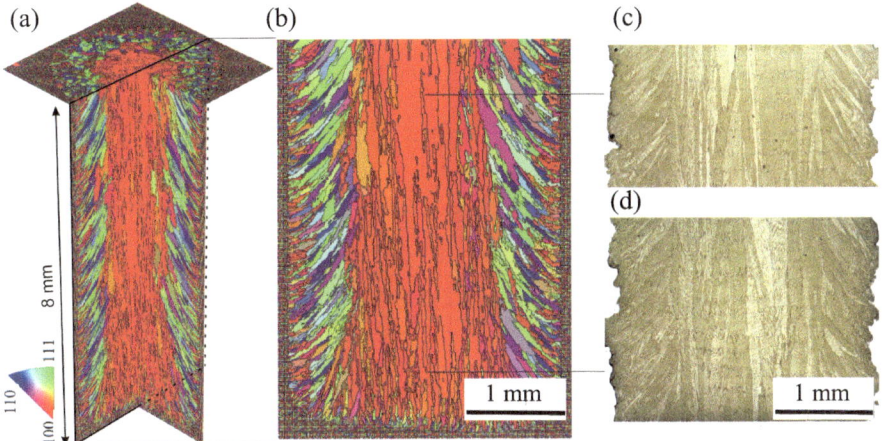

Fig. 15.6 3D view of the numerically predicted microstructure (**a**) and comparison of the simulation result (**b**) with optical micrographs (**c–d**) at different heights in the longitudinal section. Reproduced under the terms of the CC BY-NC-ND 4.0 license [28]

Subsequently, the microstructural simulation software was applied to study the effect of the PBF-EB scanning strategy on the grain structure and texture of thin tilted lattice struts [46]. The tilted lattice struts are thin cylindrical rods with an inclination of 45° and a diameter of 2 mm. The thermal field is provided by the FD solver. Different scanning strategies are applied, such as snake-hatch scanning with various degrees of rotation and contour scanning. This section uses three different scanning strategies as examples (see Fig. 15.8a). Rot180 means that the snake-hatch pattern is rotated by 180° after each layer. Uni180 is a scanning strategy in which the electron beam moves laterally from right to left while processing each layer. For the contour scanning strategy, the electron beam moves ellipse by ellipse from the inside to the outside. The spacing of the ellipses along the minor axis is 100 μm, which is the same value as the line offset in the snake-hatch strategies.

Figure 15.8c–d shows the resulting microstructure in the longitudinal section, the {100} texture represented by the orientation distribution function pole figure, and the grain orientation distribution, respectively. Different scanning strategies cause varied characteristics in various aspects, demonstrating the usefulness of the microstructural simulation software to assist PBF in microstructure customization. Figure 15.8c shows that there is a strong intensity in the {100} and near-{100} orientations, independent of the scanning strategy. However, the exact near-{100} orientations depend on the scanning strategy. In particular, a double peak is presented in the {100} pole figure of the Rot180. Furthermore, the grain orientation (or grain boundary alignment) strongly depends on the scanning strategy. For example, most of the grains in the core region of Uni180 have an inclination of approximately 60°. In contrast, the grains in the Rot180 core region and contour core region mostly grow vertically. For a more detailed discussion, the reader is referred to [46].

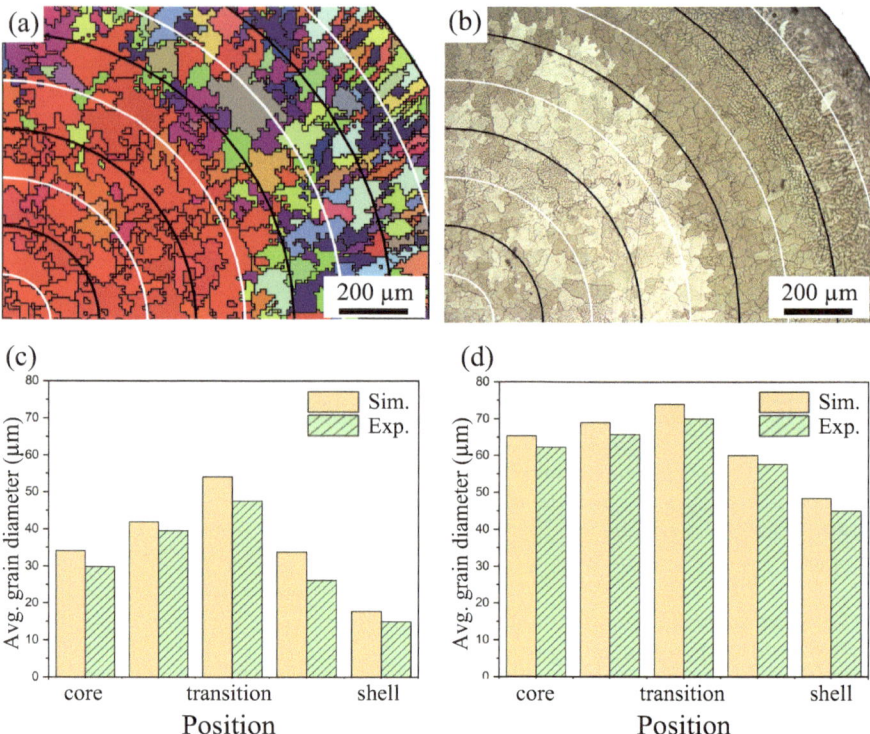

Fig. 15.7 Comparison of the simulation-predicted micrograph (**a**) with experimentally obtained optical micrograph (**b**) using a horizontal section at the upper end of the thin cylindrical rod. The white and black lines indicate the positions of the grain size measurements according to the interception and planimetric methods, respectively. The measurement results based on the interception method are shown in (**c**), and those based on the planimetric method are shown in (**d**). Reproduced under the terms of the CC BY-NC-ND 4.0 license [28]

With the multi-scale homogenization framework presented in Sect. 15.2.4, it is possible to take each grain and its material orientation into account in the calculation of the lattice stiffness. Here, it is demonstrated that the homogenized stiffness of a lattice varies when sloppily assuming the underlying material as homogeneous, or when assuming the struts material as heterogeneous with distinct texture, as it would result from the CA. The lattice at hand is displayed in Fig. 15.2. In the following, the unit cell of the lattice has the edge length 10 mm. The lattice is manufactured from IN718 with cubic symmetric properties defined by three Lamé parameters as shown in [24], Table 3. Two different scenarios are compared. In the first scenario, we assume an isotropic material defined by two Lamé parameters $\lambda^m = 110000$ MPa and $\mu^m = 80000$ MPa [32], whereby the choice of both parameters is motivated by averaging the cubic symmetric elasticity tensor of the anisotropic material. This scenario is supposed to represent the first approach to evaluate the stiffness behavior.

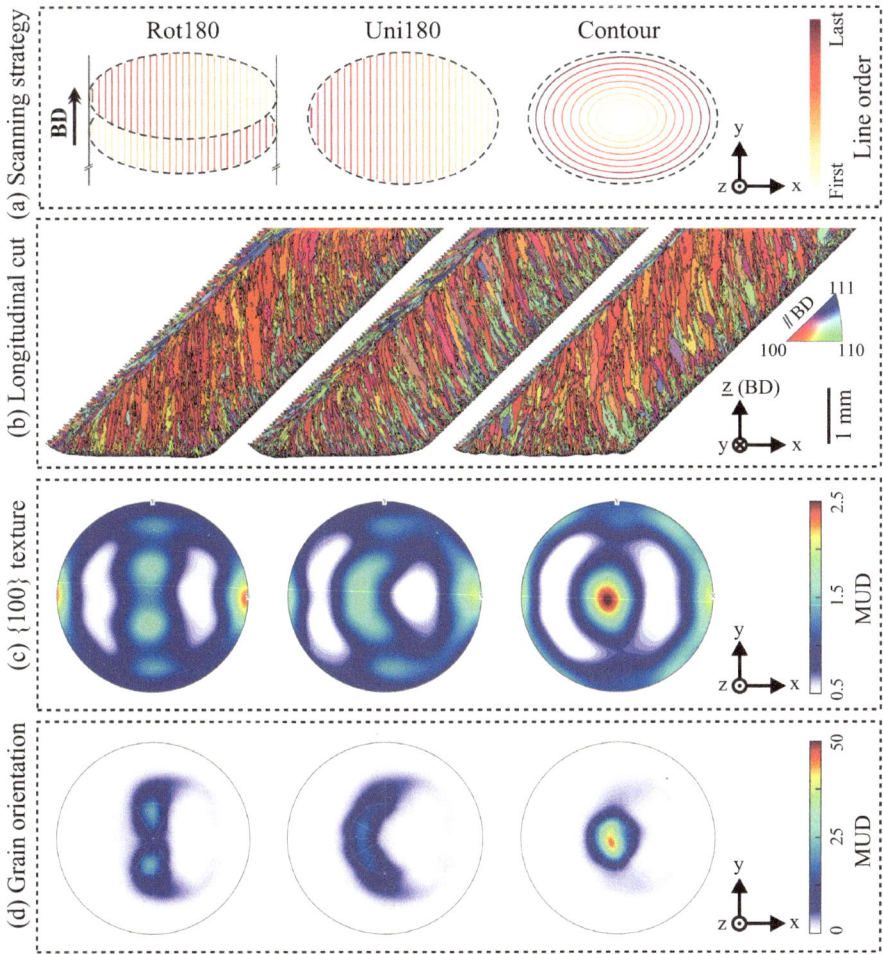

Fig. 15.8 Effect of the PBF-EB scanning strategy on the grain structure and texture of thin tilted lattice struts. Three different scanning strategies (**a**) are shown as examples. (**b–d**) show the resulting microstructure in the longitudinal section, {100} texture (**c**), and grain orientation distribution, respectively. Reproduced under the terms of the CC BY-NC-ND 4.0 license [46]

In the second scenario, we consider the cubic symmetric material behavior in each grain of the struts. Further, we take into account that the material orientations are aligned along the building direction of the lattice. For the above-presented lattice with a strut diameter of $D = 1$ mm, the cross-section is subdivided into a core with a grain size of ≈ 0.05 mm and a shell with a grain size of ≈ 0.025 mm. To compare the resulting stiffnesses, the visualization of Young's modulus surface is a useful tool [41]. The Young's modulus surface visualizes the spatial distribution of Young's modulus for uniaxial loading. Figure 15.9 shows the surfaces for the first and second scenarios we introduced, whereby the building direction is denoted by BD.

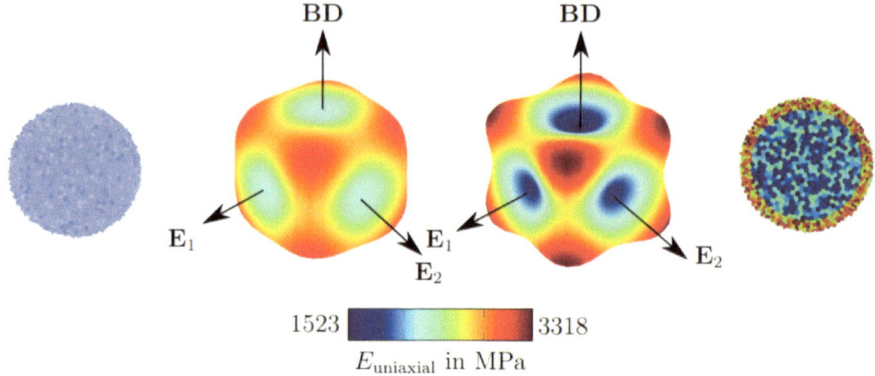

Fig. 15.9 Young's modulus surfaces depicting the spatial variation of Young's modulus for the two different scenarios introduced. On the left, the struts consist of isotropic material. The lattice displays cubic symmetric properties. On the right, we consider the crystalline structure of the struts, resulting in an orthotropic material behavior of the lattice. The cross-sections of the rods are sketched. On the left, we do not distinguish between grains, due to the presence of isotropic material, whereas on the right, we distinguish each grain. Reproduced according to the conditions of the CC BY 4.0 license [32]

In the first case, the homogenized properties result in a cubic symmetric material behavior, where the stiffness is higher in the space diagonals than in the space directions. The reason therefore lies in the topology of the lattice. The second case takes the manufacturing process into account and thus the orientation of the grains. Here, the orientation of the cubic symmetric material is aligned with the building direction. In the core, the orientation may be tilted by up to 3.5° in the shell, while values of up to 22.5° are possible, according to the CA results, compare Fig. 15.6. For the second case, the orthotropic properties reveal that the stiffness in the building direction is lower than the stiffness in the $\mathbf{E}_1$ and $\mathbf{E}_2$ directions. The resulting stiffness is thus no longer cubic symmetric. Considering the differences in the stiffness of the lattice is of crucial importance when optimizing the topology of structural parts constituting lattices, as indicated in Chap. 12 and [21, 32].

15.3.3 Inelastic Modeling of Struts

In previous sections, it was shown that the multi-scale homogenization approach enables the computation of the macroscopic material behavior by taking the manufacturing process into account. When facing structural instabilities on the RVE-scale, this procedure is no longer admissible, since the principle of scale separation is no longer fulfilled [11, 19, 34]. To properly capture the structural instabilities, computing a lattice in its full resolution is required. For the purely elastic case, this is already well-known in the literature. The elastoplastic case, however, requires the definition

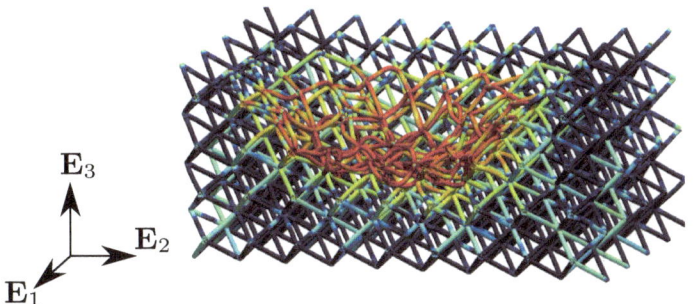

Fig. 15.10 Section through a squared energy absorber. The energy absorber is compressed in a circular area and displays a large deformation. The plastic work density, which is an indicator of the plastification of the struts, is color-coded [16]

of a yield criterion and a hardening behavior in terms of rod-specific quantities. The definition of the yield surface and hardening behavior is not trivial and depends not only on the underlying constitutive model but also on the cross-sectional geometry. To this end, we make use of multi-scale homogenization techniques in which two scales are considered: the rod-scale[1] and the smaller mesoscale—the cross-section. By integrating stresses and plastic work over the rod's cross-section, it is possible to determine the onset of the yield depending on the stress resultants as shown in [16, 17]. Since this approach describes plasticity in terms of stress resultants, we denote the first approach as the stress resultants approach (SR approach). This approach enables solving a large lattice within a few seconds to minutes. However, it requires the a priori determination of a yield criterion and a hardening behavior in terms of rod-specific quantities. Recalling that the geometry of the cross-section influences the constitutive behavior of the rod, it is evident that the yield criterion must be defined for each cross-section that is considered. For the infinite number of cross-sectional geometries resulting from the manufacturing process, this may result in a tedious task. Figure 15.10 depicts a cut of the deformed configuration of an energy absorber in which the density of plastic work, which indicates the dissipated energy, is color-coded.

A second approach, which is more versatile in the choice of the underlying constitutive behavior and geometry of the cross-section, but numerically more expensive, is the FE^2 approach, which is widely used in multi-scale homogenization. While solving the balance equations of the rod, the meso-problem is solved for a prescribed strain and curvature field and returns the energetic conjugates. Here, the problem on the meso-scale is solved while simultaneously solving the problem on the rod-scale. In contrast to the SR-approach, the FE^2 approach allows for the gradual plastification of a cross-section, whereas the SR-approach assumes the cross-section to plastify instantaneously [16–18].

---

[1] Note that the rod-scale is equivalent to the RVE-scale represented in Fig. 15.2. However, in this section, we only consider a single rod and not a network of rods building an RVE.

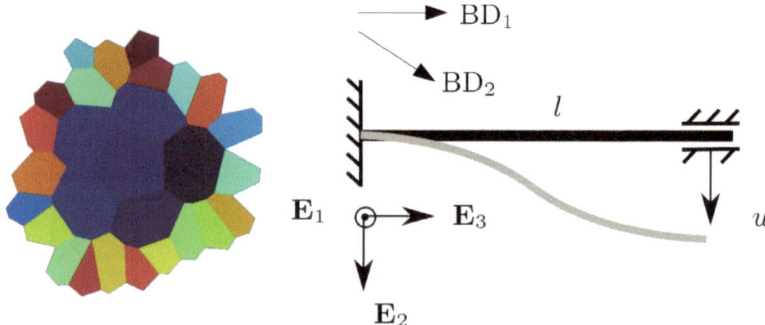

Fig. 15.11 Cross-section of a rod of length l undergoing constrained bending. The cross-section denotes grains in the core and slightly smaller grains in the shell [16]. Further, two different building directions of the rod are depicted

The FE^2 approach allows for modeling the elastoplastic response of a rod without any knowledge of the elastoplastic properties in terms of the rod-specific quantities. We make use of this property to capture the mechanic response of a rod accounting for the geometric and material specifications resulting from the additive manufacturing process. The cross-section of the rod is displayed in Fig. 15.11 (left). We distinguish between a core, where the grains have a size of ≈ 0.15 mm, and a shell, where the grains have an approximated size of ≈ 0.05 mm. The cross-section has a diameter of ≈ 0.5 mm. Each grain obeys the cubic symmetric material properties given by [23]. The inelastic behavior is modeled with the theory of small strain rate dependent crystal plasticity [25, 33]. In Fig. 15.11 on the right hand side, the considered load case is depicted. The free tip of a rod of length $l = 20$ mm is displaced by u and the rotation of the rod's tip is constrained.

The resulting forces and moments at the rod's tip are evaluated for two different cases.

The first example shows the resulting forces and moments with the rod's building direction along the rod's axis BD_1. The second case considers the building direction to be tilted by $45°$, BD_2. The change in the building direction affects both the elastic properties of the rod and its plastic properties.

The resulting forces and moments at the rod's free tip are displayed in Fig. 15.12.

A variation in the slope of the curves is visible for different entries of the forces $\mathbf{n}_0$ and moments $\mathbf{m}_0$. The values of n_{0_2} and m_{0_1} at the beginning of the yield do not vary strongly while a difference in the slope after plastification occurs for both orientations of the material. Further, the remaining entries of $\mathbf{n}_0$ and $\mathbf{m}_0$ are not equal to 0 due to the coupled stiffnesses of the strut, resulting in the non-symmetric cross-section. Although the FE^2 approach may be numerically expensive, it allows for capturing the elastoplastic behavior of additively manufactured parts.

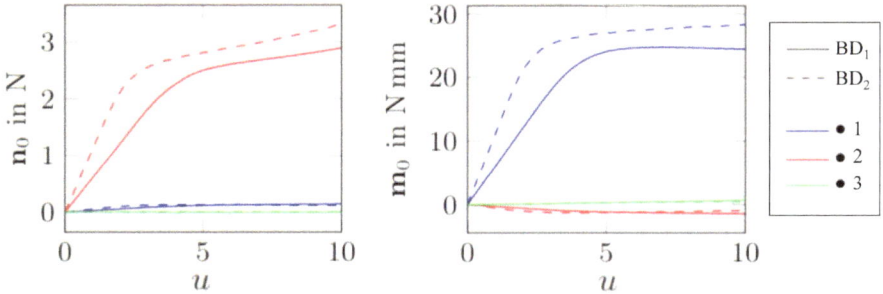

Fig. 15.12 The resulting forces and moments at the rods's tip. Two different building directions, described by BD_1 and BD_2 return different results [16]

15.4   Conclusions and Perspectives

This chapter reviews the development of a simulation chain for predicting the mechanical properties of PBF-manufactured metallic parts, i.e., from the temperature field experienced during manufacturing to the resulting microstructure and, finally, to the mechanical properties. Its usefulness is demonstrated by application examples on bulk materials and cellular structures. The numerically predicted microstructure, mainly referring to the grain morphology and texture, was validated experimentally. The predicted microstructure was used as input to define representative volume elements for meso-mechanical simulations. Macro-mechanical properties (elasticity and plasticity) were further simulated based on the meso-mechanical simulations. The conclusions of this chapter can be summarized as follows: The numerical tools applied were demonstrated to be an efficient and useful approach to gaining insights into the microstructure and mechanical properties of PBF-manufactured metallic parts. They can be used to elucidate the process-microstructure-property relationship for PBF and it is further possible to apply them to tailor the local microstructure and thus the local mechanical properties.

In future work, the following points should be considered: Further experimental validation of the simulation-predicted mechanical properties should be performed. The random powder bed should be taken into account because it has a significant influence on surface roughness and process defects such as porosity [39, 40]. These effects are important for the mechanical performance of PBF-manufactured metallic parts. Other mechanical properties such as creep and fatigue should be modeled and studied, e.g., Ni-based superalloys are typical materials for high-temperature applications, and thus creep should be of particular concern. Fatigue should be one of the most interesting mechanical properties of cellular materials.

References

1. Andersson, J.-O., Helander, T., Höglund, L., Shi, L., Sundman, B.: Thermo-calc & DICTRA, computational tools for materials science. Calphad—Comput. Coupling Phase Diagrams and Thermochemistry **26**(2), 273–312 (2002)
2. Arndt, D., Bangerth, W., Davydov, D., Heister, T., Heltai, L., Kronbichler, M., Maier, M., Pelteret, J.-P., Turcksin, B., Wells, D.: The deal. II library, version 8.5. J. Numer. Math. **25**(3) (2017)
3. Arora, A., Kumar, A., Steinmann, P.: A computational approach to obtain nonlinearly elastic constitutive relations of special cosserat rods. Comput. Methods Appl. Mech. Eng. **350**, 295–314 (2019)
4. Carslaw, H.S.: Conduction of Heat in Solids. Clarendon Press (1986)
5. Cruz, R.D.L.: Leveraging performance of 3d finite difference schemes in large scientific computing simulations (2015)
6. Ellsiepen, P., Hartmann, S.: Remarks on the interpretation of current non-linear finite element analyses as differential-algebraic equations. Int. J. Numer. Methods Eng. **51**(6), 679–707 (2001)
7. Gandin, C.-A., Rappaz, M.: A coupled finite element-cellular automaton model for the prediction of dendritic grain structures in solidification processes. Acta Metallurgica et Materialia **42**(7), 2233–2246 (1994)
8. Gandin, C.-A., Rappaz, M.: A 3d cellular automaton algorithm for the prediction of dendritic grain growth. Acta Materialia **45**(5), 2187–2195 (1997)
9. Gandin, C.A., Rappaz, M., Tintillier, R.: 3-dimensional simulation of the grain formation in investment castings. Metallurgical Mater. Trans. A **25**(3), 629–635 (1994)
10. Gandin, C.-A., Schaefer, R., Rappax, M.: Analytical and numerical predictions of dendritic grain envelopes. Acta Materialia **44**(8), 3339–3347 (1996)
11. Geers, M., Kouznetsova, V., Matouš, K., Yvonnet, J.: Homogenization Methods and Multiscale Modeling: Nonlinear Problems, pp. 1–34. Wiley (2017)
12. Grossmann, C., Roos, H.-G., Stynes, M.: Numerical Treatment of Partial Differential Equations. Springer, Berlin Heidelberg (2007)
13. Guillemot, G., Gandin, C.-A., Combeau, H., Heringer, R.: A new cellular automaton–finite element coupling scheme for alloy solidification. Modelling Simul. Mater. Sci. Eng. **12**(3), 545–556 (2004)
14. Gunasegaram, D.R., Steinbach, I.: Modelling of microstructure formation in metal additive manufacturing: Recent progress, research gaps and perspectives. Metals **11**(9), 1425 (2021)
15. Hekmatjou, H., Zeng, Z., Shen, J., Oliveira, J.P., Naffakh-Moosavy, H.: A comparative study of analytical rosenthal, finite element, and experimental approaches in laser welding of AA5456 alloy. Metals **10**(4), 436 (2020)
16. Herrnböck, L.: Numerische Methoden zur Modellierung elastoplastischer Balken und ihre Anwendung auf periodische Gitterstrukturen. Ph.D. thesis, Friedrich-Alexander-Universität Erlangen-Nürnberg, submitted (2023)
17. Herrnböck, L., Kumar, A., Steinmann, P.: Geometrically exact elastoplastic rods: determination of yield surface in terms of stress resultants. Comput. Mech. **67**(3), 723–742 (2021)
18. Herrnböck, L., Kumar, A., Steinmann, P.: Two-scale off- and online approaches to geometrically exact elastoplastic rods. Comput. Mech. **71**, 1–24 (2023)
19. Herrnböck, L., Steinmann, P.: Homogenization of fully nonlinear rod lattice structures: on the size of the RVE and micro structural instabilities. Comput. Mech. **69**(4), 947–964 (2021)
20. Hunter, L.W., Kuttler, J.R.: The enthalpy method for heat conduction problems with moving boundaries. J. Heat Transfer **111**(2), 239–242 (1989)
21. Hübner, D., Gotterbarm, M., Kergaßner, A., Köpf, J., Pobel, C., Markl, M., Mergheim, J., Steinmann, P., Körner, C., Stingl, M.: Topology optimization in additive manufacturing considering the grain structure of Inconel 718 using numerical homogenization. In: Proceedings of 7th International Conference on Additive Technologies, pp. 102–111 (2018)

22. Iserles, A.: A First Course in the Numerical Analysis of Differential Equations. Cambridge University Press (2008)
23. Kergaßner, A., Koepf, J.A., Markl, M., Körner, C., Mergheim, J., Steinmann, P.: A novel approach to predict the process-induced mechanical behavior of additively manufactured materials. J. Mater. Eng. Performance **30**(7), 5235–5246 (2021)
24. Kergaßner, A., Mergheim, J., Steinmann, P.: Modeling of additively manufactured materials using gradient-enhanced crystal plasticity. Comput. Math. Appl. **78**(7), 2338–2350 (2019)
25. Kergaßner, A.: Theorie und Numerik gradientenerweiterter Kristallplastizität. Ph.D. thesis, Friedrich-Alexander-Universität Erlangen-Nürnberg (2022)
26. Koepf, J.A., Gotterbarm, M.R., Kumara, C., Markl, M., Körner, C.: Alternative approach to modeling of nucleation and re-melting in powder bed fusion additive manufacturing. Adv. Eng. Mater. (2023)
27. Koepf, J.A., Gotterbarm, M.R., Markl, M., Körner, C.: 3d multi-layer grain structure simulation of powder bed fusion additive manufacturing. Acta Materialia **152**, 119–126 (2018)
28. Koepf, J.A., Soldner, D., Ramsperger, M., Mergheim, J., Markl, M., Körner, C.: Numerical microstructure prediction by a coupled finite element cellular automaton model for selective electron beam melting. Comput. Mater. Sci. **162**, 148–155 (2019)
29. Körner, C., Markl, M., Koepf, J.A.: Modeling and simulation of microstructure evolution for additive manufacturing of metals: a critical review. Metallurgical Mater. Trans. A **51**(10), 4970–4983 (2020)
30. Kurz, W., Giovanola, B., Trivedi, R.: Theory of microstructural development during rapid solidification. Acta Metallurgica **34**(5), 823–830 (1986)
31. Markl, M., Rausch, A.M., Küng, V.E., Körner, C.: Sample: a software suite to predict consolidation and microstructure for powder bed fusion additive manufacturing. Adv. Eng. Mater. **22**(9), 1901270 (2019)
32. Mergheim, J., Breuning, C., Burkhardt, C., Höbner, D., Köpf, J., Herrnböck, L., Yang, Z., Körner, C., Markl, M., Steinmann, P., Stingl, M.: Additive manufacturing of cellular structures: multiscale simulation and optimization. J. Manuf. Processes **95**, 275–290 (2023)
33. Miehe, C.: Numerical computation of algorithmic (consistent) tangent moduli in large-strain computational inelasticity. Comput. Methods Appl. Mech. Eng. **134**(3), 223–240 (1996)
34. Miehe, C.: Strain-driven homogenization of inelastic microstructures and composites based on an incremental variational formulation. Int. J. Numer. Methods Eng. **55**(11), 1285–1322 (2002)
35. Nye, J.: Some geometrical relations in dislocated crystals. Acta Metallurgica **1**(2), 153–162 (1953)
36. Promoppatum, P., Yao, S.-C., Pistorius, P.C., Rollett, A.D.: A comprehensive comparison of the analytical and numerical prediction of the thermal history and solidification microstructure of inconel 718 products made by laser powder-bed fusion. Engineering **3**(5), 685–694 (2017)
37. Ramsperger, M., Singer, R.F., Körner, C.: Microstructure of the nickel-base superalloy CMSX-4 fabricated by selective electron beam melting. Metallurgical Mater. Trans. A **47**(3), 1469–1480 (2016)
38. Rappaz, M., Gandin, C.-A.: Probabilistic modelling of microstructure formation in solidification processes. Acta Metallurgica et Materialia **41**(2), 345–360 (1993)
39. Rausch, A., Küng, V., Pobel, C., Markl, M., Körner, C.: Predictive simulation of process windows for powder bed fusion additive manufacturing: influence of the powder bulk density. Materials **10**(10), 1117 (2017)
40. Rausch, A.M., Markl, M., Körner, C.: Predictive simulation of process windows for powder bed fusion additive manufacturing: Influence of the powder size distribution. Comput. Math. Appl. **78**(7), 2351–2359 (2019)
41. Refai, K., Montemurro, M., Brugger, C., Saintier, N.: Determination of the effective elastic properties of titanium lattice structures. Mech. Adv. Mater. Struct. **27**(23), 1966–1982 (2020)
42. Rosenthal, D.: The theory of moving sources of heat and its application to metal treatments. J. Fluids Eng. **68**(8), 849–865 (1946)
43. Simo, J.C.: A finite strain beam formulation. The three-dimensional dynamic problem. part i. Comput. Methods Appl. Mech. Eng. **49**(1), 55–70 (1985)

44. Simo, J.C., Vu-Quoc, L.: A three-dimensional finite-strain rod model. part ii: computational aspects. Comput. Methods Appl. Mech. Eng. **58**(1), 79–116 (1986)
45. Soldner, D., Mergheim, J.: Thermal modelling of selective beam melting processes using heterogeneous time step sizes. Comput. Math. Appl. **78**(7), 2183–2196 (2019)
46. Yang, Z., Koepf, J.A., Markl, M., Körner, C.: Numerical prediction of texture and grain structure in additively manufactured lattice struts. Progr. Additive Manuf. (submitted) (2023)
47. Yang, Z., Kuesters, Y., Logvinov, R., Markl, M., Körner, C.: SAMPLE3D: a versatile numerical tool for investigating texture and grain structure of materials processed by PBF processes. In: 4th International Conference on Simulation for Additive Manufacturing (Sim-AM 2023) (submitted) (2023)

Matthias Markl completed his doctorate in 2015 in the field of simulation of electron beam powder bed fusion under the supervision of Prof. Dr.-Ing. habil. Carolin Körner. Matthias Markl then took over as head of the Numerical Simulation working group, which focuses on the simulation of metal additive manufacturing processes.

Julia Mergheim associated with the Institute of Applied Mechanics (LTM) at Friedrich-Alexander-Universität Erlangen-Nürnberg, is a researcher whose expertise encompasses Nonlinear Finite Element Method (FEM), Multiscale Material Modeling, and Fracture and Damage Mechanics.

Paul Steinmann was appointed professor at the University of Kaiserslautern in 1997. In 2007, he took over the Institute of Applied Mechanics (LTM) at Friedrich-Alexander-Universität Erlangen-Nürnberg. He is currently the Director of the UK Glasgow Computational Engineering Center (GCEC). His research interests range from ma-terial modeling to multiscale methods, multiphysics problems, non-standard con-tinua, configurational failure mechanics, biomechanics, and general advances in finite element and discretization methods.

Prof. Carolin Körner has headed the Chair of Materials Science and Technology of Metals at Friedrich-Alexander-Universität Erlangen-Nürnberg since 2011. She is primarily active in additive manufacturing (electron beam metals), casting technology (investment casting, die casting), process and microstructure simulation, and alloy development. She also heads a working group at the Central Institute for New Materials and Process Technology ZMP and Neue Materialien Fürth GmbH in Fürth.

Part V
Quality Control

Chapter 16
Geometric Measurement and Testing Technology for Additive Manufacturing

Benjamin Baumgärtner and Tino Hausotte

16.1 Introduction

16.1.1 Incremental In-Line Test Technology for Additive Production

Although powder-based additive manufacturing processes have great potential for the tool-less production of complex and individualized components with almost unlimited design freedom, they have only hesitantly been used for the flexible production of single workpieces. One reason for this is the unacceptable quality deficits regarding geometrical accuracy and component strength. In the case of selective laser melting (SLM), these deficits arise not only by component distortion during cooling, but also by fluctuations in process parameters and influencing factors such as laser energy, layer thickness, particle size, temperature history, and relative position deviations between the laser beam and the powder bed which occur during layer-by-layer production. Detecting warpage using retained samples and special test specimens is expensive and potentially inaccurate due to the correlation between part properties and their position in the building chamber, which results from spatial temperature gradient dependencies [6]. As a workaround, a layer-wise incremental in-line measurement has to be carried out. The automatic correction of process fluctuation-related component errors through the regulation of process parameters streichen should enable a significant reduction in rejects, costs, and lead times. Fringe projection, as a robust streichen optical-surface probing measurement and inspection process, is particularly suited for the incremental inspection of internal and external component structures. At present, neither the integration of optical measurement and

B. Baumgärtner · T. Hausotte (✉)
Friedrich-Alexander-Universität Erlangen-Nürnberg, Chair of Manufacturing Metrology, Nägelsbachstraß 25, 91052 Erlangen, Germany
e-mail: tino.hausotte@fmt.fau.de

© The Author(s) 2025
D. Drummer and M. Schmidt (eds.), *Progress in Powder Based Additive Manufacturing*, Springer Tracts in Additive Manufacturing,
https://doi.org/10.1007/978-3-031-78350-0_16

process control in additive manufacturing systems nor the strategies for incremental component measurement have been thoroughly investigated. In the first phase of subproject C4, the basic requirements for incremental inline inspection of SLM processes when processing materials such as Polyamide 12 (PA12) and other crystalline thermoplastics including Polyoxymethylene (POM), Polypropylene (PP), and High-Density Polyethylene (PE-HD) were analyzed. As commercial measurement systems proved to be unsuitable, a new scientific development of a measurement strategy for inline measurement systems was carried out and investigated by analyzing the specific requirements for an SLM process for polymers in cooperation with subprojects A3, B1, and B3 [10–24]. At the end of the first phase, a measurement system based on the optical measurement principle was realized [6–11, 14]. With the developed measuring system, the accuracy of powder application and the molten areas can be geometrically characterized layer by layer during the process. This enables the detection of manufacturing deviations and errors during the additive manufacturing of components and provides a new type of monitoring tool for the early termination of faulty SLM building jobs and the early start of a new building process. This has the potential to save resources such as energy, time, and material that would otherwise be wasted on completing a faulty part, and to improve the economics and sustainability of part production. However, the resources used up to the point of defect are inevitably declared as waste and disposed of as defective components. To avoid this, the process parameters would have to be recorded and a corrective measure derived, which would then be carried out during the process.

16.1.2 Layer-By-Layer Inline Inspection in Additive Manufacturing

The second phase aimed to extend the photogrammetry-based inline inspection method to include additional parameters such as laser power, melt pool geometry, and process-related temperatures, and, from this, to develop the possibility of a corrective measure. In addition, the measurement system was to be supplemented with a reference system, using the results of the preliminary work of [11] to shorten the measurement circle and to reduce manufacturing deviations. This allows a better quantification of the cause-effect relationships for manufacturing deviations, improves the understanding of the process, and facilitates parameter correction for the improvement of the reproducibility and accuracy of SLM components. Starting with geometrical post-process measurements to improve the assessment of the manufacturing deviation of retained samples and, later on, of adapted test samples [8–18], an incremental process monitoring system was finally developed. Based on adapted measurement strategies, optical coordinate measuring systems, and sensor cooling concepts from the first phase, the preliminary foundations for the selective laser melting of polymers based on optical measurement principles were established. In contrast to the integration into a DTM Sinterstation 2000 at the Chair LKT, the

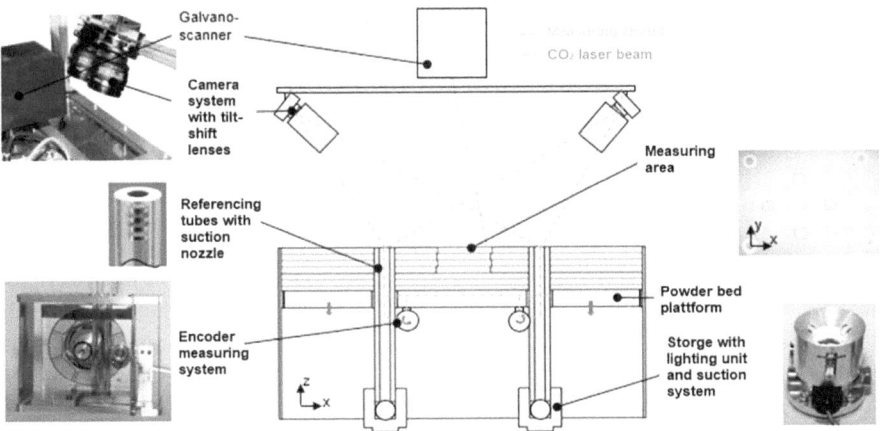

Fig. 16.1 Components of the in situ measuring system with a shortened measuring circuit

in situ measuring systems were integrated into an EOS P380 at the Bavarian Laser Center (blz) subproject B6 during the subsequent second phase [17]. The in situ measurement techniques that were developed are used to monitor the energy input of the laser, the formation of the melt pool, the lowering of the powder bed platform, and the sintered contour. This allows production deviations to be detected in the part contour on a layer-by-layer basis and corrected using a control loop.

To reduce measurement errors, a reference system was integrated into the powder bed platform to shorten the measurement loop (see Fig. 16.2). If reference marks are sintered layer by layer in the powder bed, only the last layer is a reference, which can be laterally offset from the previous layers. The four quartz glass tubes of the reference system—each of which is always held at the same height in relation to the camera system by a point contact with a vertical force and in a horizontal direction by the powder bed and the powder bed platform—are pushed through the powder bed as the powder bed is lowered layer by layer. The diffusely emitted light from the quartz tubes can be detected by the cameras of the photogrammetry system and provides a common lateral reference for all layers of the powder bed (see Fig. 16.2). The initial validation of the referencing system showed maximum standard deviations of the positions from the camera of 88 µm under process conditions (temperature: 168 °C). The camera system simultaneously captures the melted part contour and the ends of the referencing tubes. An additional illumination system consisting of three light sources integrated into the sidewalls of the process chamber provides uniform illumination of the powder bed and selective illumination to create different shadows for part contour evaluation. An evaluation of the illumination concerning metrological requirements was carried out using a separate measurement setup with two Ulbricht spheres, in collaboration with subproject B6. By coupling different laser sources, the absorption and reflection properties of the powder could be measured as a function of the temperature and aggregate state [17].

For the optical in situ imaging, two obliquely positioned high-resolution stereo cameras are used as a photogrammetry system. The cameras are each equipped with a tilt-shift lens to implement the Scheimpflug principle [27] by tilting the sensor chip and optimally capturing the powder bed surface in the depth of field. The individual measurement data sets of the melted component contour could be merged using edge detection [19] and further evaluation with the aid of the referencing tubes. With the in situ measuring system, the distances between the tubes were measured several times with and without powder application, and maximum standard deviations of 112 μm and 18 μm [21] were determined. By the end of the second phase, the lateral contour displacements due to powder application could be detected. The melt pool was also monitored using an on-axis high-speed camera. Integrating the camera directly into the optical beam path of the manufacturing system allows for recording the width of the path melted by the laser beam and a real-time evaluation.

For a comprehensive analysis of the energy input of the laser source into the powder bed layer, the laser power introduced into the powder bed is crucial in addition to the scanning speed. To detect energy losses in the optical beam path and laser power fluctuations, which can affect the layer bonding and dimensional accuracy of the manufactured components, a laser power meter was integrated into the optical beam path. It was shown that a system-related energy loss from the laser source to the powder bed of up to 60% could be expected, mainly due to material-related absorption of the ZnSe windows [15].

Based on post-process measurements with a fringe projection system, part height deviations in the order of magnitude of the layer thickness (100 μm) could be detected [7]. Two rotary encoders were integrated below the powder bed platform (see Fig. 16.2) to determine the applied layer thickness, and the resulting component height, and consider shrinkage. These roll uniformly along the low-expansion reference tubes, allowing accurate measurement of powder bed settling. After calibration with an interferometric probe, it was demonstrated that the manufacturing facility-specific powder bed lowering was approx. 2–3% smaller than the set layer height of 100 μm. The integrated measurement system also includes a thermographic camera to monitor the melt pool and several Pt100 temperature sensors to record temperature effects along the original measurement circle without referencing. Based on the synchronously acquired measurement data, and the interaction and significance of the individual process variables, the dimensional accuracy of the manufactured components was analyzed in more detail by the end of the second project phase, and strategies for the correction of manufacturing deviations were investigated [21].

For quality control in laser- and powder-based AM processes, comprehensive reference monitoring is necessary to realize reliable geometrical measurements. The workpiece contour of each layer has to be extracted by process-accompanying, preferably optical measurement methods to ensure appropriate evaluations for process control. To obtain correct geometrical measurements, optical measuring systems have to be calibrated, aligned, and focused on the melting pool, which can be seen in Fig. 16.3a after the laser energy input and Fig. 16.3b after powder application. This indicates that the camera system fulfills the requirement of stable imaging in order to reference process-related influences on the position of the melt pool.

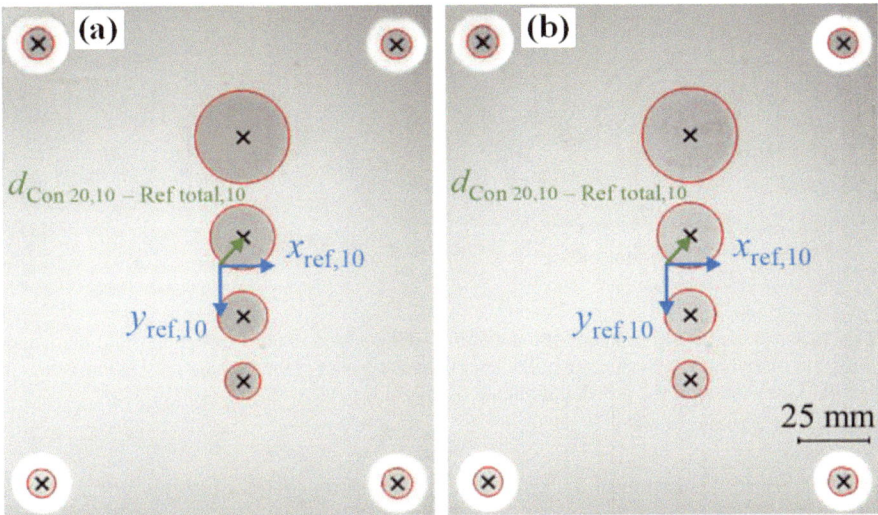

Fig. 16.2 In situ detection of reference tubes and melt pool contours **a** after laser energy input and **b** after powder application for centroid determination and reference distance evaluation [22]

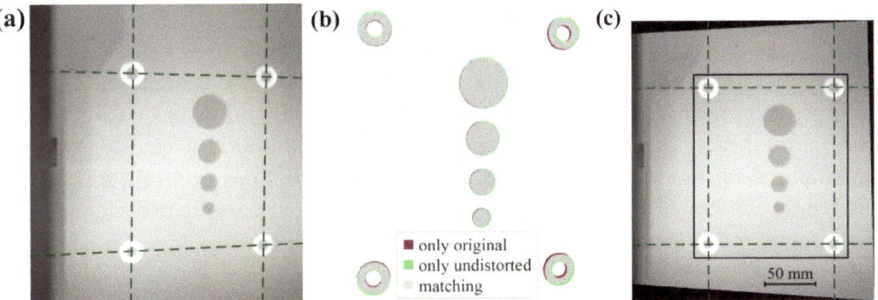

Fig. 16.3 Correction of image distortion: **a** original image I, **b** differences between original and undistorted image, and **c** corrected (undistorted and rectified) image I' with subsequent area restriction of the ROI [22]

After 2-D image acquisition, the correction of lens and perspective distortions is crucial step for subsequent image processing. Besides, illumination adaptation and filter concepts are also essential for high image quality. In [22] and [21], two methods of optical in situ verification of lateral contour displacements in AM have been presented. Therefore, an off-axis in situ monitoring system was implemented in an EOS P380 experimental LBM-P research facility.

For the optical monitoring system, monochrome, 26-megapixel CCD cameras (Genie TM NanoXL-M5100) were used and positioned off-axis according to Fig. 16.1) outside the LBM facility. For image rectification and undistortion, the extrinsic and intrinsic parameters of the camera system were determined with a

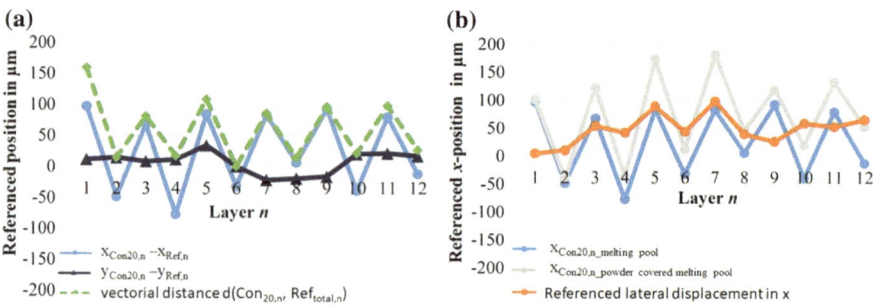

Fig. 16.4 **a** Lateral displacement of the melting pool position in the x-direction (light blue) and y-direction (dark blue). **b** Evaluation of the lateral displacement of the melting pool in the x-direction before (blue) and after the powder application (gray) [18]

checkerboard calibration. To compare the contour of the melting pool with the sliced CAD data, the images were transformed to an undistorted top view (shown in Fig. 16.3) for the camera perspective and c) for the undistorted image. To ensure uniform brightness in the image, a locally different brightness adjustment was performed using a flat-field correction.

The in situ evaluation of the melting pools and their contours after powder application was characterized by the maximum value of the first-order derivative with a Canny algorithm and the use of the referenced distance determination due to the quartz glass pipes. The melting pools in the ROI are extracted layer by layer between an illuminated reference system composed of four illuminated quartz glass pipes held lateral by the powder bed. Based on the evaluated position of the referenced melt pool contours after laser energy input, the positional stability of the components in the referenced manufacturing field can be compared layer by layer. A lateral drift of up to ±150 μm with a standard deviation sindex! of 62.1 μm was observed in the x-direction (coating direction), as shown in Fig. 16.4a, while in the y-direction the positions change less.

As the cause of the powder application cannot be assessed with this evaluation, the influence of the powder application was analyzed separately. First of all, it is necessary to correct the illumination in the building chamber, as shown in Fig. 16.5a–c. This evaluation is based on the second measurement method, in which the referenced positions of the evaluated melting contours are compared with the centroid coordinates of the translucent melting pool contours after powder application. This allows the influence of the powder application system force on the entire powder bed to be analyzed separately. During the investigations on the monitored LBM-P system, a lateral displacement between the production layers of up to 100 μm was detected, as shown in Fig. 16.4b.

However, the effect of temperature fluctuations on the measurement setup cannot be ignored, especially during the preheating, which is described in [22] in more detail. Depending on the construction material and temperature control of the AM system, the heating phase can take several hours. Due to the applied subsequently

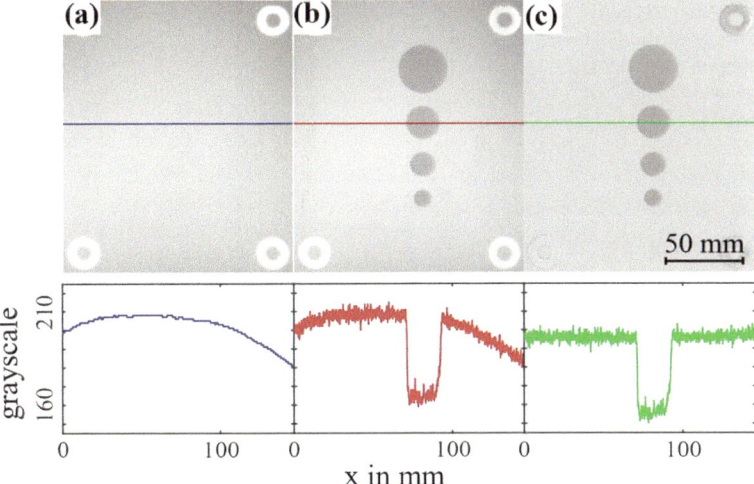

Fig. 16.5 Illumination correction and gray value curve: **a** filtered background image, **b** melting pool image after the energy input, and **c** illumination-corrected image [22]

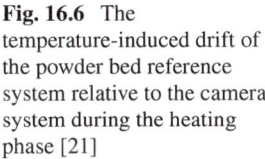

Fig. 16.6 The temperature-induced drift of the powder bed reference system relative to the camera system during the heating phase [21]

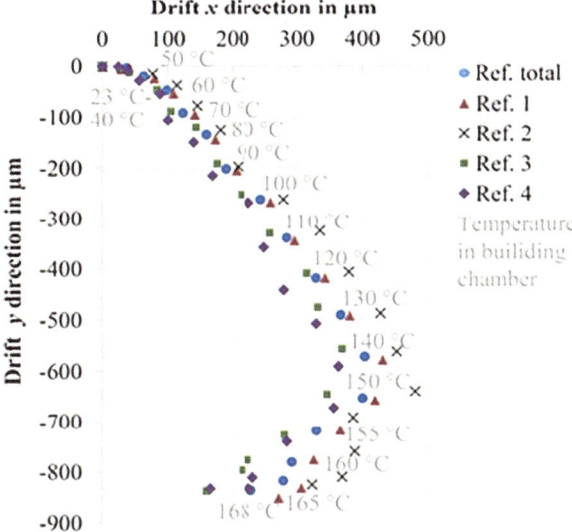

preheating of the building chamber, a temperature-induced drift of several hundred micrometers of the powder bed reference system relative to the camera system was observed [23], as shown in Fig. 16.6.

The methods of in situ contour analysis and the information generated about locally referenced contour displacements and melting pool geometry can serve as an essential database for subsequent process control during the additive manufacturing process. Although it is to be expected that the layer offset is on average greater than

Fig. 16.7 Material and measurement methods depending on roughness (average value of all specimens) [16]

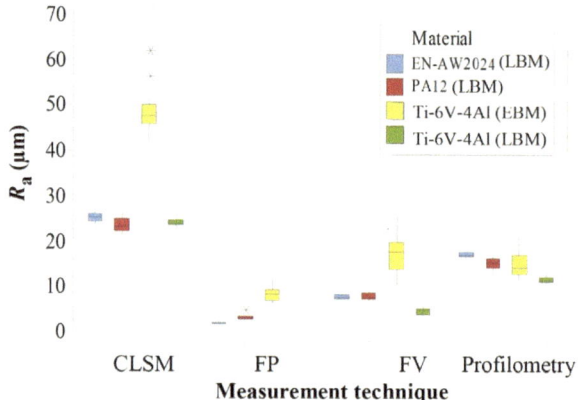

the surface roughness, the detectability in the post-process still has to be clarified by common procedures such as confocal laser scanning microscopy (CLSM), as there is no standard process description so far.

In [16], a comparison of optical measuring methods such as CLSM, fringe projection (FP), focus variation (FV), and profilometers was evaluated to determine the surface quality of powder bed-based manufactured parts. In cooperation with subprojects A5, B2, B3, and B5 the examinations were carried out with the powder materials EN-AW2024, Ti-6V-4Al, and PA12, which were processed by electron beam melting (EBM) and laser beam melting (LBM) of metals and polymers (see process parameters [16]). It was shown that the calculated material-specific surface roughness of metal and polymer parts is comparable concerning different optical and profilometer measurement methods, as indicated in Fig. 16.7. However, the material-specific roughness can vary depending on the additive manufacturing processes used. The raw data sets monitored were all processed using the same software to minimize the influence of different algorithms and users, and a subsequent fitting of the observed topographies allows for an initial assessment of the congruence of the surfaces.

Only when these conditions have been verified and the significant roughness values have been determined can a direct comparison of different surface data be made. After a brief examination of the influence of material positioning and recoating on the surface quality of the sidewalls, a qualification of application-specific surface detection methods was given. The focus variation method was found to be a suitable areal surface detection method for assessing the surface quality of additively manufactured parts, although an additional adjustment to the linear profilometer measurement is also recommended. Finally, material ratio curves and spectral density analysis of the FV analysis show the ability of this measurement system to separate individual powder particles. Together with sufficient resolution and low measurement uncertainty, this last analysis is required for a comprehensive surface evaluation of measurement systems. At the end of the second phase, a measuring system was integrated into the

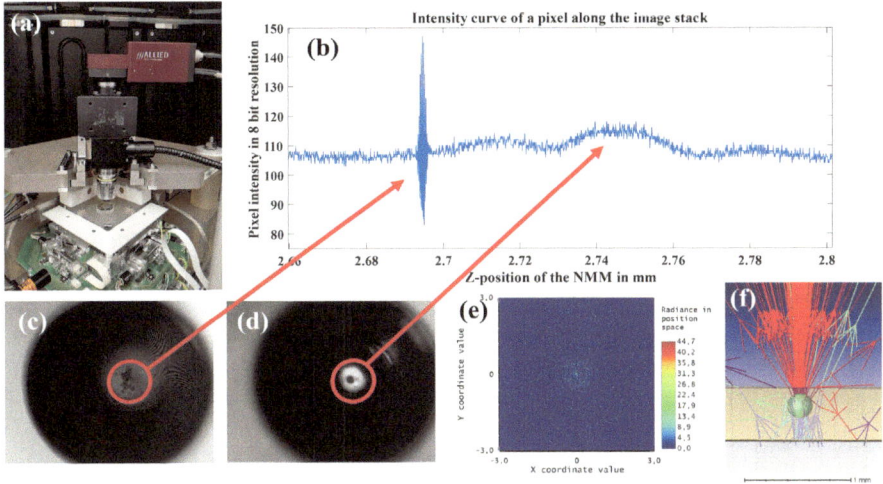

Fig. 16.8 a Illustration of the measurement setup of the white light interferometry sensor on the NMM-1. **b** Representation of the intensity curve of a central pixel along the z-axis (axial). **c** Camera image of a gas pore in fused silica with interference fringes on the sample

construction space that is capable of detecting laser and powder application-related contour displacements.

16.1.3 Traceability of the Computed Tomography Porosity Measurement

To be able to investigate the effects of the variations recorded by the developed in situ measuring techniques on the manufactured component in their entirety and to ensure end-to-end quality assurance, it is important – and hence the aim of the third phase – to compare the results with post-process measurements. In addition to the measuring techniques already investigated for checking the dimensional accuracy and form and position deviations (first phase), the methods for investigating internal defects such as pores, which are highly relevant for some mechanical properties, should be mentioned here. For a spatially resolved, non-destructive analysis, computed tomography (CT), in which a three-dimensional gray-scale data set is reconstructed from two-dimensional radiographic images, has so far been without an alternative. Segmentation algorithms can be used to measure the location and geometry of individual pores. However, the inspection of additively manufactured components with a micro-CT has so far been limited by resolution, measurement artifacts, and the lack of a possibility to specify a measurement uncertainty for defect analyses. One of the open questions is whether and to what extent a suitable parameter exists that can be used to characterize the resolving power of a CT system concerning the detectability of pores. In the field of dimensional metrology, there are current research efforts to determine the metrological structure resolution (MSR or single-sided surface resolution, smallest geometries or microstructures that can be measured dimensionally) or interface structure resolution (ISR or double surface resolution, smallest reliably resolvable interface between two surfaces) of industrial CT systems using suitable standards. In this context, approaches exist for MSR in the form of frequency analyses of multi-wave standards [1], aperiodic frequency standards [7], or by analysis of the transmission properties for rounded component edges and for ISR in the form of investigations of the minimum measurable distances between two contacting spherical surfaces. However, the cited works with their development of the measurement chain for dimensional measurements are not directly transferable to porosity measurements and, to process the proposed contents of the sub-project, a realistic X-ray simulation was necessary. The results of the AdvanCT project (EMPIR-17IND08) were used to create a virtual metrological 3D CT with the simulation software aRTist 2 of the Federal Institute for Materials Research and Testing (BAM) [20], which could be transferred to an existing metrological CT Zeiss Metrotom with adapted CT parameters for the source. Regarding the desired reference measurements for porosity standards, measurements were carried out on transparent fused silica samples (third phase) with enclosed pores using various coordinate measuring systems and optical sensors. In particular, the lateral diameter of the pores was determined using dark-field illumination. Furthermore, different strategies for the detection of multiple axially superimposed interfaces by means of white light interferometry or with the fiber-coupled, distance-modulated confocal sensor and corresponding evaluation algorithms were investigated within the framework of the DFG-funded project (HA 5915/10-1), which can be used as a principle for pore detection in optical transparent substrates. A low measurement uncertainty for measurements with different optical sensors was achieved by using the nano-coordinate measurement system NMM-1. With the help of a new type of signal evaluation based on lock-in amplification, axial control of the positioning stage is possible for the fiber-coupled, distance-modulated confocal sensor, so that surface detection can be achieved by a single lateral scan instead of axial layer-by-layer lateral scans. Due to the significant reduction in the measurement time while maintaining the known advantages of the confocal technique, it was already shown that the influence of temperature-induced drift is lower, and thus smaller measurement uncertainties were achieved [13].

The scientific approach of the sub-project C4 (third phase) was to determine the shape, volume, and position of gas pores using high-resolution optical tomography on transparent fused silica standards. The standards were then to be used for traceability of the porosity measurement by X-ray computed tomography and thus to obtain a measurement uncertainty for the optical porosity standards. Using CT simulations and the position and shape parameters obtained from the optical measurements, comparative simulations were carried out and, starting from ideal simulations, influences close to reality should be addressed. Through the possibility of changing the specimen material in a targeted manner, insights into the differences in the simulation concerning additive manufacturing materials were acquired. By measuring additively manufactured samples and reference measurements using higher-resolution synchrotron CT measurements, the influence of the CT measurement on process-typical pores was evaluated and a measurement uncertainty determined. The CT porosity analysis developed by this procedure was compared with other alternative measurement methods for porosity determination in an interlaboratory test.

16.1.4   Optical Reference Measurement

To enable the measurement uncertainty determination for porosity analysis by CT, a standard with known geometry or higher-resolution tomography is needed. The objective of the third phase was to perform an optical calibration of gas pores in fused silica. For this purpose, cylindrical fused silica specimens with gas pore inclusions were obtained. Furthermore, epoxy resins were applied between glass slides and cured. The outgassing of the solvent creates highly spherical pores, which can also be used for optical calibration. For optical calibration, the axial diameter of the gas pore was measured by white light interferometry, the equatorial diameter by dark-field illumination, and the pole areas by confocal microscopy and merged by data fusion. For the determination of the axial diameter, a Mirau objective 10x was used in connection with a sensor recording (first phase) for an NMM-1 and with a camera unit of Allied Vision (first phase) later than in connection with a monochrome camera (Imaging Development Systems GmbH IDS, unit from the third phase).

Figure 16.8a shows the measurement setup used to determine the axial diameter of a pore by white light interferometry (WLI) and a mirror placed beneath the sample as a reference plane. Figure 16.8c shows the image of the interference fringes of a gas pore close to the surface in fused silica and Fig. 16.8b the corresponding intensity curve of a pixel in the image stack when moving the sample stage along the z-axis (axial). Although an interference of the sample surface is prominent within the signal, a further evaluable interference did not occur. Regarding the law of refraction, interference was expected when there is an increase in the refraction index within the optical path, which should be the case at the bottom of the pore. A raytracing simulation of light with the software Zeemax with an adapted model of a pore in fused silica (Fig. 16.8f) also shows a strong reflection when the focal plane is in the center of the pore (e). When reaching the bottom of the pore, the simulation shows a significant loss of intensity of the reflection. It is assumed that interference is present but that due to the low reflectivity of glass and the influence of the pore shape on the beam path, the interference contrast is affected. Based on the intensity values of the 0th and the 1st order minima, a Michelson contrast of 0.02 was calculated at the upper interface of the specimen. It can be assumed that the intensity of the interference of the lower interface of the specimen must be weaker in intensity and signal-to-noise ratio, and thus leads to a loss of the measurement signal. Another problem that becomes apparent from the simulation is that for the calculation of the vertical pore diameter, probing points of both pole areas are needed. To obtain an axial pore diameter, a plan mirror (third phase) was placed below the sample and the reflection of the pore upper interface was recorded within the mirror plane.

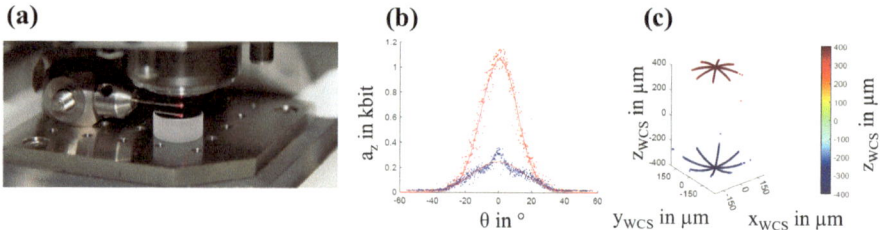

Fig. 16.9 a Image of the microsphere placed above a measuring mirror on the NMM-1. **b** Dependency of the axial response az on the surface angle θ of the real sphere (red) and the mirrored sphere (blue). **c** Measurement results of the microsphere [26]

As indicated in the further course of the intensity profile in Fig. 16.8b, there is no interference from the lower interface of the pore, the measuring mirror, or the upper interface of the pore. As a proof-of-concept of measuring the probing point of a sphere within the mirror plane, this procedure was first tested on a microsphere and using a confocal microscope, as shown in Fig. 16.9a. By using a measuring mirror, it was possible to obtain probing points (see Fig. 16.9c) for the Gaussian sphere fit of the upper pole of the stylus (red) and the lower pole (blue). The intensity of the back reflection decreased significantly but is still sufficient for fitting, which is shown in Fig. 16.9b for one measurement point. The possibility of acquiring measurement points from the underside of a microsphere can significantly reduce the uncertainty in detecting shape deviations of the sphere geometry [26].

While a transfer of the measurement process is limited by the working distance of the Mirau objective, it could be performed for the pore (see Fig. 16.8c). As shown in the further course of the intensity profile in Fig. 16.8b, there is no interference from the lower interface of the pore, the measuring mirror, or the upper interface of the pore. Since the reflectivity of fused silica is 4 % according to the literature, and the pore is shaped as two concave lenses placed on top of each other, only a low contrast of interference near the optical axis is expected. A determination of the axial diameter by WLI was not possible. To detect an axial diameter despite the low contrast, the confocal sensor mounted on the NMM-1 was used.

A gas pore ($> 500\,\mu$m diameter) in epoxy was used for this purpose. Figure 16.10a shows a measurement along the z-axis (lateral) with (orange) and without (blue) a pore in the optical path. The measurement shows that the upper and lower poles of a gas pore can be detected and a pore diameter can be determined. Similar to [26], only parts of the pole can be detected due to the limited aperture before the measurement signal is lost.

This turned out to be problematic in the measurement of gas pores in fused silica since the pore standards have significantly smaller pore diameters ($13\,\mu$m in diameter in the dark-field measurement), and thus positioning of the focal point on top of the pore failed. An extension of the confocal sensor to include dark-field illumination with a condenser, which uses the unused beam path, turned out to be technically unfeasible since the necessary condensers are too large and a raised specimen holder

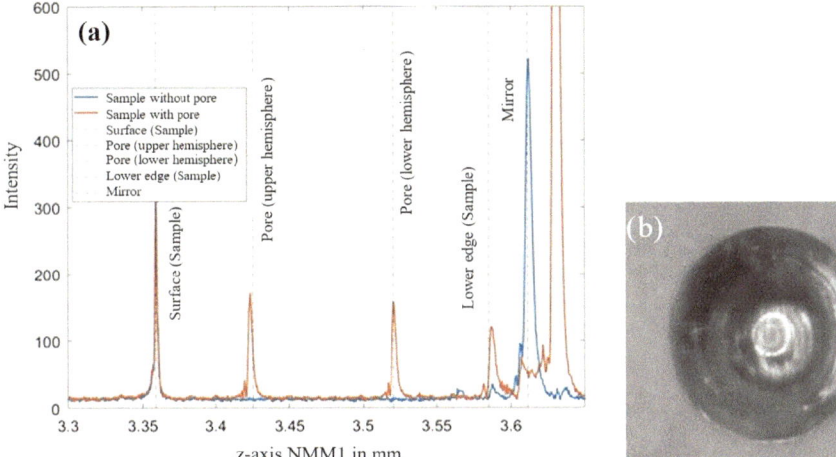

Fig. 16.10 **a** Confocal point sensor measurement of a pore (orange) and without a pore (blue) in epoxy. **b** dark-field-illumination image of a gas pore in fused silica

must supplement the positioning table. The resulting torque on the stage of the NMM-1 and additional deflections of the specimen holder lead to vibrations and distortions, which affect the measurement result. For this reason, a specimen grip was used that was structurally placed outside. Due to this construction, however, only measurements in the z-direction are possible using the NMM-1.

For the evaluation of the equatorial diameter, the dark-field illumination or alternatively, a VideoCheck IP (Werth, Germany) was used for the dark-field illumination (DF) measurements. For the determination, an edge detection algorithm with a Gaussian sphere fit was performed and a diameter was calculated.

A data fusion of all information on the position of the pore from the WLI and DF is difficult due to missing measurement points of the axial pore position. In summary, the calibration is technically feasible using confocal sensor and extension by dark-field illumination, but could not be realized in combination with the necessary measurement precision by the NMM-1 due to structural limitations. Nevertheless, the successful dark-field measurements for gas pores in fused silica for the diameter and shape deviation to a circle were determined, which can be used for CT simulations and CT measurements.

16.2 CT Simulation and Measurements

To estimate the effects of CT on porosity measurement in general, but also on material properties such as the X-ray contrast, an adaptation of the measurement process was made using the CT simulation software BAM aRTist 2.10. For this purpose, a close-to-real simulation was developed, which was based on the results of the EMPIR project AdvanCT and a realistic adaptation of the available CT measuring device. The simulation calculates projection images, which were subsequently introduced into the measurement chain instead of real CT images. Firstly, simulations of gas pores fused silica were performed and compared with real measurements. The aim was to evaluate the fundamental influence of CT on the dimensional measurement of pores and to optimize the parameters for the CT simulation.

To gain an enhanced understanding, and to avoid pores due to a correlation with the process parameters of the additive process and related specific pore shapes, CAD models of pores from the additive area based on alternative measurement techniques such as micrographs and high-resolution synchrotron CT were used and scaled in size. Due to limited access to near-process pore models, the simulation was limited to the most commonly used materials in AM, namely PA12 and Titanium-6-Aluminum-4-Vanadium (Ti64). For the specimens of AM materials, a cylindrical shape was used following ASTM 1570.11 to reduce the influence of artifacts, which can be seen in Fig. 16.11. An important question was whether there is an influence of the material-dependent contrast on the detectability of internal structures. For this purpose, a CT simulation with square cavities (25–400 µm) within a cylindrical sample was created, as shown in Fig. 16.12a as a CT cross-section of the defects. Figure 16.12b shows the corresponding contrast-to-noise ratio related to the material/cavity transition for Ti64 and PA12. For the calculation, a surface determination according to the ISO50 % [25] method was used to define the material/defect transition. Defects with a successful surface determination are drawn with a solid line, while others are depicted with dashed lines. As shown, a downscaling of the size of defects leads to a reduction of the gray value transition between the void and material, which equivalently leads to a decrease in the Contrast-to-Noise Ratio (CNR). The limit formulated in DIN EN 16016-3 for a reliable detectability of a defect of a CNR equal to 3 is undercut for the smallest defect in this simulation for Ti64, while for PA12 the limit is right after

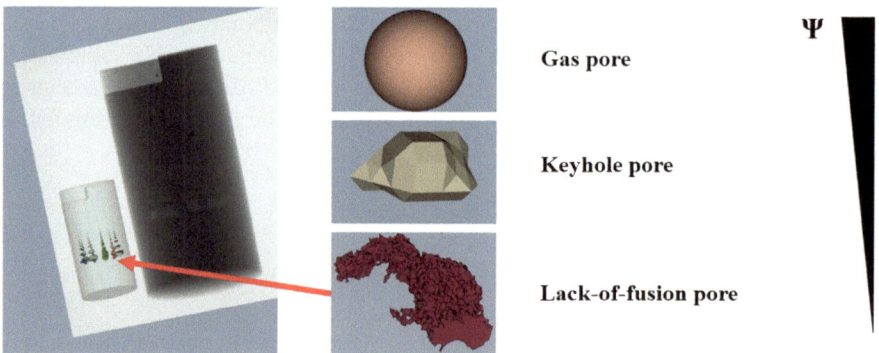

Fig. 16.11 CT simulation for a CAD model of a Ti64 sample with insert pores. The pore shape varies according to the sphericity [9]

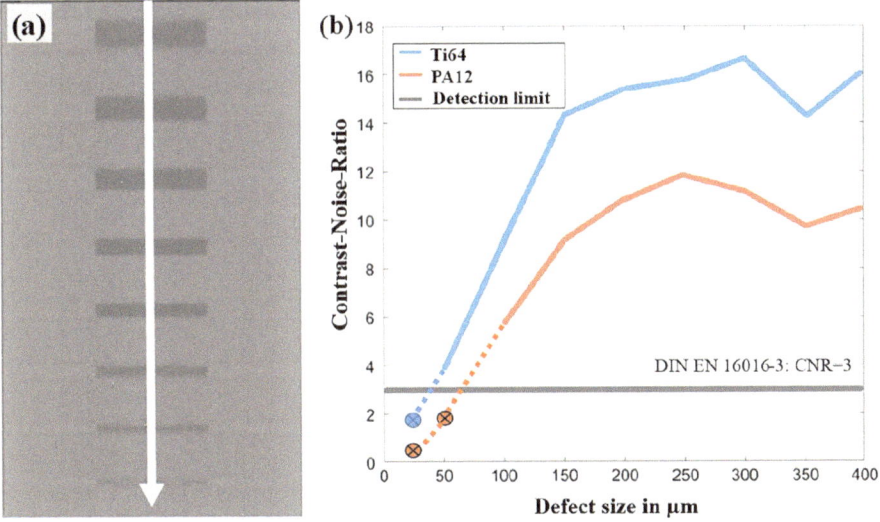

Fig. 16.12 **a** CT cross-section of a cylindric PA12 sample with square cavities inside. **b** Contrast to noise ratio for the defects in (**a**) in dependence on the size. The CNR was calculated for Ti64 (blue) and PA12 (orange)

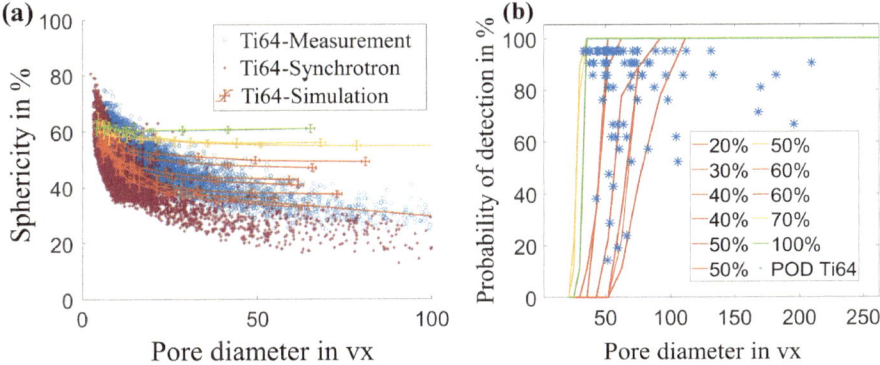

Fig. 16.13 Sphericity of pores in **a** Ti64 and **b** in PA12 for the CT measurement (blue asterisk), the synchrotron CT measurement (red/orange asterisk), and coloured solid line with error bars for the CT simulation [26]

defects smaller than 100 μm. With a voxel size of 7 μm and 8 μm for Ti64 and PA12 in the simulation, the commonly known 3-voxel-detection-criterion in diameter is only confirmed for Ti64. As can also be seen from this curve, there is already a fundamental difference in contrast between the two materials PA12 and Ti64, which has not yet been taken into account in the specified detection limit.

A comparison with real CT measurements showed a good estimation of the simulation for the lack-of-fusion pores that mainly occur in the Ti64 sample. Because of the CNR of 3 formulated in DIN EN 16016-3, this investigation questions the specification of a uniform detectability concept without a reference to the geometry. According to this, pores with a diameter of more than 50 μm should be reliably detectable, which, however, is only achieved by highly spherical pores when looking at the POD. For pores with low sphericity, detection losses is to be expected even with diameters that are 2–3 times larger. Furthermore, it must be clarified here to what extent the CT analysis depicts the recognizable pore shape. For this purpose, a comparative investigation of the pore shape was carried out with CT measurements, CT simulation, and synchrotron CT measurements, which are shown in Fig. 16.13a. Within the CT simulation, the pore shape is strongly influenced by the CT analysis. A negative offset can be observed for spherical pores, which must be interpreted as an increase in surface area based on the shape criterion. For angular pores such as lack-of-fusion, both a positive and a negative offset with a turning point at a sphericity value of 40 % can also be observed. In addition, a size-dependent rounding affects the shape of lack-of-fusion and leads to an increase in the sphericity for smaller pores. Assuming that noise and CT artifacts comprise a possible reason for this observation, a reference measurement was performed on the same sample using synchrotron CT (third phase), which provides a higher resolution, fewer artifacts, and less noise. Among the previously-mentioned classifications of pores based on their shape, a correlation between the cause of the formation and contour parameters in PBF-LB/M processes such as beam energy, layer thickness, and hatch distance can be observed, which was investigated in [5] and in [4] for Ti64. The results showed that a direct correction by adjusting the process parameters, i.e., the pore frequency and size, is possible in the contour area if the cause of formation is counteracted by a pore classification. Traceability of the pore shape within the CT analysis offers the possibility to achieve an improvement of the process control.

Although the simulation of the pore shape shows good agreement with real measurements, without direct reference it is difficult to consider the influences of the evaluation methodology. For this purpose, repeat measurements according to DIN EN ISO 15530-3 were carried out on optical pore standards. Therefore, the diameter of gas pores was measured with dark-field illumination and compared with CT measurements and CT simulations. Using the surface determination and a Gaussian fit of a sphere to calculate the diameter of highly spherical or ideal gas pores showed a size-independent offset between the DF/nominal pore size and surface determination, as shown in Fig. 16.14a [2].

A comparison of the nominal pore size and the size determined by the porosity algorithms (VGDefX, VG-Studio Max, third phase) in the CT simulation showed that good agreement between the surface determination and the algorithm could only be obtained for the pores used to optimize the thresholds of the algorithm, as shown in Fig. 16.14b. An examination of the pore shape also showed an offset in the evaluation of the sphericity of gas pores. In comparison with the adapted simulation model, a good agreement of the results could be achieved for ideal spheres in fused silica. In the CT simulation, the offset in sphericity can be understood as a direct surface enlargement of the pore. By comparison with an ideal CT simulation, the influence of noise on the projection image can be improved and thus considered

Fig. 16.14 **a** Deviation of the diameter of gas pores calculated with the surface data and CAD model in reference for CT simulation (blue) and dark-field illumination for CT measurement (orange). **b** Deviation of the pore diameter calculated with a porosity algorithm with global thresholding [2]

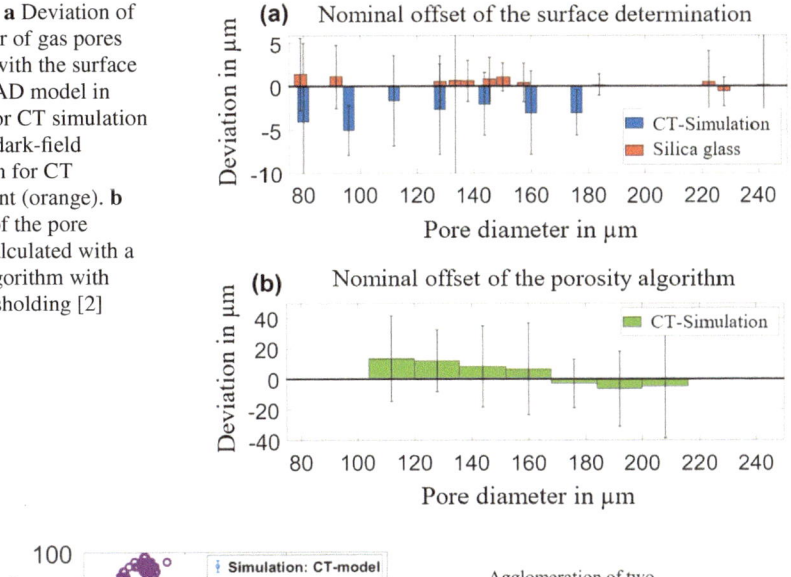

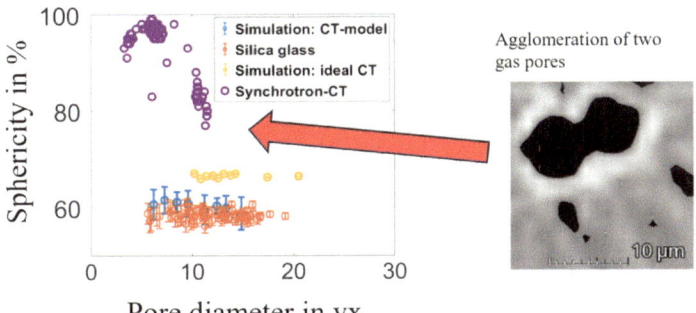

Fig. 16.15 Determination of the sphericity for high spherical pores in quartz glass with CT (orange), synchrotron CT (purple), CT simulation of the real CT (blue), and ideal CT simulation (yellow) [2]

in isolation, leading to an increase in sphericity in Fig. 16.15 and can be inferred as a cause of the surface enlargement. However, since ring artifacts can only be removed within the ideal simulation to a limited extent, the remaining offset of the sphericity cannot be completely corrected. For verification of the CT measurement, synchrotron CT measurements were performed on a part of the sample. Due to the strongly reduced noise/artifact behavior and the higher achievable resolution, the measurement can serve as a reference. The shape of the gas pores is evaluated here with 90% sphericity, which also supports the assumption of influence by noise and artifacts in the experiment. Furthermore, gas pores with significantly reduced sphericity are also detected. These are agglomerations of several pores, as shown in the CT sectional image. Due to the size-independent constant behavior of the sphericity of gas pores, a correction of systematic deviations for the selected CT parameters is achievable.

In summary, an almost realistic CT simulation was created with which the behavior of process-specific pores such as keyholes or lack-of-fusion pores could be investigated with regard to the material-dependent X-ray contrast or the position in the component. A material and shape dependency of the detectability when calculating a probability of detection (POD) was determined. A correlation of the size-dependent detectability by the Contrast-to-Noise Ratio (CNR) according to DIN EN 16016-3 revealed that these results are not included or sufficiently considered in the standard. Furthermore, the CT shows a compounding effect for keyhole and lack-of-fusion pores, which can be misjudged and lead to false conclusions about process parameters. Highly spherical gas pores are also not recognized as such and, in contrast, experience tilting. Furthermore, a comparison was made concerning the previous analysis methods such as porosity algorithms in relation to the known values of the fused silica standard, and a size-dependent deviation in the volume for gas pores was determined by using a global threshold value for the analysis.

16.3   Evaluating the CT Porosity Analysis

As already shown, CT contrast, pore size, and shape influence detectability. In addition to these physical influences on the measuring system, it must be assumed that there are also dependencies on the side of the software that is used. Since artificial intelligence (AI)- or threshold-based porosity algorithms are used for automated analysis, the detectability of planned inner structures was investigated in [3] in cooperation with subprojects B2, B3, and B5 and a workflow was derived for optimizing detectability by using porosity algorithms. Since AI-based image recognition algorithms are used, the reasons for the decision are usually not transparent and it remains unclear to what extent the result can be influenced by parameterization. Although pores are generally not limited by their size, large defects are classified as unlikely or as a planned structures and filtered out of the results. Furthermore, the detectability of planned internal defects is affected by connections to the surface, artifacts, noise, and trapped powder in PBF processes. A comparison of three commercially available porosity algorithms with optimized detectability reveals that there are large differences between the algorithms for the 2D porosity evaluation of a planned internal defect (nominal 18 % porosity in the cross-section), as shown in Fig. 16.16a. A comparison with alternative porosity measuring methods such as micrographs or the electron optical imaging method (ELO) from subproject B2 also reveals differences in validity. Regarding the evaluation of the total porosity, however, the porosity algorithm and the derived workflow showed good agreement with the measurement results from pycnometer measurements for Ti64.

In cooperation with subproject B5, the porosity in the contour area and the correlation of scanning strategy and surface roughness in Ti64 components were investigated in more detail [3]. Significant differences between porosity determination using micrographs and CT analysis in the contour area were also observed. Furthermore, the CT analysis and the porosity analysis can be used as support to understand

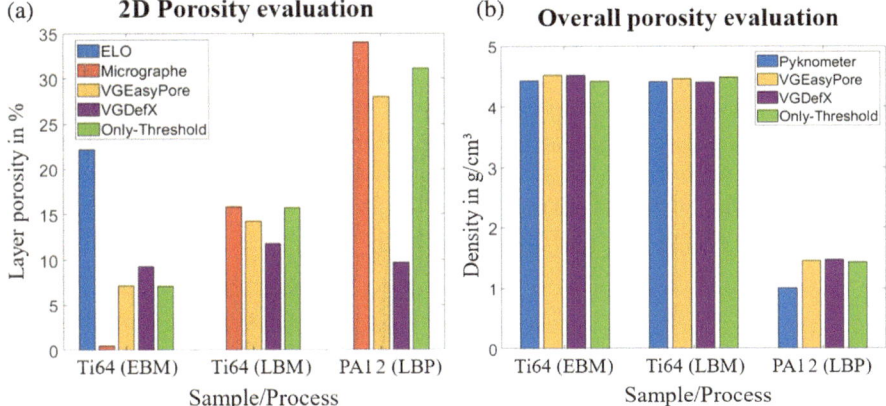

Fig. 16.16 a Comparison of 2-dimensional porosity evaluation with ELO (blue), micrographs (orange), and porosity algorithms (yellow, purple, green) for Ti64 and PA12 samples with planned internal defects. **b** Evaluation of the component density with pycnometer (blue) porosity algorithms (yellow, purple, green) [3]

the thermal influences due to the selected component geometry and the associated pore formation, as well as the material structure in Ti64 [12].

16.4 Conclusion

Throughout the CRC814, the research focus was on component testing and improving the quality of AM parts. With the use of fringe light projection as an incremental in-line test technique, it was possible to detect manufacturing defects at an early stage of the building process and save costs. A further development of this approach, which already detects manufacturing deviations such as layer offsets, was to correct them. Therefore, a camera system with a lowerable powder bed and quartz glass reference tubes that serve as stable reference points was implemented in the building chamber, and the position data correlated with process parameters. However, not only external geometric deviations represent parameters of component quality, but also internal defects such as pores and cracks. Thus, an essential aspect of the research was to develop an optically transparent porosity standard made of quartz glass for X-ray computed tomography. The contained gas pores could be detected by dark-field microscopy and confocal microscopy, and a deviation of volume and size determined for X-ray computed tomography. By using CT simulations with an almost realistic CT model of the real CT device, a transfer of the porosity measurement to additively manufactured parts could be achieved and the detectability and volume determination investigated. In this way, it was determined that, in principle, the detectability of samples made of Ti64 and PA12 has a shape dependency that has not yet been taken into account in DIN EN 16016-3. In the application of state-of-the-art porosity algorithms, size-dependent deviations in the volume determination could be assessed for all pores due to a global threshold value setting in relation to the porosity standard. Another important aspect was the comparability of already established porosity measurement methods in AM such as micrographs and Archimedean density determination in relation to different porosity algorithms. A good agreement of the algorithms in relation to the density determination was found, but significant differences in 2D comparison by micrographs.

References

1. Arenhart, F., Nardelli, V., Donatelli, G., Porath, M.: Investigation of the CT-induced random surface deviations using a multi-wave standard. In: Conference on Industrial Computed Tomography (ICT), Wels, Austria (2012)
2. Baumgärtner, B., Hausotte, T.: P19—investigation of the shape deviation for gas pores measured with x-ray computed tomography. In: SMSI 2023 Conference—Sensor and Measurement Science International, AMA Service GmbH (editor), pp. 316–317 (2023)
3. Baumgärtner, B., Rothfelder, R., Greiner, S., Breuning, C., Renner, J., Schmidt, M., Drummer, D., Körner, C., Markel, M., Hausotte, T.: Evaluation of additively manufactured internal geometrical features using the x-ray computed tomography. J. Manuf. Mater. Processing (JMMP) (2023)
4. Baumgärtner, B., Rothfelder, R., Schmidt, M., Hausotte, T.: Detektion der Einflüsse von Konturparametern auf die Randschichtporosität in ti-al6-v4 in pbf-lb/m- Prozessen mittels Röntgen-Computertomographie (2022)
5. Baumgärtner, B., Rothfelder, R., Schmidt, M., Hausotte, T.: Untersuchung der Einflüsse von Konturparametern auf die Bildung von Oberflächenporosität in ti-al6-v4-Bauteilen in pbf-lb/m-Prozessen mittels Röntgen-Computertomographie. In: 8. Industriekolloquium des Sonderforschungsbereichs 814 "Additive Fertigung" (Lehrstuhl für Kunststofftechnik LKT Friedrich-Alexander-Universität Erlangen-Nürnberg, 10. Mai 2022 - 10. Mai 2022) (2022)
6. Drummer, D., Weickmann, J.: Inkrementelle in-line Prüftechnik für die additive Fertigung. DFG Sonderforschungsbereich 814 (SFB814)—Additive Fertigung: Einrichtungsantrag Teilprojekt C4, vol. 1, pp. 353–374
7. Fleßner, M., Blauhöfer, M., Helmecke, E., Hausotte, T., Staude. A.: CT measurements of microparts: numerical uncertainty determination and structural resolution (2015)
8. Galovskyi, B., Hausotte, T.: Testing workpieces for selective laser sintering. In: American Society for Precision Engineering. Proccedings of ASPE 2015 Spring Topical Meeting Raleigh, North Carolina, pp. 89–94 (2015)
9. Galovskyi, B., Hausotte, T., Drummer, D., Harder, R.: In-line layer wise measurements for selective laser sintering process. In: XXI IMEKO World Congress of Measurement in Research and Industry, Prague, pp. 1410–1414 (2015)
10. Hartmann, W., Hausotte, T., Drummer, D., Wudy, K.: Anforderungen und Randbedingungen für den Einsatz optischer Messsysteme zur in-line- Prüfung additiv gefertigter Bauteile. RTejournal- Forum für Rapid Technologie (9) (2012)
11. Hartmann, W., Loderer, A.: Automated extraction and assessment of functional features of areal measured microstructures using a segmentation-based evaluation method. Surf. Topogr.: Metrol. Prop. **2**(4), 44001 (2014)
12. Hausotte, T.: *Nanopositionier- und Nanomessmaschinen. Geräte für hochpräzise makro bis nanoskalige Oberflächen- und Koordinatenmessungen.* Pro Business, Berlin, 2011. Zugl.: Ilmenau, Techn. Univ., Habil.-Schr

13. Hausotte, T., Gröschl, A., Schaude, J.: High-speed focal-distance-modulated fiber-coupled confocal sensor for coordinate measuring systems. Appl. Opt. **57**(14), 3907–3914 (2018)
14. Hausotte, T., Hartmann, W., Timmermann, M., Galovskyi, B.: Optische Messsysteme zur inline-Prüfung im additiven Fertigungsprozess. Nürnberg (2012)
15. Heinl, M., Galovskyi, B., Bayer, F., Laumer, T., Hausotte, T.: Influence of laser power fluctuations on the quality of additive manufactured workpieces. In: Wissenschaftliche Gesellschaft Lasertechnik e.V. (editor). Lasers in Manufacturing (LiM) (2017)
16. Heinl, M., Greiner, S., Wudy, K., Pobel, C., Rasch, M., Huber, F., Papke, T., Merklein, M., Schmidt, M., Körner, C., Drummer, D., Hausotte, T.: Measuring procedures for surface evaluation of additively manufactured powder bed-based polymer and metal parts. Meas. Sci. Technol. **31**(9), 1–14 (2020)
17. Heinl, M., Laumer, T., Bayer, F., Hausotte, T.: Temperature-dependent optical material properties of polymer powders regarding in-situ measurement techniques in additive manufacturing. Polym. Testing 378–383 (2018)
18. Heinl, M., Loderer, A., Galovskyi, B., Hausotte, T.: Vereinfachte Qualitätsbeurteilung in jeder Raumrichtung additive manufacturing—Qualitätssicherung additiv gefertigter Bauteile anhand von Test-Artefakten. wt Werkstattstechnik—Online, WA119R, pp. 799–803 (2016)
19. Heinl, M., Schmitt, F., Hausotte, T.: In-situ contour detection of additive manufactured workpieces. In: ScienceDirect—Procedia CIRP 74, pp. 664–668. Elsevier (2018)
20. Helmecke, E., Fleßner, M., Staude, A., Hausotte, T.: Numerical measurement uncertainty determination for dimensional measurements of microparts with CT. The e-J. Nondestructive Testing (2016)
21. Lerchen, M.: Messverfahren für die pulverbettbasierte additive Fertigung zur Sicherstellung der Konformität mit geometrischen Produktspezifikationen. Ph.D. thesis, Friedrich-Alexander-Universität Erlangen-Nürnberg (2017). https://opus4.kobv.de/opus4-fau/files/17422/Lerchen_Diss_MB_377.pdf
22. Lerchen, M., Hornung, J., Zou, Y., Hausotte, T.: Methods and procedure of referenced in situ control of lateral contour displacements in additive manufacturing. J. Sens. Sens. Syst. **10**(2), 219–232 (2021)
23. Lerchen, M., Schinn, J., Hausotte, T.: Referencing of powder bed for in situ detection of lateral layer displacements in additive manufacturing. J. Sens. Sens. Syst. **10**(2), 247–259 (2021)
24. Loderer, A., Galovskyi, B., Hartmann, W., Hausotte, T.: Qualifying measuring systems by using six sigma. KEM **637**, 37–43 (2015)
25. Otsu, N.: A threshold selection method from gray-level histograms. IEEE Trans. Syst. Man Cybern. **9**, 62–66 (1979)
26. Schaude, J., Baumgärtner, B., Hausotte, T.: Bidirectional confocal measurement of a microsphere. Appl Opt. **60**(17), 8890–8895 (2021)
27. Scheimpflug, T.: Improved method and apparatus for the systematic alteration or distortion of plane pictures and images by means of lenses and mirrors for photography and for other purposes. GB Patent N0 1196, Intellectual Property Office, Newport, South Wales (1904)

Tino Hausotte has headed the Institute of Manufacturing Metrology at Friedrich-Alexander-Universität Erlangen-Nürnberg since 2011. The Institute works in the fields of meso-, micro-, and nanoscale as well as cross-scale surface, contour, form, and coordinate metrology. This objective is complemented by work on the characterization of measuring system properties, the determination of measurement uncertainty, the assurance of measuring equipment capability, and measuring equipment management. In the field of additive manufacturing, he focuses, in particular, on component analysis using computer tomography.